T0275958

Advances in Intelligent Systems and Computing

Volume 407

Series editor

Janusz Kacprzyk, Polish Academy of Sciences, Warsaw, Poland
e-mail: kacprzyk@ibspan.waw.pl

About this Series

The series "Advances in Intelligent Systems and Computing" contains publications on theory, applications, and design methods of Intelligent Systems and Intelligent Computing. Virtually all disciplines such as engineering, natural sciences, computer and information science, ICT, economics, business, e-commerce, environment, healthcare, life science are covered. The list of topics spans all the areas of modern intelligent systems and computing.

The publications within "Advances in Intelligent Systems and Computing" are primarily textbooks and proceedings of important conferences, symposia and congresses. They cover significant recent developments in the field, both of a foundational and applicable character. An important characteristic feature of the series is the short publication time and world-wide distribution. This permits a rapid and broad dissemination of research results.

Advisory Board

Chairman

Nikhil R. Pal, Indian Statistical Institute, Kolkata, India
e-mail: nikhil@isical.ac.in

Members

Rafael Bello, Universidad Central "Marta Abreu" de Las Villas, Santa Clara, Cuba
e-mail: rbellop@uclv.edu.cu

Emilio S. Corchado, University of Salamanca, Salamanca, Spain
e-mail: escorchado@usal.es

Hani Hagras, University of Essex, Colchester, UK
e-mail: hani@essex.ac.uk

László T. Kóczy, Széchenyi István University, Győr, Hungary
e-mail: koczy@sze.hu

Vladik Kreinovich, University of Texas at El Paso, El Paso, USA
e-mail: vladik@utep.edu

Chin-Teng Lin, National Chiao Tung University, Hsinchu, Taiwan
e-mail: ctlin@mail.nctu.edu.tw

Jie Lu, University of Technology, Sydney, Australia
e-mail: Jie.Lu@uts.edu.au

Patricia Melin, Tijuana Institute of Technology, Tijuana, Mexico
e-mail: epmelin@hafsamx.org

Nadia Nedjah, State University of Rio de Janeiro, Rio de Janeiro, Brazil
e-mail: nadia@eng.uerj.br

Ngoc Thanh Nguyen, Wroclaw University of Technology, Wroclaw, Poland
e-mail: Ngoc-Thanh.Nguyen@pwr.edu.pl

Jun Wang, The Chinese University of Hong Kong, Shatin, Hong Kong
e-mail: jwang@mae.cuhk.edu.hk

More information about this series at http://www.springer.com/series/11156

Tarek Gaber · Aboul Ella Hassanien
Nashwa El-Bendary · Nilanjan Dey
Editors

The 1st International Conference on Advanced Intelligent System and Informatics (AISI2015), November 28–30, 2015, Beni Suef, Egypt

 Springer

Editors
Tarek Gaber
Faculty of Computers and Informatics
Suez Canal University
Ismailia
Egypt

Aboul Ella Hassanien
Faculty of Computers and Information
Cairo University
Giza
Egypt

Nashwa El-Bendary
Arab Academy for Science
Giza
Egypt

Nilanjan Dey
Bengal College of Engineering and
   Technology
Durgapur, West Bengal
India

ISSN 2194-5357   ISSN 2194-5365   (electronic)
Advances in Intelligent Systems and Computing
ISBN 978-3-319-26688-6   ISBN 978-3-319-26690-9   (eBook)
DOI 10.1007/978-3-319-26690-9

Library of Congress Control Number: 2015955384

Springer Cham Heidelberg New York Dordrecht London

Printed on acid-free paper

Springer International Publishing AG Switzerland is part of Springer Science+Business Media
(www.springer.com)

Preface

This edited volume of *Advances in Intelligent Systems and Computing* contains the accepted papers presented at the main track of the 1st International Conference on Advanced Intelligent System and Informatics (AISI2015), November 28–30, 2015, Beni Suef, Egypt.

The aim of AISI2015 is to provide an internationally respected forum for academics, researchers, analysts, industry consultants, and practitioners in the fields involved to discuss and exchange recent progress in the area of informatics and intelligent systems technologies and applications. The conference has three major tracks, namely Intelligent Systems, Intelligent Robotics Systems, and Informatics.

The plenary lectures, tutorials, and the progress and special reports bridged the gap between the different fields of machine learning and informatics, making it possible for nonexperts in a given area to gain insight into new areas. Also included among the speakers were several young scientists, namely postdocs and students, who brought new perspectives to their fields. We expect that the future AISI conferences will be as stimulating as this most recent one was, as indicated by the contributions presented in this proceedings volume.

The first edition of AISI2015 was organized by the Scientific Research Group in Egypt (SRGE) jointly by Faculty of Computers and Information, Beni Suef University, Beni Suef, Egypt. The conference was organized under the patronage of Prof. Amin Lotfy, President of Beni Suef University. The conference received 85 papers from 29 countries and accepted 43 papers.

The organization of the AISI2015 conference was entirely voluntary. The reviewing process of AISI2015 required huge effort from the International Technical Program Committee. Each paper has been reviewed by at least three reviewers. Therefore, we would like to thank all the members of this committee for their contribution to the success of the AISI2015 conference. Also, we would like to express our sincere gratitude and appreciation to the host of AISI2015, Beni Suef

University, Beni Suef, Egypt, and to the publisher, Springer, for their hard work and support in the organization of the conference. Last but not least, we would like to thank all the authors of this proceedings for their high-quality contributions.

The friendly and welcoming attitude of the AISI2015 conference supporters and contributors made this event a success!

Egypt   Tarek Gaber
Egypt   Aboul Ella Hassanien
Egypt   Nashwa El-Bendary
India   Nilanjan Dey
September 2015

Contents

Part III   Swarms Optimization and Applications

Part IV   Hybrid Intelligent Systems

Part I
Intelligent Systems and Informatics (I)

Automatic Rules Generation Approach for Data Cleaning in Medical Applications

Asmaa S. Abdo, Rashed K. Salem and Hatem M. Abdul-Kader

Abstract Data quality is considered crucial challenge in emerging big data scenarios. Data mining techniques can be reutilized efficiently in data cleaning process. Recent studies have shown that databases are often suffered from inconsistent data issues, which ought to be resolved in the cleaning process. In this paper, we introduce an automated approach for dependably generating rules from databases themselves, in order to detect data inconsistency problems from large databases. The proposed approach employs confidence and lift measures with integrity constraints, in order to guarantee that generated rules are minimal, non-redundant and precise. The proposed approach is validated against several datasets from healthcare domain. We experimentally demonstrate that our approach outperform significant enhancement over existing approaches.

Keywords Data quality · Data mining · Data inconsistency · Data cleaning · Electronic medical records (EMR)

1 Introduction

Data is most important value in today's economy [1]. The value of data highly depends on its degree of quality. Henceforward, existence of inconsistency issues in data intensely decreases their assessment, making them misinformed, or even harmful.

A.S. Abdo (✉) · R.K. Salem · H.M. Abdul-Kader
Information System Department, Faculty of Computers and Information Menoufia
University, Menoufia, Egypt
e-mail: asm2009_asm2009@yahoo.com

R.K. Salem
e-mail: rdkhalil@yahoo.com

© Springer International Publishing Switzerland 2016
T. Gaber et al. (eds.), *The 1st International Conference on Advanced Intelligent System and Informatics (AISI2015), November 28–30, 2015, Beni Suef, Egypt,*
Advances in Intelligent Systems and Computing 407, DOI 10.1007/978-3-319-26690-9_1

The quality of data is an increasingly pervasive problem, as data in real world databases quickly degenerates over time and effects the results of the mining. Such poor data quality often emerges due to violations of integrity constraints, which results in incorrect statistics, and ultimately wasting of time and money [2, 3].

Companies lose billions of dollars annually due to poor data quality [4]. As a result, detecting inconsistent data is very important task in the data cleaning process. Doubtless, ensuring high quality dependable data is a competitive advantage to all businesses, which requires accurate data cleaning solutions [5, 6].

Data cleaning refers to the process of maintaining corrupted and/or inaccurate records. This process is mandatory in data management cycle before mining and analyzing data [1].

However, most of the existing data cleaning techniques in the literature focus on record matching [4], which match master cleaned records with a probably imprecise records. It is still necessary to tackle the problem of data inconsistency with the help of data themselves and without the need for external master copy of data. During resolving data inconsistencies, several integrity constraints are ensured, e.g., Functional Dependencies (FDs) and Conditional Functional Dependencies (CFDs) [7, 8].

Medical application domain is one of the most critical applications that suffers from inconsistent and dirty data issues. As ensuring data quality of electronic medical records is very important in healthcare data management purposes, whereas critical decisions are based on patient status apprised from medical records [9–11]. Herein, we are interested to generate data quality rules, which then used for resolving data inconsistencies in such medical databases.

Indeed, the main contribution of this paper is to propose an approach for generating dependable data cleaning rules. Such discovered rules are exploited not only for detecting inconsistent data, but also for correcting them. The proposed approach utilizes data mining techniques for discovering dependable rules, it bases mainly on frequent closed-patterns and their associated generators to their closure that speed up rules generation process. The experimental results conducted over medical datasets verify the effectiveness and accuracy of the proposed approach against CCFD-ZartMNR algorithm [12].

This paper is organized as follows: Sect. 2 discusses related works. In Sect. 3 present data quality in medical applications. Section 4 introduces the proposed approach for generating dependable data cleaning rules. Section 5 discusses the experimental study and results conducted for different medical datasets. Finally, Sect. 6 concludes the proposed work and highlight future trends.

2 Related Work

Despite the urgent need for precise and dependable techniques for enhancing data quality and data cleaning problems, there is not vital solution up to now to these problems. There has been little discussion and analysis about enhancing data inconsistency. However, most of recent work focus on record matching and duplicate detection [13].

Database and data quality researchers have discussed variety of integrity constraints based on Functional Dependencies (FD) [7, 14–16]. These constraints are developed essentially for schema design, but are often not able to embed the semantic of data for data cleaning.

Other researchers focuses on extension of FD, they have proposed what is so-called Conditional Functional Dependencies (CFD) and Conditional Inclusion Dependencies (CID). Such integrity constraints are employed for capturing errors in data [8, 17, 18].

Moreover, several data quality techniques are proposed to clean missy tuples in databases [19]. Statistical inference approaches are studied in [20]. These approaches tackle missing values in order to enhance quality of data. From technological part, there are several open source tools which developed for handling messy data [21].

Besides, data transformation methods such as commercial ETL (Extract, Transformation and Loading) tools are developed for data cleaning [22].

The usage of editing rules in combination with master data is discussed in [23]. Such rules are able to find certain fixes by updating input tuples with master data. However, this approach requires defining editing rules manually for both relations, i.e., master relation and input relation, which is very expensive and time-consuming.

Furthermore, lots of work are proposed in the literature relying on domain specific similarity and matching operators, such works include record matching, record linkage, duplicate detection, and merge purge, [13, 24, 25]. These approaches define two functions; namely match and merge [26]. While match function identifies duplication of records, the merge function combines the two duplicated records into one.

From the literature, we reutilized CFD as a special case of association rules [8, 12]. The relationship between minimal constant CFD and item set mining, association rule as similar relationship to CFD that work on minimal non redundant rules [27].

3 Data Quality in Medical Applications

In era of Electronic Medical Records (EMR), which are objects of knowledge about patients medical and clinical data [28]. Healthcare stakeholders and providers of health care services need such information and knowledge not only at the point of service but also at the point of clinical treatment decisions for improving health care quality [29, 30]. They require such knowledge with precise quality to maximize the benefit of decision-making process [31]. However, maintaining exact and reliable information about diseases treated is based on precise data stored about patient [32]. Therefore, clinical and health service research aimed at accurate, reliable and

Table 1 Records from thyroid (hypothyroid) dataset

Patient number	Pregnant	Hypopituitary	On antithyroid medication	TBG measured
98	f	f	t	f
162	f	f	f	f
164	f	f	f	f
175	f	f	t	f
183	f	f	f	f
214	f	f	t	f
218	f	f	f	f
261	f	f	f	f
742	f	f	t	f
1253	f	f	f	f

complete statistical information about the uses of health care services within a community.

Example 1 Consider a sample from thyroid (hypothyroid) dataset. The following relation schema R is about patient data given in Table 1. Technicians know that these data suffer from inconsistency. The proposed approach aims to discover dependable data quality rules for detecting such inconsistencies.

4   Proposed Approach

The main purpose of the proposed approach is to discover interest minimal non re-dundant constant CFD rules that cover all set of rules. In other words, the discovered rules are minimal and complete with respect to specified support and confidence thresholds. Such discovered rules are employed for dirty data identification and treating data inconsistencies. The proposed approach relies on generating closed frequent patterns and their associated generators according. The associated generators are defined as the closure of closed frequent patterns from all set of frequent patterns.

Given an instance r of a relation schema R, support s, and confidence c thresholds, the proposed approach discovers Interest-based Constant Conditional Functional Dependencies (ICCFDs), abbreviated ICCFD-Miner approach. The discovered ICCFDs ensure finding interest minimal non-redundant dependable Constant Conditional Functional Dependencies (CCFDs) rules with constant patterns in r.

The flow of two main steps for generating minimal non redundant rules is shown in Fig. 1. The proposed approach is detailed as follows:

Input: Dataset and two predefined thresholds, i.e., minimum support (minsup), minimum confidence (minconf), are the input to the ICCFD-Miner approach.

Fig. 1 The proposed
ICCFD-Miner approach

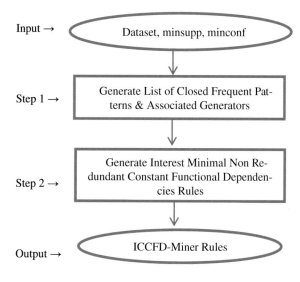

Fig. 2 Search space domain

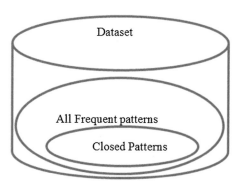

Step 1: Given user defined minimum support threshold, the list of closed frequent patterns and their associated generators to their closure can be generated. In
order to minimize the search space and saving the time of rules generation, The
proposed approach utilize such generated closed frequent patterns and their associated generators instead of working on generating all frequent patterns [12]. The
search space domain is indicated in Fig. 2.

Let us define Support CFD φ: $(X \rightarrow A)$, as the number of records in the dataset
that contain XUA to the total number of records in the database. Support threshold
based on the idea that values that occur together frequently have more evidence to
validate that they are correlated. Support of a CFD φ: $(X \rightarrow A)$ where X generator
pattern and A is (closed/generator) [33], defined as

$$\text{support}(X \rightarrow A) = \frac{\text{Number of tuples containning values in X and A}}{\text{Total number of tuples in relation}} \qquad (1)$$

Step 2: Given user defined confidence threshold. The set of interest minimum non redundant constant conditional functional dependencies data quality rules are generated. While the literature utilizes only support and confidence for generating such rules, the proposed approach consider interest measure into account for generating more dependable and reliable rules. The form of rules for each frequent generator pattern X finds its proper supersets A from set of frequent closed patterns. Then, from X and A add rule antecedent (Generator) $\rightarrow$ consequence (closed/generator) as φ: $X \rightarrow A$.

Let us define Confidence CFD as the number of records in the dataset that satisfy CFD divided by number of records that satisfy left hand side of rule.

$$\text{confidence}(X \rightarrow A) = \frac{\text{support}(\phi)}{\text{support}(x)} \qquad (2)$$

Confidence measures reliability of rule, since the value of confidence is real number between 0 and 1.0 [34]. The pitfall of Confidence is that ignores support of right hand side of rules. As consequence, we add data quality measure called Interest (Lift) which generates more dependent rules when defined it as greater than one.

Let us define *Lift CFD* as measuring the degree of compatibility of left hand side and right hand side of rules as, i.e., occurrence of both left hand side and right hand side [35]. We set here Lift value > 1 to obtain dependent ICCFD-Miner rules. For example lift of this CFD rule ϕ: $(X \rightarrow A)$ is

$$\text{lift}(X \rightarrow A) = \frac{\text{confidence}(\phi)}{\text{support}(A)} \qquad (3)$$

This approach optimizes the process of the rules generation compared with the most related methods. Let us addressing an example of generated rule form from dataset utilized in Sect. 3 as follows:

ϕ: pregnant = f, hypopituitary = f $\rightarrow$ on antithyroid medication = f, TBG measured = f

[support = 0.98 (3679/3772) confidence = 0.99 lift > 1].

"If she is not pregnant and not has hypopituitary then it must not take any antithyroid medication and its TBG measured value should be false".

This consider very important rule to physicians when determining drug to patient. Moreover, such rules are used as critical review to decision maker which determine total number of patient that take/not take antithyroid medication. Tuples that violate this generated rule are shown in Table 2. We highlight detected error values in data.

Table 2 Records that violate generating rule

Patient number	Pregnant	Hypopituitary	On antithyroid medication	TBG measured
98	f	f	t	f
162	f	f	f	f
164	f	f	f	f
175	f	f	t	f
183	f	f	f	f
214	f	f	t	f
218	f	f	f	f
261	f	f	f	f
742	f	f	t	f
1253	f	f	f	f

5 Experimental Study

Using two real-life datasets, we evaluate our proposed approach described in Sect. 4. In order to assess their performance in current domain of real life application especially in critical application such as medical applications. The exploited datasets have large amount of information about patients and their status. The proposed approach used for generating dependable rules in these datasets in order to enhance their data quality. The cleaned data become on demand data access for decision maker in healthcare systems, enabling accurate decisions based on their exact quality data.

We evaluate the following factors on the efficiency and the accuracy of ICCFD-Miner rules produced such threshold support (*sup*), confidence (*conf*), size of sample relation r (the number of instances in r), arity of relation r (the number of columns in r), and time complexity.

5.1 Experimental Setting

The experiments are conducted using two real-life datasets about diseases taken from the UCI machine learning repository (http://archive.ics.uci.edu/ml/) namely, Thyroid (hypothyroid), primary-tumor. Table 3 shows the number of attributes and the number of instances for each dataset.

The proposed approach is implemented using java (JDK1.7). The implementation is tested on machine equipped with Intel(R) Pentium(R) Dual CPUT3400 @ 2.16 GHz 2.17 GHz processor with 2.00 GB of memory running on windows 7 operating system. The proposed approach runs mainly in main memory. Each experiment is repeated at least five times and the average reported here.

Table 3 Datasets description

Dataset name	Arity (Number of columns)	Size (Number of instances)
Thyroid(hypothyroid)	30	3772
primary-tumor	18	339

5.2 Experimental Results

Now, we show and discuss the results on real world dataset described in previous section. Note that we aim to evaluate the effectiveness of rules generation of the proposed approach against CCFD-ZartMNR algorithm [12]. Experiments show that proposed approach always produce less number of rules but more accurate. The generated rules are interest-based minimal and non-redundant. Experiments also show that ICCFD-Miner outperforms the other algorithm with respect to time for rule generation.

Experiment-1: In this experiment, rules are generated from thyroid (hypothyroid) dataset. This data set contains 30 attributes, 3772 records of patient data describing patient information about the hypothyroid diagnoses data.

By varying values of support (sup) and confidence (conf) thresholds as shown in Fig. 3, we notice that the proposed approach always generate accurate interest minimum non redundant rules compared to CCFD-ZartMNR algorithm. For example in Fig. 3, at minimum support = 0.97 and minimum confidence = 0.99 the number of rules generated from the proposed ICCFD-Miner approach = 85 rules compared to output generated rules from CCFD_ZartMNR = 220 rules. Results from Fig. 4 shows that the proposed approach generates rules at different sup, conf values in less time compared to CCFD-ZartMNR algorithm. For example in Fig. 4, at minimum support = 0.97 and minimum confidence = 0.99 response time of the proposed ICCFD-Miner approach = 312 ms compared to response time of existing approach CCFD_ZartMNR = 368 ms.

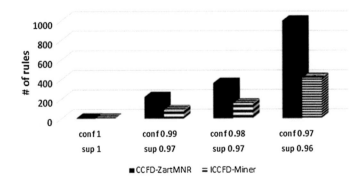

Fig. 3 Total number of rules generated of thyroid (hypothyroid) dataset

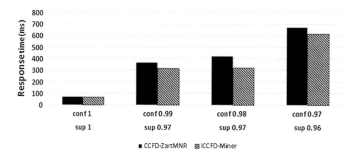

Fig. 4 Response time measure about thyroid (hypothyroid) dataset

Experiment-2: The experimental results conducted over primary-tumor disease dataset are shown in Figs. 5 and 6. This data set contains 18 attributes, 339 records describing patient information about the on primary-tumor disease diagnoses data.

Figures 5 and 6 validate the efficiency of the proposed approach against CCFD-ZartMNR algorithm in both number of rules generated and response time measure.

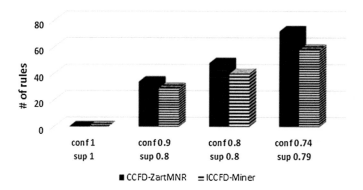

Fig. 5 Total number of rules generated of primary-tumor dataset

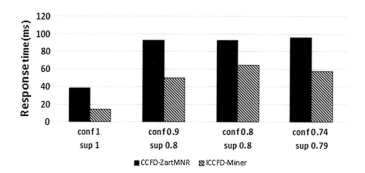

Fig. 6 Response time measure about primary-tumor dataset

Finally, we believe that the proposed approach, i.e., ICCFD-Miner, outperforms CCFD_ZartMNR due to its implying lift measure when generating dependable rules. Furthermore, the proposed approach focus on closed patterns with their associated generators, i.e., supersets of closed patterns, as a search space for generating the more accurate and reliable rules.

6 Conclusions

In this paper, we have presented ICCFD-Miner approach that discovers precise data quality rules for resolving data inconsistency problem. The proposed approach yields a promising method for detecting semantic data inconsistency. The main target of the proposed approach is keeping the database in consistent state. Generated rules are exploited as data cleaning solution to resolve inconsistency problem in current application domains. ICCFD-Miner relies on lift measure in addition to support and confidence measures for generating dependable minimal and non-redundant rules. The ICCFD-Miner is validated and evaluated over two real life datasets from medical domain. The experimental results confirm the effectiveness and usefulness of the proposed approach against CCFD_ZartMNR algorithm. The proposed approach perform well across several dimensions such as effectiveness, accuracy of number of rules generated, and run time. Finally, we plan to investigate a technique for fixing errors autonomously with generated rules from ICCFD-Miner.

References

1. Mezzanzanica, M., Boselli, R., Cesarini, M., Mercorio, F.: Automatic synthesis of data cleansing activities (2011)
2. Li, J., Liu, J., Toivonen, H., Yong, J.: Effective pruning for the discovery of conditional functional dependencies. Comput. J. **56**, 378–392 (2013)
3. Yakout, M., Elmagarmid, A.K., Neville, J.: Ranking for data repairs. In: Proceeding—International Conference Data Engineering, pp. 23–28 (2010)
4. Fan, W., Li, J., Ma, S., Tang, N., Yu, W.: Interaction between record matching and data repairing. In: Proceeding Journal of Data and Information Quality (JDIQ) vol. 4(4), p. 16 (2014)
5. Wang, J., Tang, N.: Towards dependable data repairing with fixing rules. In: SIGMOD Conference, pp. 457–468 (2014)
6. Fan, W., Geerts, F.: Foundations of data quality management. Synth. Lect. Data Manage. **4**, 1–217 (2012)
7. Liu, J., Li, J., Liu, C., Chen, Y.: Discover dependencies from data—a review. IEEE Trans. Knowl. Data Eng. **24**, 251–264 (2012)
8. Vo, L.T.H., Cao, J., Rahayu, W.: Discovering conditional functional dependencies. Conf. Res. Pract. Inf. Technol. Ser. **115**, 143–152 (2011)
9. Rodríguez, C.C.G., Riveill, M., Antipolis, S.: e-Health monitoring applications : what about data quality ? (2010)

10. Mans, R.S., van der A., Wil M.P., Vanwersch, R.J.: Data Quality Issues. Process Mining in Healthcare, pp. 79–88. Springer, Berlin (2015)

11. Kazley, A.S., Diana, M.L., Ford, E.W., Menachemi, N.: Is electronic health record use associated with patient satisfaction in hospitals? Health Care Manage. Rev. **37**, 23–30 (2012)

12. Kalyani, D.D.: Mining constant conditional functional dependencies for improving data quality. **74**, 12–20 (2013)

13. Bharambe, D., Jain, S., Jain, A.: A survey : detection of duplicate record. **2**, (2012)

14. Cong, G., Fan, W., Geerts, F., Jia, X., Ma, S.: Improving data quality: consistency and accuracy. In: Proceeding 33rd International Conference Very Large Data Bases, pp. 315–326. Vienna, Au (2007)

15. Hartmann, S., Kirchberg, M., Link, S.: Design by example for SQL table definitions with functional dependencies. VLDB J. **21**, 121–144 (2012)

16. Yao, H., Hamilton, H.J.: Mining functional dependencies from data. Data Min. Knowl. Discov. **16**, 197–219 (2008)

17. Bohannon, P., Fan, W., Geerts, F., Jia, X., Kementsietsidis, A.: Conditional functional dependencies for data cleaning. In: Proceeding—International Conference Data Engineering, pp. 746–755 (2007)

18. Bauckmann, J., Abedjan, Z., Leser, U., Müller, H., Naumann, F.: Discovering conditional inclusion dependencies. In: 21st ACM International Conference on Information and Knowledge Management, pp. 2094–2098. (2012)

19. Fan, W., Geerts, F.: Capturing missing tuples and missing values. In: Proceeding 29th ACM SIGACT-SIGMOD-SIGART Symposium Principle of Database System, pp. 169–178 (2010)

20. Mayfield, C., Neville, J., Prabhakar, S.: ERACER: a database approach for statistical inference and data cleaning. In: Proceeding ACM SIGMOD International Conference Management Data, pp. 75–86 (2010)

21. Larsson, P.: Evaluation of open source data cleaning tools : open refine and data wrangler. (2013)

22. Vassiliadis, P., Simitsis, A.: Extraction, transformation, and loading. Encycl. Database Syst. 1095–1101 (2009)

23. Fan, W., Li, J., Ma, S., Tang, N., Yu, W.: Towards certain fixes with editing rules and master data. VLDB J. **21**, 213–238 (2012)

24. Fan, W., Gao, H., Jia, X., Li, J., Ma, S.: Dynamic constraints for record matching. VLDB J. **20**, 495–520 (2011)

25. Reiter, J.: Data quality and record linkage techniques. J. Am. Stat. Assoc. **103**(482), 881 (2008)

26. Benjelloun, O., Garcia-Molina, H., Menestrina, D., Su, Q., Whang, S.E., Widom, J.: Swoosh: a generic approach to entity resolution. VLDB J. **18**, 255–276 (2009)

27. Zaki, M.J.: Mining non-redundant association rules. Data Min. Knowl. Discov. **9**, 223–248 (2004)

28. Chang, I.-C., Li, Y.-C., Wu, T.-Y., Yen, D.C.: Electronic medical record quality and its impact on user satisfaction—Healthcare providers' point of view. Gov. Inf. Q. **29**, 235–242 (2012)

29. Weiskopf, N.G., Weng, C.: Methods and dimensions of electronic health record data quality assessment: enabling reuse for clinical research. J. Am. Med. Inform. Assoc. 144–151 (2012)

30. Groves, P., Kayyali, B., Knott, D., Van Kuiken, S.: The " Big Data " Revolution in Healthcare. McKinsey, New York (2013)

31. Kush, R.D., Ph.D., Helton, E., Rockhold, F.W., Hardison, C.D.: Electronic health records, medical research, and the tower of Babel. 16–18 (2008)

32. Koh, H.C., Tan, G.: Data mining applications in healthcare. J. Healthc. Inf. Manage. **19**, 64–72 (2005)

33. Chiang, F., Miller, R.J.: Discovering data quality rules. In: Proceeding VLDB Endowment, pp. 1166–1177 (2008)

34. Medina, R., Nourine, L.: A unified hierarchy for functional dependencies, conditional functional dependencies and association rules. In: LNAI, Lecture Notes Computer Science (including Subseries Lecture Notes Artifical Intelligent Lecture Notes Bioinformatics). vol. 5548, pp. 98–113 (2009)
35. Hussein, N., Alashqur, A., Sowan, B.: Using the interestingness measure lift to generate association rules. J. Adv. Comput. Sci. Technol. **4**, 156 (2015)

Action Classification Using Weighted Directional Wavelet LBP Histograms

Maryam N. Al-Berry, Mohammed A.-M. Salem, Hala M. Ebeid, Ashraf S. Hussein and Mohamed F. Tolba

Abstract The wavelet transform is one of the widely used transforms that proved to be very powerful in many applications, as it has strong localization ability in both frequency and space. In this paper, the 3D Stationary Wavelet Transform (SWT) is combined with a Local Binary Pattern (LBP) histogram to represent and describe the human actions in video sequences. A global representation is obtained and described using Hu invariant moments and a weighted LBP histogram is presented to describe the local structures in the wavelet representation. The directional and multi-scale information encoded in the wavelet coefficients is utilized to obtain a robust description that combine global and local descriptions in a unified feature vector. This unified vector is used to train a standard classifier. The performance of the proposed descriptor is verified using the KTH dataset and achieved high accuracy compared to existing state-of-the-art methods.

Keywords 3D stationary wavelet · Action recognition · Local binary pattern · Hu invariant moments · Spatio-temporal space Multi-resolution analysis

M.N. Al-Berry (✉) · M.A.-M. Salem · H.M. Ebeid · M.F. Tolba
Faculty of Computer and Information Sciences, Ain Shams University, Cairo, Egypt
e-mail: maryam_nabil@cis.asu.edu.eg

M.A.-M. Salem
e-mail: salem@cis.asu.edu.eg

H.M. Ebeid
e-mail: halam@cis.asu.edu.eg

M.F. Tolba
e-mail: fahmytolba@gmail.com

A.S. Hussein
Faculty of Computing Studies, Arab Open University-Headquarters, 13033,
Kuwait, Kuwait
e-mail: ashrafh@acm.org

© Springer International Publishing Switzerland 2016
T. Gaber et al. (eds.), *The 1st International Conference on Advanced Intelligent System and Informatics (AISI2015), November 28-30, 2015, Beni Suef, Egypt.,*
Advances in Intelligent Systems and Computing 407, DOI 10.1007/978-3-319-26690-9_2

1 Introduction

Wavelets [1] and multi-scale techniques have been widely used in many image processing and computer vision applications [2, 3]. The most important property of wavelet analysis is the ability to analyze functions at different scales. Wavelets also provide a basis for describing signals accurately [4]. These properties of wavelet analysis have led to very successful applications within the fields of signal processing, image processing and computer vision. One of the fields that benefit from wavelets is the field of human action recognition [3, 5, 6].

In this way, wavelets have been used for spatio-temporal representation and description of human actions. In spatio-temporal methods, the action is represented by fusing spatial and temporal information into a single template. Spatio-temporal techniques can be divided into global and local techniques [7]. The advantages of both global and local representation techniques can be combined to obtain robust representations for human actions [8]. The 3D Wavelet Transforms [2, 6, 9] have been proposed and used to process video sequences as a 3D volume having time as the third dimension. The wavelet coefficients are then used to obtain global [3, 5, 10] or local [6] representations for the performed actions.

In this paper, a new method for action representation and description is proposed based on the wavelet analysis. The proposed method aims at extracting discriminative features and increasing the classification accuracy. The power of the 3D Stationary Wavelet Transform (SWT) is used to highlight spatio-temporal variations in the video volume and provides a multi-scale directional representation for human actions. These representations are described using a new proposed descriptor based on the concept of the Local Binary Pattern (LBP) [11], which is called the weighted directional wavelet LBP histogram. The proposed local descriptor is combined with Hu invariant moments [12] that are extracted from the Multi-scale Motion Energy Images (*MMEI*) [5] to get benefit from the global information contained in the wavelet representation. The performance of the proposed method is verified through several experiments using KTH dataset [13]. The rest of the paper is organized as follows: Sect. 2 reviews the background of the field and the state-of-the-art methods. The proposed representation and description method is illustrated in Sect. 3. In Sect. 4, the proposed method is applied, and the performance evaluation and comparison is performed. Finally, the work is concluded, and possible directions for improvement are pointed out in Sect. 5.

2 Related Work

The wavelet transforms [1, 14, 15] are being used in different fields such as signal, image and video compression as they provide the basis to describe signal features easily. The Decimated bi-orthogonal Wavelet Transform (DWT) [14] is probably the most widely used wavelet transform as it exhibits better properties than the

Continuous Wavelet Transform (CWT). It has much less scales than the CWT as only dyadic scales are considered. DWT is also very computationally efficient, since it requires O(N) operations for N samples of data and can be easily extended to any dimension by separable products of the scaling and wavelet functions [16].

When dealing with images in the spatial domain, two-dimensional DWT is used to decompose an image into four sub-images; an approximation and three details. The approximation A is obtained by applying the low-pass filter to the two spatial dimensions of the image and represents the image at a coarse resolution. The three detailed images are the horizontal detail H, the vertical detail V and the diagonal detail D [2]. Wavelet analysis has been used in the context of 2D spatio-temporal action recognition. For example, Siddiqi et al. [17] have used 2D DWT to extract features from the video sequence. The most important features were then selected using a Step Wise Linear Discriminant Analysis (SWLDA). Sharma et al. [18] have represented short actions using Motion History Images (MHI) [19]. They used two dimensional, 3 level dyadic wavelet transforms and modified their proposed representation to be invariant to translation and scale, but they concluded that the directional sub-bands alone were not efficient for action classification. In [20], Sharma and Kumar have described the histograms of MHIs by orthogonal Legendre moments and modeled the wavelet sub-bands by Generalized Gaussian Density (GGD). They have found that the directional information encapsulated in the wavelet sub-band enhances the classification accuracy.

Wavelet transforms have also 3D extension. Rapantzikos et al. have used the 3D wavelet transform to keep the computational complexity low while representing dynamic events. They have used wavelet coefficients to extract salient features. They treated the video sequence as a spatio-temporal volume, and local saliency measures are generated for each visual unit (voxel). Shao et al. [21] have applied a transform based technique to extract discriminative features from video sequences. They have shown that the wavelet transform gave promising results on action recognition.

While the decimated wavelet transform has been successful in many fields, it has the disadvantage of translation variance, so to overcome this drawback and maintain shift invariance, the Stationary Wavelet Transform (SWT) was developed. It has been proposed in the literature under different names such as; un-decimated wavelet transform [4], redundant wavelet [22] and shift-invariant wavelet transform [23]. The SWT gives a better approximation than the DWT since it is linear, redundant and shift-invariant [2]. In, the 3D SWT has been proposed and used for spatio-temporal motion detection. A 3D SWT is applied to each set of frames, and then the analysis is performed along the x-direction, the y-direction and the t-dimension of the time varying data [2], which is formed using the input frames.

Based on this 3D SWT, the authors of [10] have proposed two spatio-temporal human action representations. The proposed representations were combined with Hu invariant moments as features for action classification. Results obtained using the public dataset were promising and provide a good step towards better enhancements. They have proposed a directional global wavelet-based representation of natural human actions [3] that utilizes the 3D SWT [2] to encode the directional spatio-temporal characteristics of the motion.

Wavelets have also been used in combination with local descriptors such as Local Binary Patterns (LBPs) [11, 24] to describe texture images. LBPs are texture descriptors that have proved to be robust and computationally efficient in many applications [21, 25, 26]. The most important characteristic of LBPs is that it is not only computationally efficient, but also invariant to gray level changes caused by illumination variations [11]. The original LBP operator labels the pixels of an image by thresholding the 3×3 neighborhood of each pixel with the center value and considering the result as a binary number. The LBP was extended to operate on a circular neighborhood set of radius R. This produces 2^P different binary patterns for the P pixels constituting the neighborhood. Uniform LBP was defined as the pattern that contains at most 2 transitions for 1 to 0 or from 0 to 1. These patterns describe the essential patterns that can be found in texture. The disadvantage of LBP is lack of directional information.

In this paper, the 3D SWT is combined with LBPs and employed in the field of human action recognition. A new descriptor is proposed for describing human actions represented by the wavelet coefficients. The new descriptor combines the directional information contained in the wavelet coefficients in a weighted manner using the entropy value. The new local descriptor is expected to provide discriminative local features for the human actions.

3   Proposed Action Representation and Description

In this section, the proposed approach for extracting features is described. The proposed approach uses the power of the 3D SWT to detect multi-scale directional spatio-temporal changes and describes these changes using a weighted Local Binary Pattern (LBP) histogram. The weighted directional wavelet LBP is fused with moments to maintain global relationships. The proposed method is illustrated in Fig. 1.

3.1   Weighted Directional Wavelet Local Binary Pattern Histogram

The proposed descriptor is obtained by treating the video sequence as a 3D volume; considering time the third dimension. First, the video sequence is divided into disjoint blocks of frames and these blocks are supplied to the 3D SWT. The number of frames in one block depends on the intended number of wavelet analysis levels (resolutions). The number of frames is obtained by $L = 2^J$, where J is the number of resolutions. The 3D SWT produces the spatio-temporal detail coefficients

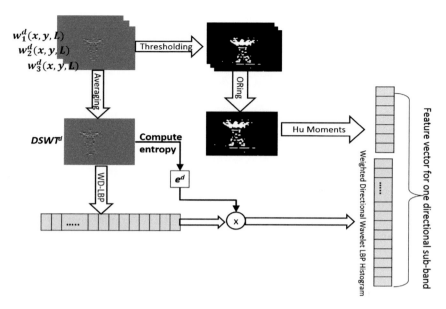

Fig. 1 Illustration of the proposed method

$w_j^d(x, y, t)$, where (x, y) are the spatial coordinates of the frames, t is the time, $j = 1, 2, \ldots, J$ is the resolution (scale) and d is the sub-band orientation, and the approximation coefficient $c_j(x, y, t)$ computed by the associated scaling function as described in [2].

The detail sub-bands $w_j^d(x, y, t), d = (5, 6, 7)$ are used for representing the action as they contain highlighted motion detected in the temporal changes that happened along the t-axis. Sub-band 4 also highlights temporal changes along with global spatial approximation. It is not used here because the focus is on investigating the effect of using local details only. The resulting coefficients are used to obtain a directional multi-scale stationary wavelet representation for the action by averaging the wavelet coefficients of the obtained resolutions at time $t = L$. This results in three Directional multi-scale Stationary Wavelet Templates $DSWT^d, d = (5, 6, 7)$, for the three detail sub-bands.

$$DSWT^d(x, y) = \frac{\sum_{j=1}^{3} w_j^d(x, y, L)}{J}, d = (5, 6, 7) \qquad (1)$$

Each directional stationary wavelet template is treated as texture images and the Local Binary Pattern is used to extract local features from it. The Directional Wavelet—LBP (DW-LBP) is computed as follows:

$$DW - LBP^d_{(P,R)} = \sum_{p=0}^{p-1} S(f_p^d - f_c^d)2^p \tag{2}$$

where (P,R) denotes a neighborhood of P equally spaced sampling points on a circle of radius R, $S(z)$ is the thresholding function

$$S(z) = \begin{cases} 1, & z \geq 0 \\ 0, & z < 0 \end{cases} \tag{3}$$

f_c^d is the center pixel in the defined neighborhood in sub-band d, f_p^d, $p = 0, 1, \cdots, P-1$ are the pixels in the neighborhood. The normalized histogram of the obtained Directional Wavelet—Local Binary Pattern ($DW - LBP$) is computed and the entropy value [27] for each directional wavelet template (e^d) is computed and is used to give a weight for the computed histogram by multiplying it by the histogram values.

3.2 Global Description Using Invariant Moments

To obtain global description, the coefficients of the three selected sub-bands are thresholded to produce motion energy images. The motion images obtained at time $t = L$ of the J different scales are fused into three directional multi-scale motion energy images ($MMEI^d(x, y, t)$) in the same way proposed in [3]. These sub-band motion energy images encode the directional motion energy during the processed L frames. These coefficients of the different scales are threshold to obtain motion images $O_j^d(x, y, t)$

$$O_j^d(x, y, t) = \begin{cases} 1 & \text{if } w_j^d(x, y, t) > \tau \text{ for some } j, d \\ 0 & \text{otherwise} \end{cases} \tag{4}$$

The threshold value τ is computed automatically using Otsu's thresholding technique [27]. This technique is completely based on the computation of the image histogram and maximizes the inter-class variance [27]. By fusing the motion images obtained from the different resolutions for each sub-band, multi-scale motion energy images ($MMEI^d(x, y, t)$) is obtained. These motion energy images are described using the seven Hu moments [12]. The Hu moments are known to be good global shape descriptors and are translation, scale, mirroring and rotation invariant [27].

To compute the set of invariant moments, the central moment of order $p + q (\mu_{pq})$ is computed by:

$$\mu_{pq} = \sum_x \sum_y (x - \bar{x})^p (y - \bar{y})^q MMEI^d(x, y) \tag{5}$$

where, $\bar{x} = \frac{M_{10}}{M_{00}}$, $\bar{y} = \frac{M_{01}}{M_{00}}$, and M_{ij} are the raw image moments calculated by:

$$M_{ij} = \sum_x \sum_y x^i y^j MMEI^d(x, y) \tag{6}$$

The set of seven invariant moments are then computed as described in [27] and constitutes the second part of the final feature vector.

4 Experimental Results

This section describes the performed experiments and demonstrates the achieved results. First, the used datasets are described, then the experimental results are presented.

4.1 Human Action Dataset

The performance evaluation of the proposed method is carried out using the KTH dataset [13]. This dataset contains samples for six actions: "boxing", "hand-clapping", "hand-waving", "jogging", "running" and "walking". These actions are performed by 25 different subjects in four different scenarios. The first scenario s1 is "outdoors", s2 is "outdoors with scale variation", s3 is "outdoors with different clothes" and s4 is "indoors". Samples of the "boxing" action in the four different scenarios are shown in Fig. 2.

4.2 Experiments and Results

A simple K-Nearest Neighbor (KNN) classifier with $(K = 1)$ was used for testing the accuracy of the proposed features. The used KNN classifier uses Euclidean

Fig. 2 Samples of the "boxing" action in the four different scenarios

distance. Leave-one-out cross-validation is used to calculate the classification accuracy. The descriptor is computed in (8,5) neighborhood, i.e., the radius of the circular neighborhood is 5 and 8 sampling points are used on the circle. These values for the radius and the sampling points were chosen empirically. The number of bins in the weighted histogram of DW-LBP is 59, as only uniform patterns are considered. The final feature vector length is 66 for each sub-band after combination with the seven Hu invariant moments.

The four different scenarios were first examined separately. In another experiment, the overall performance is measured by using the samples of the four scenarios. Two classification schemes were used in the evaluation. First, the features of the three directional sub-bands are combined in a single feature vector and this feature vector is used to train the classifier. In the other scheme, three classifiers are trained; each one using one feature vector of the three sub-bands and the class assigned to the test pattern is selected by majority voting between the three classifiers.

Table 1 lists the classification accuracy obtained for the four scenarios and the overall accuracy using the two classification schemes. In the case of the combined feature vector (CFV), scale variations and variations in clothes didn't affect the classification accuracy. Using voting the second scenario recorded less accuracy than the combined feature vector. It can be seen that the proposed method recorded a high classification accuracy consistently using the two classification schemes.

Another experiment was carried out using the gait actions only. The proposed method achieved a high accuracy of 97.33 % using the combined feature vector while gait actions are 100 % accurately classified using the voting scheme. The results are shown in Table 2.

Table 3 shows a comparison between the proposed method and existing state-of-the-art methods. The proposed method outperformed existing techniques, achieving higher classification accuracy. All compared techniques used leave-one-out cross validation.

Table 1 Accuracy using combined feature vector versus voting between directional feature vectors

Scenario	Combined feature vector %	Voting %
s1	94.67	95.33
s2	94.67	92.00
s3	94.67	94.67
s4	96.00	96.00
Overall	**95.00**	**94.50**

Table 2 Accuracy using combined feature vector versus voting for gait actions

Scenario	Combined feature vector %	Voting %
s1	97.33	100
s2	96.00	100
s3	98.67	100
s4	97.33	100
Overall	**96.00**	**100**

Method	Accuracy %
Table 3 Performance comparison with existing techniques	
Proposed method (CFV)	**95.00**
Proposed method (voting)	**94.50**
Kong et al. (2011) [28]	88.81
Bregonzio (2012) [29]	94.33
Gupta et al. (2013) [30] (gait actions Only)	95.10
Proposed method (gait actions only)(CFV)	**96.00**
Proposed method (gait actions only)(voting)	**100**

5   Conclusions

In this paper, a new human action descriptor is proposed. The proposed descriptor combines the power of wavelets in highlighting directional variations and the simplicity and robustness of local binary patterns in describing local structures. The new local descriptor is based on the 3D stationary wavelet transform that highlights spatio-temporal variations representing human actions in video sequences. A weighted combination of directional features extracted from the wavelet coefficients is fused with global moments and used for describing actions. The proposed features are tested on a public dataset, and their accuracy is verified using a standard classifier. The proposed method outperformed state-of-the-art methods achieving 95 % average classification accuracy for all actions using combined feature vector and an average accuracy of 100 % for gait actions using voting.

Future work may include block -based processing of the wavelet multi-scale templates and extending the new descriptor to the 3D domain.

References

1. Ahad, M., Tan, J., Kim, H., Ishikawa, S.: Motion history image: its variants and applications. Mach. Vis. Appl. **23**, 255–281 (2012)
2. Al-Berry, M.N., Ebied, H.M., Hussein, A.S., Tolba, M.F.: Human action recognition via multi-scale 3D stationary wavelet analysis. In: Hybrid Intelligent Systems 2014 (HIS'14). Kuwait (2014)
3. Al-berry, M.N., Salem, M.A.-M., Ebeid, H.M., Hussein, A.S., Tolba, M.F.: Action recognition using stationary wavelet-based motion images. In: IEEE conference o n Intelligent systems 14, pp. 743–753. Warasaw, Poland: Springer International Publishing (2014)
4. Al-Berry, M.N., Salem, M.A.-M., Hussein, A.S., Tolba, M.F.: Spatio-Temporal motion detection for intelligent surveillance applications. Int. J. Comput. Methods, **11**(1) (2014)
5. Al-Berry, M.N., Salem, M.A.-M., Mousher, H.E., Hussein, A.S., Tolba, M.F.: Directional stationary wavelet-based representation for human action classification. In: Advanced Machine Learning Technologies and Applications(AMLTA2014), pp. 309–320. Cairo: Springer International Publishing (2014)
6. Bradley, A.P.: Shift-invariance in the discrete wavelet transform. In: Proceeding of VIIth Digital Image Computing: Techniques and Applications, 1, pp. 29–38. Sydney (2003)

7. Bregonzio, M., Xiang, T., Gong, S.: Fusing appearance and distribution information of interest points for action recognition. Pattern Recogn. **45**, 1220–1234 (2012)
8. Daubechies, I.: Recent results in wavelet applications. J. Electron. Imaging **7**(4), 719–724 (1998)
9. Daubechies, I.: Ten Lectures on Wavelets. Industrial and Applied Mathematics Publishing Company, Philadelphia (1992)
10. Fadili, J., Starck J.L.: Numerical issues when using wavelets. In: Robert, M.A. (ed.) Encyclopedia of Complexity and Systems Science, vol. 3, pp. 6352–6368. Springer, New York (2009)
11. Fowler, J.E.: The redundant discrete wavelet transform and additive noise. Signal Process. Lett. **12**(9), 629–632 (2005)
12. Gonzalez, R.C., Woods, R.E.: Digital Image Processing, 3rd edn. Printice Hall, Upper Saddle River (2008)
13. Gupta, J.P., Singh, N., Dixit, P., Semwal, V.B., Dubey, S.R.: Human activity recognition using gait pattern. Int. J. Comput. Vision Image Process. **3**(3), 31–53 (2013)
14. Hu, M.-K.: Visual pattern recognition by moment invariants. IEEE Trans. Inf. Theory **8**(2), 179–187 (1962)
15. Kong, Y., Zhang, X., Hu, W., Jia, Y.: Adaptive learning codebook for action recognition. Pattern Recogn. Lett. **32**, 1178–1186 (2011)
16. Liu, S., Liu, J., Zhang, T., Lu, H.: Human action recognition in videos using hybrid features. In: Advances in Multimedia Modeling, Lecture Notes in Computer Science, vol. 5916, pp. 411–421 (2010)
17. Mallat, S., Hwang, W.: Singularity detection and processing with wavelets. IEEE Trans. Inf. Theory **38**(2), 617–643 (1992)
18. Nanni, L., Barnham, S., Lumini, A.: Combining different local binary pattern variants to boost performance. Expert Syst. Appl. **38**, 6209–6216 (2011)
19. Pietikainen, M.: Computer vision using local binary patterns, Computational Imaging and Vision, vol. 40. Springer, Berlin (2011)
20. Poppe, R.: A survey on vision-based human action recognition. Image Vis. Comput. **28**, 976–990 (2010)
21. Rapantzikos, K., Tsapatsoulis, N., Avrithis, Y., Kollias, S.: Spatiotemporal saliency for video classification. Sig. Process. Image Commun. **24**, 557–571 (2009)
22. Salem, M.: Multiresolution image segmentation. Ph.D. Thesis, Department of Computer Science, Humboldt-Universitaet zu Berlin. Berlin, Germany (2008)
23. Schüldt, C., Laptev, I., Caputo, B. Recognizing human actions: a local SVM approach. In: International Conference on Pattern Recognition (ICPR2004). vol. 3, pp. 32–36. IEEE (2004)
24. Shao, L., Gao, R., Liu, Y., Zhang, H.: Transform based spatio-temporal descriptors for human action recognition. Neurocomputing **74**, 962–973 (2011)
25. Sharma, A., Kumar, D.K. Moments and wavelets for classification of human gestures represented by spatio-temporal templates, pp. 215–226 (2004)
26. Sharma, A., Kumar, D.K., Kumar, S., McLachlan, N.: Wavelet directional histograms for classification of human gestures represented by spatio-temporal templates. In: 10th International Multimedia Modeling Conference MMM'04, pp. 57–63. IEEE (2004)
27. Siddiqi, M.H., Ali, R., Rana, M.S.,Hong, E.-K., Kim, E.s., Lee, S.: Video-based human activity recognition using multilevel wavelet decomposition and stepwise linear discriminant analysis. Sensors **14**(4), 6370–6392 (2014)
28. Song, K.-C., Yan, Y.-H., Chen, W.-H., Zhang, X.: Research and perspective on local binary pattern. Acta Automatica Sinica **39**(6) (2013)
29. Stark, J.-L., Fadili, J., Murtagh, F.: The undecimated wavelet decomposition and its reconstruction. IEEE Trans. Image Process. **16**(2) (2007)
30. Zhao, Y., Jia, W., Hu, R.-X., Min, H.: Completed robust local binary pattern for texture classification. Neurocomputing **106**, 68–76 (2013)

Markerless Tracking for Augmented Reality Using Different Classifiers

Faten A. Khalifa, Noura A. Semary, Hatem M. El-Sayed
and Mohiy M. Hadhoud

Abstract Augmented reality (AR) is the combination of a real scene viewed by the user and a virtual scene generated by the computer that augments the scene with additional information. The user of an AR application should feel that the augmented object is a part of the real world. One of the factors that greatly affect this condition is the tracking technique used. In this paper, an augmented reality application is adopted with markerless tracking as a classification task. ORB algorithm is used for feature detection and the FREAK algorithm is used for feature description. The classifiers used for the tracking task are KNN, Random Forest, Extremely Randomized Trees, SVM and Bayes classifier. The performance of each classifier used is evaluated in terms of speed and efficiency. It has been observed that KNN outperforms other classifiers including Random Forest with different number of trees.

Keywords Augmented reality · Markerless tracking · Classification

F.A. Khalifa (✉) · N.A. Semary · H.M. El-Sayed · M.M. Hadhoud
Faculty of Computers and Information, Menofia University, Menofia, Egypt
e-mail: faten_a_khalifa@ci.menofia.edu.eg; faten_a_khalifa@yahoo.com

N.A. Semary
e-mail: noura.samri@ci.menofia.edu.eg

H.M. El-Sayed
e-mail: hatem6803@yahoo.com

M.M. Hadhoud
e-mail: mmhadhoud@yahoo.com

© Springer International Publishing Switzerland 2016
T. Gaber et al. (eds.), *The 1st International Conference on Advanced Intelligent
System and Informatics (AISI2015), November 28–30, 2015, Beni Suef, Egypt,*
Advances in Intelligent Systems and Computing 407, DOI 10.1007/978-3-319-26690-9_3

1 Introduction

Augmented Reality (AR) [1] is one part of the general area of mixed reality as shown in Fig. 1. Both virtual environments (virtual reality) (completely synthetic) and augmented virtuality, in which real objects are added to virtual ones, replace the surrounding environment by a virtual one. On the other hand, AR changes the way we view the world by providing local virtuality. It blurs the line between what's real and what's computer-generated by enhancing what we see, hear, feel and smell. Augmented reality adds graphics, sounds, haptic feedback and smell to the real world as it exists. This enhances the user's sensory perception of the virtual world they are seeing or interacting with. Today, AR is used in many applications and fields like medical, manufacturing, annotated environment, entertainment, education, training, design and assembly.

The AR enabling technologies include displays, tracking, registration, user interface and interaction, networks, databases, 3D modeling and other hardware used. Tracking is following the user's and virtual object's movements by means of special device or techniques. Compared to virtual reality, AR tracking devices must have higher accuracy, wider input variety and bandwidth, and longer ranges. Tracking is usually easier in indoor settings than in outdoor settings. Pose tracking for AR must deliver full six degrees of freedom (6DoF) and be accurate, robust and run in real-time.

The aim of this paper is to investigate the performance of different types of classifiers for markerless tracking of augmented reality. The classifiers included in the evaluation are Random Forest [2], Extremely Randomized Trees [3], K Nearest Neighbor (KNN) [4], Support Vector Machines (SVM) [5] and Bayes classifier [6].

The organization of the remaining content is as follows: Related work is presented in Sect. 2 and an overview of the classifiers used is in Sect. 3. Section 4 presents the methodology and the implementation details. In Sect. 5, the experimental results are discussed. Finally, the conclusion is reported in Sect. 6.

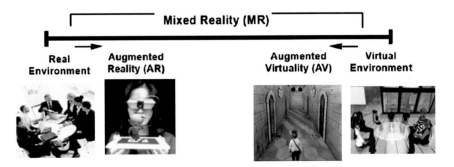

Fig. 1 Mixed reality

2 Related Work

In [7], Barandiaran et al. presented two approaches for markerless tracking, recursive tracking and tracking by detection. They used the Random Forest classifier for tracking by detection. They concluded that recursive tracking is very unstable and unsuitable for fast camera movement. On the other hand, tracking by detection is more robust and faster. A real-time camera tracking for markerless AR was introduced in [8]. It was based on planar tracking. They also suggested a method for re-detecting features to maintain registration of virtual objects, when the tracked features disappear. In [9], the authors proposed a markerless camera tracking approach for augmented reality on unprepared tabletop environments. They detected image features of the scene which are tracked frame-to-frame by computing optical flow. They used a synchronized multithreaded manner to achieve real-time performance.

A lot of work has been done in multiple target detection. Wagner et al. [10] presented a method for real-time pose estimation and tracking of multiple planar targets on mobile phones. They have split up the system into detection and tracking subsystems that work at different frame rates. This makes the system faster. In [11], a 3D tracking method that supports several hundreds of pre-trained planar targets was proposed. It was targeted at a digilog book application. Parallel processing was used as two threads, one for detection and the other for frame-to-frame tracking. The vocabulary tree-based searching method was used to cover large data sets. A method that is able to track several 3D objects for augmented reality was introduced in [12]. It detects new targets over multiple frames, but without real-time guarantees. Our work evaluates different classifiers for the task of markerless tracking within an augmented reality application.

3 Theoretical Background

Random Forest [2] is an ensemble classifier of randomized decision trees. The main principle behind ensemble methods is that a group of weak classifiers can come together to form a strong classifier. Training the Random Forest is based on randomness. A different subset of the training data features is randomly selected using bootstrap sampling, with replacement, to train each tree. A randomly selected subset of variables is used to split each node. When a new input is entered into the system, it is run down all of the trees. Class assignment is made by the number of votes from all of the trees and for regression the average of the results is used. The Forest chooses the classification having the most votes over all the trees. Random Forests can be used for classification or regression.

Extremely Randomized Trees [3] (also named Extra Trees) algorithm is considered an extension to Random Forest. In Extremely Randomized Trees, randomness goes one step further in the way splits are computed. Extremely

Randomized Trees algorithm is the same as Random Forest, but there are two differences. First, Extremely Randomized Trees don't apply the bagging procedure to construct a set of the training samples for each tree. The same input training set is used to train all trees. Second, Extremely Randomized Trees select a node split a randomly, whereas Random Forest finds the optimal split (by variable index and variable splitting value) among random subset of variables.

K Nearest Neighbor (KNN) [4] is a very simple and efficient algorithm that is used in many areas. It assumes that the data is in a feature space. It is a non-parametric as it does not make any assumptions on the underlying data distribution. It is also a lazy algorithm as it does not use the training data points to do any generalization. In other words, there is no explicit training phase or it is very minimal. This means the training phase is pretty fast. It stores all available cases and classifies new cases based on a similarity measure (e.g., distance functions). A case is classified by a majority vote of its neighbors, with the case being assigned to the class most common amongst its K nearest neighbors measured by a distance function.

Support Vector Machine (SVM, also named Support Vector Networks) [5] is based on the concept of decision planes that define decision boundaries. It finds and constructs the best hyperplane that separates cases of different classes in a multi-dimensional space. To find an optimal hyperplane, SVM employs an iterative training algorithm, which is used to minimize an error function. It supports both regression and classification tasks and can handle multiple continuous and categorical variables. It can perform linear classification as well as non-linear classification.

Bayes classifier [6] is a supervised learning algorithm that is based on the Bayes' theorem and is particularly suited with the high dimensional inputs. It assumes that the values of all features are independent of each other, given the class variable. It starts with calculating the prior probabilities of the data objects. To classify a new object, the likelihood of each object is calculated. According to the Bayes' theorem, a conditional probability (also called posterior probability) is formed by combining both the prior and the likelihood. The final classification decision is produced as the class which achieves the largest posterior probability.

4 Methodology

4.1 Camera Calibration

Camera calibration is the estimation of the internal (intrinsic) parameters of a camera. It is an important step in order to correct any optical distortion artifacts. Equation (1) shows the intrinsic matrix (also called camera matrix) containing 5 intrinsic parameters. cx and cy represent the coordinates of the principal point, which would be ideally at the image center. ∞x and ∞y represent the focal length and s is the skew coefficient. Intrinsic parameters are specific to a camera, so once calculated; it can be stored for future purposes. There are many approaches used to calibrate a camera. One of them is to take a number of images of a planar pattern by the targeted camera from different distances and points of view [13]. The pattern used in our case is a chessboard.

$$A = \begin{bmatrix} \alpha_x & s & c_x \\ 0 & \alpha_y & c_y \\ 0 & 0 & 1 \end{bmatrix} \qquad (1)$$

4.2 Camera Pose Estimation

Camera pose estimation is the problem of determining the geometric transformation between the world coordinate system and the camera coordinate system. This transformation is represented by a 3×4 matrix, consisting of a 3×3 rotation matrix and a translation vector as shown in Eq. (2). They are called the external (extrinsic) camera parameters. Pose estimation is obtained from 2D-3D correspondences using a solution to Perspective-n-Point (PnP) problem [14]. It is about the estimation of the pose of a calibrated camera, given n ($n \geq 3$) 3D reference points in the object framework and their corresponding 2D projections. This estimation should be accurate enough for correct augmentation.

$$[R|t] = \begin{bmatrix} R_{11} & R_{12} & R_{13} & t_x \\ R_{21} & R_{22} & R_{23} & t_y \\ R_{31} & R_{32} & R_{33} & t_z \end{bmatrix} \qquad (2)$$

The Projection of a 3D point in the world coordinate system to a 2D point in the camera coordinate system is obtained by Eq. (3). Both points are expressed in homogeneous coordinates according to a pinhole calibrated camera model.

$$\begin{bmatrix} u \\ v \\ 1 \end{bmatrix} = A.[R|t]. \begin{bmatrix} x_w \\ y_w \\ z_w \\ 1 \end{bmatrix} \qquad (3)$$

4.3 System Framework

The flowchart of the system framework is shown in Fig. 2. The first step is camera calibration which is done offline and only once. The next step which is also offline is building a model of the pattern to be tracked. This is known as tracking by detection or model-based tracking. There is another type of tracking called

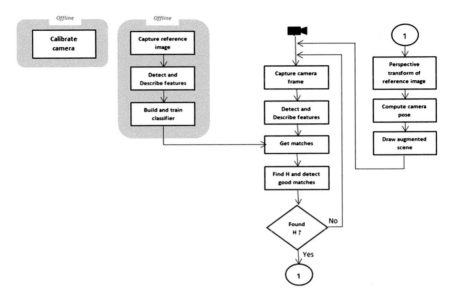

Fig. 2 Flowchart of augmented reality

recursive tracking that depends on frame-to-frame tracking such as [15]. It always faces the problem of accumulated error. Model-based tracking does not suffer from this problem. In our work, the model consists of descriptors of point features. A reference image of the pattern is taken. Then, point features are detected in this reference image using ORB (Oriented FAST and Rotated BRIEF) detector [16]. It is a fast and a robust feature detector. It uses FAST to detect keypoints and then it applies the Harris corner detector for selecting only the strongest features. It uses scale pyramid for scale invariance and moments for rotation invariance. The detected features are then described using FREAK (Fast Retina Keypoint) descriptor [17]. It computes a cascade of binary strings by efficiently comparing image intensities over a retinal sampling pattern. Different numbers of features are examined in our experiments. The classifier is build and trained with the descriptors of the extracted features.

For the tracking process, a camera is captured and the features are detected and described as in the model building step. Those features are matched against the features of the reference image using the classifier. The RANSAC (Random Sample Consensus) [18] algorithm is used to keep only the inliers and reject all outliers. From the remaining good matches, an accurate homography (H) is estimated. Then, the camera pose could be calculated as explained previously. Once the camera pose is estimated, the augmented scene is drawn.

4.4 Implementation Details

OpenCV [19] implementations for the feature detector, descriptor and all types of classifiers are used. All the experiments have been carried out on a 2.13 GHz/3 MB cache Intel® Core™ i3 330 M, 2 GB RAM, x64 Windows 7 machine. The code is compiled by the Microsoft Visual C++ 2010 Express with OpenCV-2.4.9. CMake-3.1.0-rc3-win32-x86 is used for enabling OpenGL with OpenCV. A 5MP camera of SAMSUNG GT-I9003 mobile phone is used with resolution of 2560 × 1920.

5 Experimental Results

In this section, the results of the performance of different classifiers are discussed. First, the resulting average reprojection error of our calibration task was 0.31 pixels. The classifiers used are Bayes classifier, SVM, KNN (k = 1), Extremely Randomized Trees and Random Forest with 18 trees. Random Forests with different number of trees are also tested, as shown later. The performance criteria used are the training time, the frame rate and the reprojection error. The pattern to be tracked and screen shots of our augmented reality system are shown in Fig. 3. The green box is the augmented object which is correctly aligned to the pattern in the scene.

As shown in Fig. 4, the training time of each classifier with different number of feature is reported. As it is known, KNN classifier doesn't have a training stage, so it is always equal to 0. For the rest classifiers, the training time increases whenever the number of features increases. Both Random Forest and Extremely Randomized Trees take less than 1 s training time for all the number of features. On the other hand, the training time of SVM and Bayes classifier highly increases. It approaches quarter a minute with 1000 features.

Fig. 3 Screen shots of augmented reality

Fig. 4 Training time in
seconds

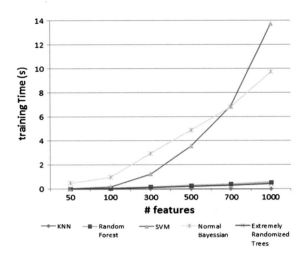

Fig. 5 Frame rate

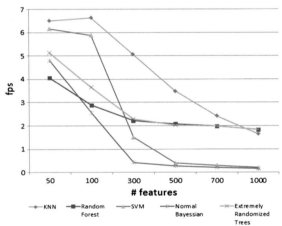

Figure 5 presents the frame rate obtained with different classifiers. It shows that the highest frame rated is obtained with the KNN classifier. Frame rates of SVM and Bayes classifier descend with higher number of features.

Average reprojection error results are shown in Fig. 6. In all cases, Random Forest was the worst classifier in terms of accuracy. KNN, SVM and Bayes classifier are very close to each other and are better than Bayes classifier. The best accuracy was obtained by KNN classifier with 100 features as an error of 1.2 pixels.

Radom Forest with different number of trees was tested with the same number of features used before. The training time for each number of trees is shown in Fig. 7. As expected, the training time increases whenever the number of trees and the number of features increase.

According to the frame rate, Random Forest with 40 trees was the slowest one as shown in Fig. 8. All numbers of trees descend to very close values of frame rate with 700 and 1000 features.

Fig. 6 Average reprojection error in pixels

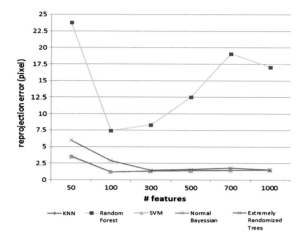

Fig. 7 Training time of random forest in seconds

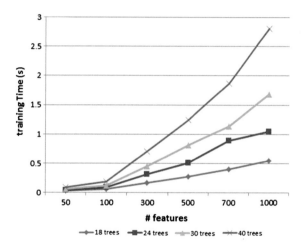

Fig. 8 Frame rate of random forest

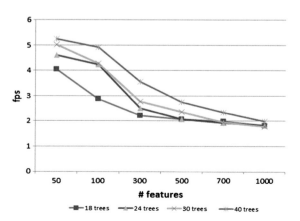

Fig. 9 Average reprojection
error of random forest in
pixels

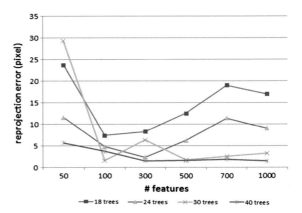

Figure 9 presents the average reprojection error of Random Forest with different number of trees. Random Forest with 18 trees was the worst one, whereas 40 trees was the best one in most cases. The best accuracy was obtained by 40 trees with 1000 features.

6 Conclusion

In this paper, we have presented an evaluation of different classifiers for markerless tracking. We have implemented a complete augmented reality system using machine learning techniques to solve the problem of keypoint matching. Five classifiers have been tested with different number of extracted features. The results showed that KNN outperforms other classifiers in terms of training time, frame rate and reprojection error. According to the number of extracted features, KNN with 100 features was the best combination. Moreover, Random Forests with different numbers of trees were included in the evaluation. Random Forest with 30 trees and 100 extracted features was better than the others. But also KNN outperform Random Forest with different number of trees. As a future work, the system is to be enhances in order to be real-time.

References

1. Martínez, H., Skournetou, D., Hyppölä, J., Laukkanen, S., Heikkilä, A.: Drivers and bottlenecks in the adoption of augmented reality applications. J. Multimedia Theory Appl. **2**(1), 27–44 (2014)
2. Breiman, L.: Random forests. Mach. Learn. **45**(1), 5–32 (2001)
3. Geurts, P., Ernst, D., Wehenkel, L.: Extremely randomized trees. Mach. Learn. **36**(1), 3–42 (2006)

4. Cover, T., Hart, P.: Nearest neighbor pattern classification. IEEE Trans. Inf. Theory **13**(1), 21–27 (1967)
5. Cortes, C., Vapnik, V.: Support vector networks. Mach. Learn. **20**(3), 273–297 (1995)
6. Rish, I.: An empirical study of the naive bayes classifier. In: International Joint Conference on Artificial Intelligence vol. 3(22), pp. 41–46 (2001)
7. Barandiaran, I., Paloc, C., Grana, M.: Real-time optical markerless tracking for augmented reality applications. J. Real-Time Image Proc. **5**(2), 129–138 (2010)
8. Lee, A., Lee, J., Lee, S., Choi, J.: Markerless augmented reality system based on planar object tracking. In: Proceeding of Frontiers of Computer Vision, pp. 1–4 (2011)
9. Lee, T., Hollerer, T.: Multithreaded hybrid feature tracking for markerless augmented reality. IEEE Trans. Visual Comput. Graphics **15**(3), 355–368 (2009)
10. Wagner, D., Schmalstieg, D., Bischof, H.: Multiple target detection and tracking with guaranteed framerates on mobile phones. In: Proceedings of the 8th IEEE International Symposium on Mixed and Augmented Reality (ISMAR'09), pp. 57–64 (2009)
11. Kim, K., Lepetit, V., Woo, W.: Scalable real-time planar targets tracking for digilog books. Visual Comput. **26**(6–8), 1145–1154 (2010)
12. Park, Y., Lepetit, V., Woo, W.: Multiple 3D object tracking for augmented reality. In: Proceedings of the 7th IEEE and ACM International Symposium on Mixed and Augmented Reality (ISMAR'08), pp. 117–120 (2008)
13. Zhang, Z.: A flexible new technique for camera calibration. IEEE Trans. Pattern Anal. Mach. Intell. **22**(11), 1330–1334 (2000)
14. Lepetit, V., Moreno-Noguer, F., Fua, P.: EPnP: an accurate O(n) solution to the pnp problem. Int. J. Comput. Vision **81**(2), 155–166 (2009)
15. Davison, A., Mayol, W., Murray, D.: Real-time localisation and mapping with wearable active vision. In: Proceeding 2nd IEEE and ACM International Symposium on Mixed and Augmented Reality (ISMAR'03), (2003)
16. Rublee, E., Rabaud, V., Konolige, K., Bradski, G.: ORB: an efficient alternative to SIFT or SURF. In: Proceedings of the IEEE International Conference on Computer Vision, pp. 2564–2571 (2011)
17. Alahi, A., Ortiz, R., Vandergheynst, P.: Freak: fast retina keypoint. In: Proceedings of the 2012 IEEE Conference on Computer Vision and Pattern Recognition (CVPR '12), pp. 510–517 (2012)
18. Fischler, M.A., Bolles, R.C.: Random sample consensus: a paradigm for model fitting with applications to image analysis and automated cartography. Commun. ACM **24**(6), 381–395 (1981)
19. Bradski, G.: The opencv library. Doctor Dobbs J. **25**(11), 120–126 (2000)

An Experimental Comparison Between Seven Classification Algorithms for Activity Recognition

Salwa O. Slim, Ayman Atia and Mostafa-Sami M. Mostafa

Abstract The daily activities recognition is one of the most important areas that attract the attention of researchers. Automatic classification of activities of daily living (ADL) can be used to promote healthier lifestyle, though it can be challenging when it comes to intellectual disability personals, the elderly, or children. Thus developing a technique to recognize activities with high quality is critical for such applications. In this work, seven algorithms are developed and evaluated for classification of everyday activities like climbing the stairs, drinking water, getting up from bed, pouring water, sitting down on a chair, standing up from a chair, and walking. Algorithms of concern are K-nearest Neighbor, Artificial Neural Network, and Naïve Bayes, Dynamic Time Warping, $1 recognizer, Support Vector Machine, and a novel classifier (D$1). We explore different algorithm activities with regard to recognizing everyday activities. We also present a technique based on $1 and DTW to enhance the recognition accuracy of ADL. Our result show that we can achieve up to 83 % accuracy for seven different activities.

Keywords ADL · K-nearest neighbor · Artificial neural network · Naïve Bayes · Dynamic time warping · $1 recognizer · Support vector machine

1 Introduction

The ability to recognize human's daily life activities provides good opportunities for application in many areas such as health care [1, 2]. This area is in continuous development and researchers encounter new challenges every day. The importance of improving the health care motivated a lot companies to develop different applications to track the activities of personals with health problems and try to help

S.O. Slim (✉) · A. Atia · M.-S.M. Mostafa
Computer Science Department, Faculty of Computers and Information,
Helwan University, Cairo Governorate, Egypt
e-mail: eng.s.osama@gmail.com

© Springer International Publishing Switzerland 2016   37
T. Gaber et al. (eds.), *The 1st International Conference on Advanced Intelligent System and Informatics (AISI2015), November 28–30, 2015, Beni Suef, Egypt,*
Advances in Intelligent Systems and Computing 407, DOI 10.1007/978-3-319-26690-9_4

them improve. In recent years, many researchers have turned to find an effective way to recognize daily physical activities using wearable sensors. In particular, techniques for machine learning have been used to recognize the daily activities such as walking and lying [3, 4].

Activities recognition is not limited to recognize ADL but can be used to identify the abnormal behavior of personals suffering from disabilities such as children with autism [5], or adults suffering from intellectual disability [6]. Another application used activities recognition to detect abnormal plan landing [7], or abnormal motor vehicles driving [8].

The purpose of this paper is presenting a comparative study between seven classifier algorithms applied to ADL dataset (6 known algorithms and a novel classifier is called D$1). The six known algorithms are Artificial Neural Network (ANN), K-Nearest Neighbor (KNN), Naïve Bayes (NB), $1 Recognize, Dynamic Time Warping (DTW), and Support Vector Machines (SVM). We analyze and present the accuracy of each algorithm, and change the size of training data to compare the algorithms in order to identify the algorithm with the highest accuracy. The contributions are (1) providing a comparison between seven classifier algorithms regarding accuracy, and (2) developing a technique based on filtering $1 training data using DTW classifier.

The rest of the paper is as follows: first we review related work in Sect. 2. Section 3 explains the classification methods. The experimental methodology and results are explained in Sect. 4. Section 5 discusses the values obtained and conclusions.

2   Related Work

There are many comparative studies on classification algorithms in different applications [9] and datasets like the study of Naïve Bayes (NB) and HMM on hand gesture [10], the paper [11] applied two different hand datasets to compare classification algorithms (k-Nearest Neighbor, and Naïve Bayes, Artificial Neural Network, Support Vector Machines).

Machine learning algorithms can be used in many fields like, classifying human physical activity from on-body accelerometers [12, 13] and mobile phone [14]. Lei Gao focuses on comparing and analyzing the following classifiers: the Naïve Bayes classifier, the Decision Tree classifier, the Artificial Neural Networks classifier, the K-Nearest Neighbor classifier and The Support Vector Machines classifier to achieve higher recognition accuracy with signal processing algorithms [15].

Jacob O compares between $1, DTW, and Rubine classifier on user-supplied gestures, and the authors said $1 Accuracy is 97 % with only 1 loaded template and 99\% accuracy with 3+ loaded templates [16] but the problem in that dataset is that it has very low variations, therefore it is easily results in high accuracy and this has been proven in the experiment section.

Barbara proposes a comparison and classification procedure for an automatic recognition system for simple ADLs (Climbing the stairs, Drinking from a glass, Getting up from the bed, Pouring water in a glass, Sitting down on a chair, Standing up from a chair, Walking) by using a hybrid distance metric based on the integration of Mahalanobis distance and Dynamic Time Warping (DTW) [17].

By using Jacob DTW and $1 classifier [16] and applying them on Barbara dataset we found that DTW and $1 accuracy are not as high as Jacob mentioned in his paper. Therefore, in the sake of increasing the accuracy of $1 and DTW by integrate between them (which called D$1). Experiment section shows applying Barbara's Model Builder on D$1.

3   Classification Methods

3.1   Naïve Bayes Classification

It is statistical Classifier [18]. It assumes the features are conditionally independent [19]. There are k classes $C_1, C_2 \ldots C_k$ with features $X = X_1, X_2 \ldots X_k$. The maximum posterior $P(C_k|X)$ is calculated in order to classify those classes. The posterior can be derived from Bayes' theorem.

$$P(C_k|X) = \frac{P(C_k)P(X|C_k)}{P(X)} \qquad (1)$$

or

$$Posterior = \frac{prior \times likelihood}{evidence} \qquad (2)$$

where

- $P(C_k|X)$ is the probability of the class C_k being X feature (*Posterior*)
- $P(C_k)$ is the probability of the class C_k (*prior*)
- $P(X|C_k)$ is the probability of X feature belong to the class C_k (*likelihood*)
- $P(X)$ is measurement of the probability of feature (*evidence*)

3.2   K-Nearest Neighbor Classification

K-NN is a supervised learning algorithm, where the user partitions the given data into a number of classes. There are two phases—training and testing phase—in the algorithm. The training phase contains the features and the classes' labels of the training data. New data are classified based on the K value. When K value is not equal one, we use voting criteria to classify the data [20]. The algorithm uses the distance measurement to assigns the object to the most frequently labeled class. The distance between test object and each training objects is calculated using the appropriate K value [21]. The K value effects on the accuracy of the algorithm.

3.3 Neural Network

The multilayer perception (MLP) is a feed-forward neural network [22]. The algorithm is used extensively in regression and classification. The neural network consists of three layers. Those layers are: input layer, hidden layers, and output layer. The hidden layer maybe consists of one or more layers. Each layer consists of one or more neurons, which directly linked with neurons from the previous and the next layer except for neurons in the input and output layers. Each neural network has several inputs and several outputs. The neural takes the output values from the neurons which exist in the previous layer as input and then passes the response to several neurons in the next layer. The values retrieved from the previous layer are summed up with certain weights and the bias term. The input for each neuron is calculated individually. The activation function is used to transform the summation value. That function isn't stable in different neurons.

3.4 Support Vector Machine

SVM is based on statistical learning theory. The goal of the classifier is determining the location of Hyperplane [23]. Hyperplane separates objects of different classes by drawing separating lines among the objects. SVM classifies the data by creating Hyperplanes in a multidimensional space.

3.5 Dynamic Time Warping

DTW is significant in time series classification because of its non-linear mapping capability. It finds an optimal match between two sequences by allowing a non-linear mapping of one sequence to another, and minimizing the distance between them.

Suppose we have two time series X and Y, of length n and m respectively, where: $X = x_1, x_2 \ldots x_i \ldots x_n$ and $Y = y_1, y_2 \ldots y_j \ldots y_m$ to align two sequences using DTW we make an n-by-m matrix where the (ith, jth) element of the matrix contains the distance $d(x_i, y_j)$ between the two points x_i and y_j (With Euclidean distance, $d(x_i, y_j) = (x_i - y_j)^2$). Each matrix element (i,j) corresponds to the alignment between the points x_i and y_j.

A warping path W is a neighboring set of matrix elements that determines a mapping between x and y. The kth element of W is defined as $w_k = (i,j)_k$ therefore we have: $W = w_1, w_2 \ldots w_k$ and $\max(m,n) \leq K < (m+n-1)$.

The warping path has several constraints [24]:

- Boundary conditions: the beginning and ending points of warping path should be the first and the last points of the path matrix.
- Continuity: the path advances one step at a time.
- Monotonicity: the points in W have to be monotonically spaced in time.

The experimental section display the comparison between re-sample and non-resample the points of the path.

3.6 $1 Gesture Recognition

The significant purpose for design $1 is recognize the speedy prototyping of gesture-based user interface [16]. $1 is an instance-based nearest neighbor classifier with a Euclidean Distance measurement. A user's gesture displays as a set of raw points, and we should determine which set of previously recorded template points it most closely matches. The raw points, whether those of training gesture or testing gesture, are initially processed the same. There are four steps to process raw points, those steps are the following:

1. Re-sample: it converts the gesture path to fixed points (N).
2. Rotate: rotate the points (results of previous step) to angle zero.
3. Scale: scale the gesture into reference square.
4. Translate: translate the gesture to the original position (0, 0).

Candidate points are then scored against each set of template points T_i over a series of angular adjustments to C that finds its optimal angular alignment to T_i.

3.7 D$1 Recognition

The significant purpose for developing D$1 is increasing the accuracy of $1 and DTW with low execution time. The accuracy of $1 is lower than DTW but the $1 speed is better than DTW speed. Therefore, we developed a novel technique depending on DTW and $1. The Fig. 1 displays the D$1 architecture.

In D$1 algorithm; first the training data was passed to DTW, then the DTW measures the distance between the testing data and each sample in the training data. The R sample(s) which has the lowest distance is selected to be used as training templates to the $1. Finally $1 recognizes the testing activity data.

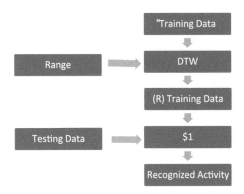

Fig. 1 D$1 architecture

4 Experimental Methodology and Results

To test the classifier algorithms we used ADL dataset [25]. The dataset are collected by a single tri-axial accelerometer attached to the right-wrist of the volunteer. It is composed of 700 trials for 8 motions which are collected from 16 volunteers (11 men and 5 women with age range from 19 to 83). It was previously used in the paper [17]. A C++ application was developed to apply seven classification algorithms on ADL dataset for the comparison study.

The common step in all algorithms is the re-sample phase (re-sample the points of the gesture to N points) in order to fix the number of points in the training and testing activity templates to easily compare between them except for DTW algorithm because it use the original points. The settings and parameters of the classifiers is an important aspect to take into consideration. Thus, we set N = 150 in the re-sample phase, k = 1 in the K-NN algorithm.

The Fig. 2 displays the accuracy of the algorithms. The accuracy is calculated by using 1 and 3 samples for each activity in the training data.

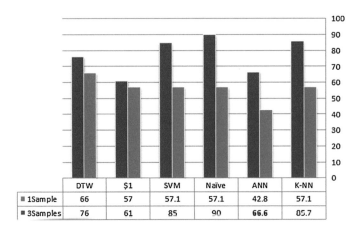

	DTW	$1	SVM	Naïve	ANN	K-NN
■ 1Sample	66	57	57.1	57.1	42.8	57.1
■ 3Samples	76	61	85	90	66.6	05.7

Fig. 2 A comparison between different training data sizes

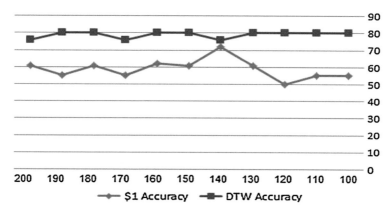

Fig. 3 DTW and $1 accuracy

DTW algorithm is the only one from the six implemented algorithms doesn't resample the gesture. Therefore we applied a resample phase on DTW and the error rate with using original point (non-Resample) is 24 % but in the resample phase is 20 %.

The Fig. 3 shows the effectiveness of DTW and $1 accuracy when we change the number of points in the resample phase. X-axis is the number of resample points (N = 100, 110, 120 ... 200).

$1 has the best accuracy when N = 140 but in DTW is the worst. We try to increase the accuracy of $1 by integrate $1 and DTW. Table 1 shows the comparisons between $1 and D$1.

The D$1 accuracy didn't increase when the training data size increases but it is better than $1. Table 2 displays the percentage of detecting each activity by

Table 1 A comparison between $1 and D$1

# Training data	$1 (%)	D$1 (%)
1	57	61
4	61	81
9	61	76
14	57	66.6

Table 2 The accuracy of activity recognition

Activity	DTW (%)	$1 (%)	D$1 (%)
Climbing the stairs	100	33.3	100
Drinking from a glass	66.6	100	100
Getting up from the bed	0	0	0
Pouring water in a glass	100	66.6	100
Sitting down on a chair	100	33.3	100
Standing up from a chair	66	100	100
Walking	100	100	100

Table 3 A comparison between Hindi-digits and unistroke

Dataset	Hindi-digits		Unistroke	
	DTW (%)	$1 (%)	DTW (%)	$1 (%)
1 sample	50	36	95	90
20 sample	62	45	100	100

Fig. 4 A comparison between recognition accuracy of D$1 and DTW with Mahalanobis distance

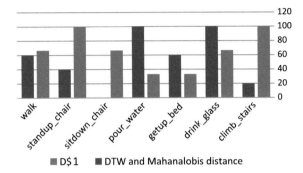

■ D$ 1   ■ DTW and Mahanalobis distance

calculating true positive (TP). TP is the number of occurrences of the activity that were correctly labeled.

Dataset properties affect the algorithm quality. As mentioned in the related work section, the authors of paper [16] applied Unistroke dataset on the $1 and DTW algorithms which are described in section of methodology. Jacob said that "$1 accuracy obtains 97 % accuracy with using 1 loaded template (as training) and 99 % accuracy with 3+ loaded templates. These results were nearly identical to DTW". Those algorithms are applied in Hindi-digits dataset collected from 10 volunteers and each one has 10 tries.

Table 3 shows a comparison between the results of DTW and $1 [16] applied on the Hindi-digits and Unistroke datasets on different training data size.

Barbara applied model builder and integrated Mahalanobis distance and DTW to recognize the activities on-line and they found DTW didn't improve the quality of the algorithm. We used Barbara's model builder and our D$1 as a classifier in on-line phase. The Fig. 4 shows a comparison between our experiment and Barbara experiment based on DTW.

5 Discussion and Conclusion

The size of the training data affects the accuracy and the speed of the algorithm. The Fig. 2 displays the accuracy of all algorithms when the training samples are 1 and 3 for each activity. The accuracy of all algorithms is low in 1 training sample but DTW is the best one. Using 3 training sample increased the algorithms accuracy but the least one is $1 and that opposite to what the paper [16] said.

In the sake of improving the accuracy of DTW by adding a resample phase, we found that DTW with resample phase is better than non-resample. Therefore we added the resample phase to DTW which is used in D$1. The Fig. 3 shows the impact of resample phase on the accuracy of DTW and $1. The best points number in $1 is N = 140 unlike DTW as it depends on the number of original points.

Table 1 shows the accuracy of $1 and D$1 in different training data sizes. D$1 is better than $1 in all cases. Table 2 illustrated that D$1 classifier is the best to detect all Activities with high percentage. That classifier used resample phase with N = 150 and R = 2. The size of training data is 28(4 for each activity) and testing data is 21(3 for each activity). All activities are detected with percentage 100 % using D$1 except for getting-up-from-the-bed activity. The percentage of getting-up-from-the-bed activity is zero in all classifier. Therefore, we exchanged the getting-up-from-the-bed activity with combing-hair activity and we found that the accuracy of D$1 is 100 % in all activities. Observing Fig. 4 and Table 2, we will find that getting-up-from-the-bed activity is detected but with low accuracy. D$1 accuracy is better than Barbara's model accuracy. D$1 detected all activities but Barbara's model can't detect sitdown_chair activity. D$1 confused between pour_water and drink_glass activities but Barbara's model recognized those activities with percentage 100 %. D$1 recognized climb_stairs and standup_chair activities with percentage 100 % but Barbara's model recognized those activities in the range between 20 to 40 %.

The training data size effects on the accuracy of any algorithm. DTW has shown the highest accuracy when using a 1 training data. The accuracy of $1 is lower than DTW in different training data size. We tried to integrate DTW and $1 to enhance the $1 accuracy by developing D$1. D$1 is better than $1. Using the resample phase in DTW enhance the accuracy of recognition. In the future work, we will integrate between $1 and K-NN and improve the resample phase by detecting curves and edges in the templates.

References

1. Parkk, J., Ermes, M., Korpipaa, P., Mantyjarvi, J., Peltola, J., Korhonen, I.: Activity classification using realistic data from wearable sensors. IEEE Trans. Inf. Technol. Biomed. **10**(1), 119–128 (2006)
2. Suryadevara, N.K., Mukhopadhyay, S.C.: ADLs recognition of an elderly person and wellness determination, Smart Sensors, Measurement and Instrumentation, vol. 14, pp. 111–137. Springer (2015)
3. Yang, J.Y., Wang, J.S., Chen, Y.P.: Using acceleration measurements for activity recognition: an effective learning algorithm for constructing neural classifiers. Pattern Recognit. Lett. **29**(16), 2213–2220 (2008) (Elsevier)
4. Atallah, L., Lo, B., King, R., Yang, G.Z.: Sensor positioning for activity recognition using wearable accelerometers. IEEE Trans. Biomed. Circuits Syst. **5**(4), 320-329 (2011)
5. Min, C.H., Tewfik, A.H.: Automatic characterization and detection of behavioral patterns using linear predictive coding of accelerometer sensor data. In: The Proceedings of the Annual

International Conference of the IEEE Engineering in Medicine and Biology Society (EMBC), pp. 220–223 (2010)

6. Zouba, N., Brémond, F., Thonnat, M., Anfosso, A., Pascual, E., Mallea, P., Guerin, O.: A computer system to monitor older adults at home: preliminary results. Gerontechnology J. **8**(3), 129–139 (2009)

7. Vaswani, N., Roy-Chowdhury, A.K., Chellappa, R.: "Shape Activity": a continuous-state HMM for moving/deforming shapes with application to abnormal activity detection. IEEE Trans. Image Process. **14**(10), 1603–1616 (2005)

8. Pradeep, T.R., Karthikamani, R.: Abnormal behavior detection in intelligent transport system for intelligent driving. Int. J. Softw. Hardw. Res. Eng. (IJSHRE). **2**(5) (2014)

9. Bhavsar, H., Ganatra, A.: A comparative study of training algorithms for supervised machine learning. Int. J. Soft Comput. Eng. (IJSCE). **2**(4), 2231–2307 (2012)

10. Avilés-Arriaga, H., Sucar-Succar, L., Mendoza-Durán, C., Pineda-Cortés, L.: A comparison of dynamic naive bayesian classifiers and hidden markov models for gesture recognition. J. Appl. Res. Technol. **9**(1), 81–102 (2011)

11. Trigueiros, P., Ribeiro, F., Reis, L.P.: A comparison of machine learning algorithms applied to hand gesture recognition. In: Proceedings of the 7th Iberian IEEE Conference on Information 273 Systems and Technologies (CISTI), pp. 1–6 (2012)

12. Mannini, A., Sabatini, A.: Machine learning methods for classifying human physical activity from on-body accelerometers. Sensors **10**(2), 1154–75 (2010)

13. Moiz, F., Natoo, P., Derakhshani, R., Leon-Salas, W.D.: A comparative study of classification methods for gesture recognition using a 3-axis accelerometer. In: The 2011 International Joint Conference on IEEE Neural Networks (IJCNN), pp. 2479–2486 (2011)

14. Ayu, M.A., Ismail, S.A., Abdul Marin, A.F., Mantoro, T.: A comparison study of classifier algorithms for mobile-phone's accelerometer based activity recognition. Elsevier-International Symposium on Robotics and Intelligent Sensors, vol. 41, pp. 224–229 (2012)

15. Gao, L., Bourke, A.K., Nelson, J.: A comparison of classifiers for activity recognition using multiple accelerometer-based sensors. In: 2012 IEEE 11th International Conference on Cybernetic Intelligent Systems (CIS), pp. 149–153 (2012)

16. Wobbrock, J.O., Wilson, A.D., Li, Y.: Gestures without libraries, toolkits or training: a $1 recognizer for user interface prototypes. In: Proceedings of the 20th Annual ACM Symposium on User Interface Software and Technology, pp. 159–168 (2007)

17. Bruno, B., Mastrogiovanni, F., Sgorbissa, A., Vernazza, T.: Analysis of human behavior recognition algorithms based on acceleration data. In: IEEE International Conference on Robotics and Automation (ICRA), pp. 1602–1607 (2013)

18. Duda, R.O., Hart, P.E.: Pattern Classification and Scene Analysis. Wiley, New York (1973)

19. Noaman, H.M., Elmougy, S., Ghoneim, A., Hamza, T.: Naive Bayes classifier based Arabic document categorization. In: The Proceedings of the 7th International Conference on Informatics and Systems (INFOS), pp. 1–5 (2010)

20. Förster, K., Monteleone, S., Calatroni, A., Roggen, D., Tröster, G.: Incremental KNN classifier exploiting correct-error teacher for activity recognition. In: 9th International Conference on IEEE-Machine Learning and Applications (ICMLA), pp. 445–450 (2010)

21. Dash, M., Liu, H.: Feature selection for classification. Intell. Data Anal. **1**(1), 131–156 (1997)

22. Friedman, N., Geiger, D., Goldszmidt, M.: Bayesian network classifiers. Mach. Learn. **29**, 131–163 (1997)

23. Burges, C.J.: A tutorial on support vector machines for pattern recognition. Data Mining Knowl. Discov. **2**(2), 121–167 (1998)

24. Jeong, Y.S., Jeong, M.K., Omitaomu, O.A.: Weighted dynamic time warping for time series classification. Pattern Recognit. **44**(9), 2231–2240 (2011)

25. ADL recognition with wrist-worm accelerometer data set. https://archiveics.uci.edu/ml/datasets/dataset+for+ADL+Recognition+with+Wrist-worm+Accelerometer

Machine Learning Based Classification Approach for Predicting Students Performance in Blended Learning

Celia González Nespereira, Esraa Elhariri, Nashwa El-Bendary, Ana Fernández Vilas and Rebeca P. Díaz Redondo

Abstract Nowadays, recognizing and predicting students learning achievement introduces a significant challenge, especially in blended learning environments, where online (web-based electronic interaction) and offline (direct face-to-face interaction in classrooms) learning are combined. This paper presents a Machine Learning (ML) based classification approach for students learning achievement behavior in Higher Education. In the proposed approach, Random Forests (RF) and Support Vector Machines (SVM) classification algorithms are being applied for developing prediction models in order to discover the underlying relationship between students past course interactions with Learning Management Systems (LMS) and their tendency to pass/fail. In this paper, we considered daily students interaction events, based on time series, with a number of Moodle LMS modules as the leading characteristics to observe students learning performance. The dataset used for experiments is constructed based on anonymized real data samples traced from web-log files of students access behavior concerning different modules in a Moodle online LMS throughout two academic years. Experimental results showed that the proposed RF classification system has outperformed the typical SVMs classification algorithm.

Keywords Learning analytics · Blended learning · Features extraction · Learning management systems (LMS) · Moodle · Random forests (RF) · Support vector machines (SVMS)

C.G. Nespereira · N. El-Bendary (✉) · A.F. Vilas · R.P.D. Redondo
I&C Laboratory AtlantTIC Research Center, University of Vigo, 36310 Vigo, Spain
e-mail: nashwa.elbendary@ieee.org

E. Elhariri
Faculty of Computers and Information, Fayoum University, Fayoum, Egypt

N. El-Bendary
Arab Academy for Science,Technology, and Maritime Transport, Cairo, Egypt

E. Elhariri · N. El-Bendary
Scientific Research Group in Egypt (SRGE), Cairo, Egypt
URL: http://www.egyptscience.net

© Springer International Publishing Switzerland 2016   47
T. Gaber et al. (eds.), *The 1st International Conference on Advanced Intelligent System and Informatics (AISI2015), November 28–30, 2015, Beni Suef, Egypt,*
Advances in Intelligent Systems and Computing 407, DOI 10.1007/978-3-319-26690-9_5

1 Introduction

Literature concerning blended learning proved that there are various definitions that describe it. However, among them, blended learning can be defined as a mixture of direct face-to-face instruction in classrooms, as a traditional learning media, and electronic online interaction [1, 2].

Moreover, Learning Management Systems (LMS) offer an extended variety of channels and modules to enable smooth information sharing and connection among course participants. Also, LMS facilitate a great range of activities in order to let instructors share information with students, develop content materials, design assignments and tests, take part in discussions, and enable collaborative learning with forums, chats, file storage areas, news services, etc. [3].

Recently, educational activities datasets, such as web-log files traced from Learning Management Systems (LMS), are increasingly used to analyze students learning behaviors via utilizing Learning Analytics (LA) and Educational Data Mining (EDM) techniques [4–7]. A variety of studies using web-log data to predict students performance has been conducted. Related studies have indicated that blended learning contributes to increased interactions between students and faculty. That is, the strengths and weaknesses of the online learning are complemented by the corresponding weaknesses and strengths of the traditional face-to-face learning and vice versa [1]. Observations obtained from analyzing such datasets provide different factors that can be employed for predicting students learning performance based on their access and interactions with the LMS. Accordingly, early-warnings could be triggered for students in order to help increasing their retention rate and facilitating their learning motivation.

For this study, we collected and analyzed real anonymized data samples traced from web-log files of Higher Education students access behavior, throughout two academic years, concerning different modules of a second year blended course in the Moodle online LMS of Telecommunication Engineering Degree at the University of Vigo. Different types of LMS modules and corresponding events have been considered for the dataset. In this experiment the daily number of interactions of each student with the Moodle LMS platform has been obtained. That has been achieved via recording the number of access events by each student for each of the course Moodle modules and store them in the database each day. The data recorded for the blended course has been obtained during a total duration of eighteen weeks, fourteen weeks between September and January, two weeks without classes because of the Christmas holidays, and two weeks of exams period.

The rest of this article is organized as follows. Section 2 presents the core concept of RF and SVMs classification algorithms. Section 3 describes the different phases of the proposed classification and prediction approach of students learning performance. Section 4 discusses the obtained experimental results. Finally, Sect. 5 presents conclusions and discusses future work.

2 Core Concepts

2.1 Support Vector Machines (SVMs)

One of the most widely used machine learning techniques for classification and regression of high dimensional datasets with excellent results is support vector machines [8–10].

Classification problem can be solved by SVMs via trying to find an optimal separating hyperplane between different classes [9–11]. The main aim of SVMs is that it seeks to maximize the margin around the separating hyperplane between a positive class from a negative class [8–10].

Any classification problem can be defined as follows: Given a training dataset with n samples $\{(x_1, y_1), (x_2, y_2), \ldots, (x_n, y_n)\}$, where x_i is a feature vector in a v-dimensional feature space and with labels $y_i \in -1, 1$ belonging to either of two linearly separable classes C_1 and C_2. Geometrically, Finding an optimal hyperplane with the maximal margin to separate two classes requires to solve the optimization problem, as shown in Eqs. (1) and (2).

$$maximize \sum_{i=1}^{n} \alpha_i - \frac{1}{2} \sum_{i,j=1}^{n} \alpha_i \alpha_j y_i y_j . K(x_i, x_j) \tag{1}$$

$$Subject - to : \sum_{i=1}^{n} \alpha_i y_i, 0 \leq \alpha_i \leq C \tag{2}$$

where, α_i is the weight assigned to the training sample x_i. If $\alpha_i > 0$, x_i is called a support vector. C is a regulation parameter used to trade-off the training accuracy and the model complexity. K is a kernel function, which is used to measure the similarity between two samples.

2.2 Random Forests (RF)

The Random Forests (RF) is one of the best classification and regression techniques. It has the ability to classify large dataset with excellent accuracy as it generates an ensemble of decision trees. Ensemble methods main principle is to group weak learners together to build a strong learner [12, 13]. The input is entered at the top and as it traverses down the tree, the original data is sampled in random, but with replacement into smaller and smaller sets. The class of sample is determined using random forests trees, which are of an arbitrary number. Algorithm 1 shows the steps of the Random Forests classification algorithm [12]:

1: Draw N_{tree} bootstrap samples from the original data
2: For each of the bootstrap samples, grow an un-pruned classification or regression tree
3: At each internal node, rather than choosing the best split among all predictors, randomly select m_{try} of the M predictors and determine the best split using only those predictors
4: Save tree as is, alongside those built thus far (No cost complexity pruning)
5: Predict new data by aggregating the predictions of the N_{tree} trees

Algorithm 1: Random Forests Classification Algorithm

The predictions of the Random Forests classifier are taken to be the majority votes of the predictions of all trees for classification. On the other hand, for regression, they are taken to be the average of the predictions of the all trees, as shown in Eq. (3) [12, 13]:

$$S = \frac{1}{K} \sum_{K=1}^{K} K^{th} \tag{3}$$

where S is a random forests prediction, Kth is a tree response, and k is the index runs over the individual trees in the forest.

3 The Proposed Classification System

In particular, the proposed framework is capable of classifying students status fail or pass. The proposed classification approach consists of three phases; namely, pre-processing, feature extraction, and classification phases. Fig. 1 shows the general structure of the proposed approach.

3.1 Pre-processing Phase

The dataset used for experiments in this research is constructed via anonymizing real data samples recorded from web-log files of bachelor degree students in Universidade de Vigo. The traced data in the dataset records access events concerning different modules of Moodle online LMS. Pre-processing phase can be achieved via applying the following steps:

1. segment users log file into 3 control points.
2. ignore non-student users log information.
3. ignore withdrawal students log information.
4. segment students into two groups based on grades (G1 from 0 to less than 6 and G2 above 6)

Fig. 1 Architecture of the proposed classification approach

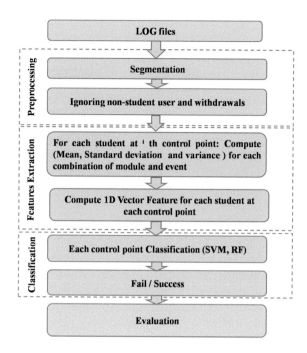

3.2 Feature Extraction Phase

Features extracted from the students log data involve mean, standard deviation for each module-event combination. During feature extraction phase, the proposed approach implements steps shown in Algorithm 2.

1: **for** Each control point **do**
2:    **for** Each student **do**
3:       **for** Each Module-Event combination **do**
4:          Compute features like (Mean, Standard deviation and variance)
5:       **end for**
6:       Form a 1D feature vector via combining all features of all combinations
7:    **end for**
8:    Ignore all Zeros features
9:    Sort (Mean, Standard deviation and variance), in order to measure frequency regardless event type
10: **end for**

Algorithm 2: Feature Extraction Algorithm

Figure 2 shows the position of each control/prediction point. Based on the specific characteristics of the course, the professor placed these three control points.

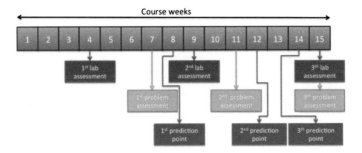

Fig. 2 3 Control Points

3.3 Classification Phase

At the final phase of the system, the proposed system applied both SVMs with RBF kernel function and RFs using 10 folds cross-validation for classification of student status (Fail or Pass). RF is a special bagging of decision TREEs and it has been recognized as a popular and accurate method to investigate the non-linear relations between variables to provide high accuracy compared to other methods [P16]. Also, RF has several advantages as it handles very large data sets, provides an estimate of variables importance, handles missing data, has a built-in method for balancing unbalanced data set, and it runs relatively fast [14].

SVMs inputs are training dataset feature vectors with their corresponding classes, while the outputs are the status of each student in the testing dataset. For RF, the inputs are number of trees, training dataset feature vectors and their corresponding classes, while the outputs are the status of each student in the testing dataset.

4 Experimental Results

For this study, real anonymized data samples, via web-log files of Higher Education students access behavior, has been collected and analyzed throughout two academic years. The traced data focused on students access behavior concerning different modules of a second year blended course in the Moodle online LMS of Telecommunication Engineering Degree at the University of Vigo. This is a blended course, given during fourteen weeks between September and January, plus two weeks without classes because of the Christmas holidays and two weeks of exams period (total eighteen weeks). We obtained the data from two academic years: 12/13 and 13/14. Figure 3 describes the pass/fail/withdraw distribution of students throughout the two academic years 2012/2013 and 2013/2014.

The data source is the Moodle database, where is stored all the information about what the students do in the platform, in events of different types. Different types of LMS modules and corresponding events are shown in Fig. 4, respectively.

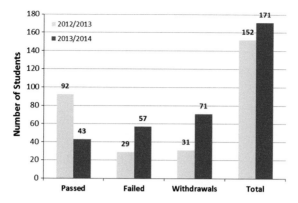

Fig. 3 Pass/fail/withdraw distribution of students

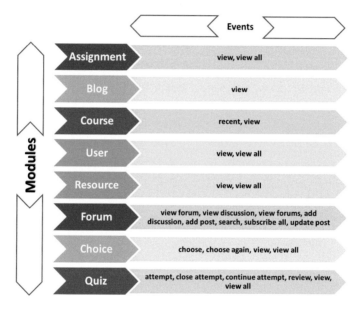

Fig. 4 Corresponding events for each module

In this experiment we want to obtain some features which describe the frequency of interactions of each student with the e-learning platform. In order to do this we compute features like (Mean, Standard deviation and variance) for each module-event combination.

- **Accuracy using SVMs:**

  1. For Control Point 1 ranges from 58 to 62.45 %.
  2. For Control Point 2 ranges from 66 to 71.05 %.
  3. For Control Point 3 ranges from 59.45 to 63.59 %.

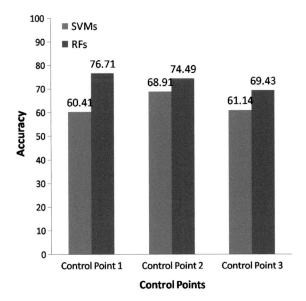

Fig. 5 The accuracy of different algorithms applied for classification using 10 folds cross-validation

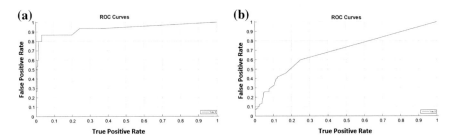

Fig. 6 ROC curve and area under curve for control point 1. **a** Accuracy $= 77.29\%$, AUC $= 0.9330$ using RFs. **b** Accuracy $= 61.57\%$, AUC $= 0.6910$ using SVMs

- **Accuracy using RF:**

1. For Control Point 1 ranges from 75 to 78.1 %.
2. For Control Point 2 ranges from 73 to 76.32 %.
3. For Control Point 3 ranges from 67.28 to 71.89 %.

Figure 5 shows classification accuracies obtained via applying the proposed classification approaches, with SVMs and RFs using 10-fold cross-validation (Note: mean of six times run).

Figure 6 shows ROC curve for the best feature using SVMs with RBF kernel and RFs for the first control point. Figure 7 shows ROC curve for the best feature using SVMs with RBF kernel and RFs for the second control point. Figure 8 shows ROC curve for the best feature using SVMs with RBF kernel and RFs for the third control

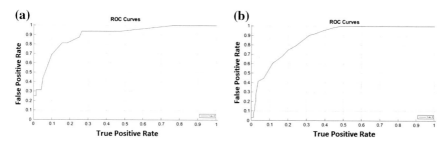

Fig. 7 ROC curve and area under curve for control point 2. **a** Accuracy = 75.88 %, AUC = 0.8902 using RFs. **b** Accuracy = 71.73 %, AUC = 0.8731 using SVMs

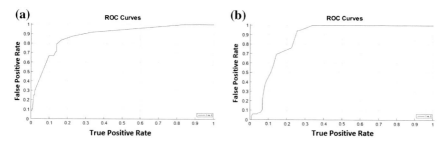

Fig. 8 ROC curve and area under curve for control point 3. **a** Accuracy = 71.89 %, AUC = 0.8113 using RFs. **b** Accuracy = 63.59 %, AUC = 0.8670 using SVMs

point. It's noticed that the obtained classification accuracy is not very high. This is because of the dataset on its current status has many unpredictable cases such as students with high access ratio and failed/withdrawn or students with low access ratio and succeeded. After investigating the whole dataset multiple times, we figured out that the best indicator, which will be considered for our future research work, would be the length-of-stay of each student module-event combination. But unfortunately, currently this is not available due to the problem of the Moodle system. The problem is that the logout event is only store when the student close the Moodle session. That means, if a student enter to the platform and then close the browser without close session, Moodle doesnt store any event. In the future, we will consider these modifications to Moodle system.

5 Conclusions and Future Work

In this article, a comparative study between Support Vector Machine (SVMs) and Random Forests (RF) classifiers for the problem of classifying student status has been presented. The proposed system has three main stages; pre-processing, feature extraction and status classification. The proposed system applies SVMs and RF using 10 folds cross-validation for classification purpose. Obtained experimental results showed that the proposed RF classification system has outperformed SVMs classification algorithm. For future work, performance enhancement will be achieved via considering log out information for each event in order to compute the event period. We are planning to estimate the length of the duration by tracking the time stamp of the user in all the Moodle events. Moreover, Optimization and features selection techniques will be considered, in order to improve system performance.

Acknowledgments This work is funded by: the European Regional Development Fund (ERDF) and the Galician Regional Government under agreement for funding the Atlantic Research Center for Information and Communication Technologies (AtlantTIC); the Spanish Government and the European Regional Development Fund (ERDF) under project TACTICA; and the Spanish Ministry of Economy and Competitiveness under the National Science Program (TEC2014-54335-C4-3-R). This work is also partially funded by the European Commission under the Erasmus Mundus GreenIT project (3772227-1-2012-ES-ERA MUNDUS-EMA21). The authors also thank GRADIANT for its computing support and the University of Vigo for its e-learning service for their support.

References

1. Kim, J.H., Park, Y., Song, J., Jo, I.-H., Predicting students' learning performance by using online behavior patterns in blended learning environments: comparison of two cases on linear and non-linear model. In: The International Conference on Educational Data Mining (EDM 2014), London, United Kingdom (2014)
2. Liebowitz, J., Frank, M.: Knowledge Management and E-learning. CRC Press (2010)
3. Romero, C., Ventura, S., Garca, E.: Data mining in course management systems: moodle case study and tutorial. Comput. Educ. **51**, 368–384 (2008)
4. Romero, C., Ventura, S.: Data mining in education. Wiley Interdiscip. Rev. Data Min. Knowl. Discov. **3**, 12–27 (2013)
5. Romero, C., Ventura, S.: Educational data mining: a review of the state of the art. IEEE Trans. Syst. Man Cybern. Part C: Appl. Rev. **40**(6), 601–618 (2010)
6. van Barneveld, A., Arnold, K.E., Campbell, J.P.: Analytics in higher education: establishing a common language. Educause Learn. Initiat. **1**, 1–11 (2012)
7. Bienkowski, M., Feng, M., Means, B.: Enhancing teaching and learning through educational data mining and learning analytics: an issue brief. Office of Educational Technology, U.S. Department of Education (2012)
8. Wu, Q., Zhou, D.-X.: Analysis of support vector machine classification. J. Comput. Anal. Appl. **8**, 99–119 (2006)
9. Zawbaa, H.M., El-Bendary, N., Hassanien, A.E., Abraham, A.: SVM-based soccer video summarization system. In: Proceedings of the Third IEEE World Congress on Nature and Biologically Inspired Computing (NaBIC2011), Salamanca, Spain, pp. 7–11 (2011)
10. Zawbaa, H.M., El-Bendary, N., Hassanien, A.E., Kim, T.H.: Machine learning-based soccer video summarization system. In: Proceedings of the Multimedia, Computer Graphics and Broadcasting FGIT-MulGraB (2), Jeju Island, Korea, vol. 263, pp. 19–28. Springer (2011)
11. Suralkar, S.R., Karode, A.H., Pawade, P.W.: Texture image classification using support vector machine. Int. J. Comput. Appl. Technol. **3**(1), 71–75 (2012)
12. Kulkarni, V.Y., Sinha, P.K.: Efficient learning of random forest classifier using disjoint partitioning approach. In: Proceedings of the World Congress on Engineering, vol. 2 (2013)
13. Bosch, A., Zisserman, A., Munoz, X.: Image classification using random forests and ferns. In: IEEE 11th International Conference on Computer Vision (2007)
14. Nisbet, R., Elder, J., Miner, G.: Handbook of Statistical Analysis and Data Mining Applications. Academic Press (2009)

Detection of Dead Stained Microscopic Cells Based on Color Intensity and Contrast

Taras Kotyk, Nilanjan Dey, Amira S. Ashour,
Cornelia Victoria Anghel Drugarin, Tarek Gaber,
Aboul Ella Hassanien and Vaclav Snasel

Abstract Apoptosis is an imperative constituent of various processes including proper progression and functioning of the immune system, embryonic development as well as chemical-induced cell death. Improper apoptosis is a reason in numerous human/animal's conditions involving ischemic damage, neurodegenerative diseases, autoimmune disorders and various types of cancer. An outstanding feature of neurodegenerative diseases is the loss of specific neuronal populations. Thus, the detection of the dead cells is a necessity. This paper proposes a novel algorithm to achieve the dead cells detection based on color intensity and contrast changes and aims for fully automatic apoptosis detection based on image analysis method. A stained cultures images using Caspase stain of albino rats hippocampus

T. Kotyk
Ivano-Frankivsk National Medical University, Ivano-Frankivsk, Ukraine
e-mail: taras1390@gmail.com

N. Dey
Department of CSE, Bengal College of Engineering and Technology, Durgapur, India
e-mail: neelanjan.dey@gmail.com

A.S. Ashour
Department of Electronics and Electrical Communications Engineering,
Faculty of Engineering, Tanta University, Tanta, Egypt
e-mail: amirasashour@yahoo.com

A.S. Ashour
CIT College, Taif University, Al Huwaya, Saudi Arabia

C.V. Anghel Drugarin
Department of Electrical and Informatics Engineering, Faculty of Engineering
and Management, Eftimie Murgu University of Resita, Resita, Romania
e-mail: c.anghel@uem.ro

T. Gaber (✉) · V. Snasel
IT4Innovation, VSB-TU of Ostrava, Ostrava, Czech Republic
e-mail: tmgaber@gmail.com
URL: http://www.egyptscience.net

V. Snasel
e-mail: vaclav.snasel@vsb.cz

© Springer International Publishing Switzerland 2016
T. Gaber et al. (eds.), *The 1st International Conference on Advanced Intelligent System and Informatics (AISI2015), November 28–30, 2015, Beni Suef, Egypt*,
Advances in Intelligent Systems and Computing 407, DOI 10.1007/978-3-319-26690-9_6

57

specimens using light microscope (total 21 images) were used to evaluate the system performance. The results proved that the proposed system is efficient as it achieved high accuracy (98.89 ± 0.76 %) and specificity (99.36 ± 0.63 %) and good mean sensitivity level of (72.34 ± 19.85 %).

Keywords Microscopic imaging · Dead cell detection · Apoptosis · Color intensity · Color contrast

1 Introduction

Widespread neurodegenerative diseases such as Huntington's disease, Alzheimer's disease, Parkinson's disease and front temporal dementias are prominently due to a philosophical loss of neurons [1]. Simultaneously, other diseases such as diabetes mellitus (DM) involves increased cell death [2]. Thus, cell death is considered the main cause of the progressive neurons loss and leads to various diseases. There are several modes of cell death including necrosis, apoptosis, and autophagy. For example, apoptosis occurs during the growth and aging as a homeostatic mechanism that preserve the cell populations in the tissues. In addition, it occurs as a defence mechanism as in the immune reactions or damaged cells by disease [3]. Typically, the mode of cell death referred to as apoptosis that defined by several techniques such as biochemistry, and imaging-based (exclusively upon morphological criteria). Apoptosis refers to the pathway resulting in cell shrinkage and karyorhexis, while necrosis illustrates finding dead cells in the histological specimen, regardless of the involved pathway [4, 5]. There are an extensive variety of stimuli and conditions that cause apoptosis such as both pathological and physiological. It arises in an orderly, step-wise behavior starting with a series of biochemical changes that causes characteristic alterations in the cell prior to its death. The apoptosis is characterized by two foremost markers: (i) the morphological features such as reduction in cell size, nuclear fragmentation and chromatin condensation resulting in apoptotic bodies, (ii) the DNA cleavage. Thus, its process

T. Gaber
Faculty of Computers and Informatics, Suez Canal University, Ismailia, Egypt

A.E. Hassanien
Faculty of Computers and Information, Cairo University, Cairo, Egypt
e-mail: aboitcairo@gmail.com
URL: http://www.egyptscience.net

V. Snasel
FEECS, Department of Computer Science, VSB-TU of Ostrava, Ostrava, Czech Republic

T. Gaber · A.E. Hassanien
Scientific Research Group in Egypt (SRGE), Cairo, Egypt

consists of cell shrinking, DNA degradation, membrane blebbing, and apoptotic bodies' formation [6]. Consequently, apoptosis detection (programmed cell death) is essential to recognize the underlying mechanism of the cell progress that has a significant role in disease progression and diagnosis.

Since, microscopy imaging techniques are occupied by the researcher and scientists to facilitate the extensive range visualization of biological features and processes in the cell structure. Light and electron microscopy have recognized a variety of morphological changes that arise during apoptosis [7]. Different types of stains can be used to prepare the specimen under investigation. Originally, the apoptosis can be detected using various assays that include absorbance measurements/colorimetric stains to measure the apoptotic level. These procedures involve gathering the specimen at each time-instant for measurement and cell monitoring. Thus, this process is not practicable and multiple samples may be required, as manual assessment is time consuming, costly and subjective. This necessitates the use of automated systems based image analysis that used in various medical and biological applications [8–11]. Therefore, image analysis is essential for the apoptosis detection as cell shrinkage can be visible by light microscopy [12] to be detected. Cell shrinkage refers to cells with reduced size that causes the cytoplasm to be dense and more tightly packed organelles. These characteristic features can be used during image analysis for the apoptosis detection, thus requires staining as the apoptosis involves single cells or small clusters of cells. On the histological examination, the apoptosis is often energy-dependent process that engages the activation of a group of cysteine proteases called "Caspases" [13]. Thus, Caspase stain for light microscopic specimens can be used as the apoptotic cell appears as a round or oval mass with dense brown nuclear chromatin fragments and dark eosinophilic cytoplasm.

In this paper, a full automatic apoptosis detection method is proposed using image analysis scheme based on color intensity and contrast change that result due to the apoptosis existence. Then, albino rat hippocampus specimens are used for system performance evaluation. The structure of the remaining sections is as follows: Sect. 2 introduces a literature review, while Sect. 3 includes the proposed study algorithms used methodology. Section 4 includes the results and discussion, followed by the conclusion in Sect. 5.

2   Literature Survey

Research studies continue to focus on the cell cycle machinery analysis that control cell cycle arrest and apoptosis. Thus, the domain of apoptosis research has been moving forward its early detection in cells, tissues or organs. A number of methods have been improved to study apoptosis in individual cells. There are extensive varieties of assay and imaging techniques are available for the apoptosis investigation. Each assay has advantages/disadvantages that may make it suitable for one application but unsuitable for another application.

Hacker [7] identified the different morphological changes that arise during apoptosis using Light and electron microscopy. Consistent with Galluzzi et al. [14], cell death can be classified into four diverse types based upon morphological characteristics: apoptosis (Type 1), autophagy (Type 2), necrosis (oncosis, Type 3), and mitotic catastrophe. Huh et al. [15] suggested that the apoptosis detection can be performed using the mitosis (cell division) detection method. In [16], the authors proposed a cell area detection method based on the optical principle of phase-contrast microscopy for adherent cells that detects apoptosis candidates. This technique engaged changes in the cell morphology and image intensity during apoptosis. The results proved an achieved accuracy of about 90 % in terms of average precision and recall.

Generally, dead cells cannot be distinguished from living cells based merely on size as addressed in several parametric analytic methods including speckle fluctuation in time-lapse images [17]. In 2012 [18], an image analysis method was presented to automatically extract deterministic information of a single cell. The authors used an Ultrahigh-resolution optical coherence tomography (UR-OCT) system to image single-cell basal cell carcinoma (BCC) in three dimensions and differentiate between live and dead BCC cells using both the morphological recognition as well as parametric analysis for the signal average, cellular density, average dynamic range, and cell volume. Based on these parameters, a significant of P-value less than 0.05 was achieved. In the same year, Scherf et al. [19] introduced a method to quantify changes in colonies morphology of the embryonic stem cells under various environmental conditions. While, Adiga and colleagues [20] classified the macrophages infection state via segmentation and morphological quantification in amplitude contrast bright field images. A fully automated system of high-throughput bright field microscopy data to track single-cell genealogies was proposed in [21]. The robustness of the cell detection with fast computation time supported the analysis. The authors used a manually inspected test set of bright field images besides statistically on the full time lapse experiment to evaluate their method. Their results proved that this method surpassed the standard approach with an overall cell detection accuracy of at least 82 %. A multilevel threshold based gray scale image approach was used for image analysis as in [22].

This survey established that using morphological changes based on the cells color intensity and contrast is mainly used for dead cell detection. Consequently, the current proposed system designs a novel algorithm based color intensity and contrast for apoptosis detection as follows.

3 Proposed System

This work proposes a narrative algorithm to detect the dead cells based image processing and analysis to extract and quantify cells and patterns. As, image analysis includes the conversion of features and objects in the image into quantitative information about these measured features and attributes.

Fig. 1 Proposed system for
the apoptosis detection

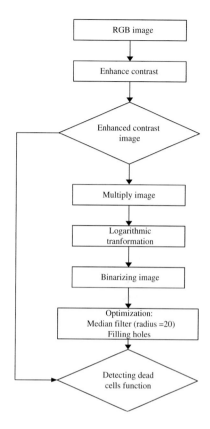

The proposed system, as illustrated in Fig. 1, consists of 6 stages. In the first
stage, the data were collected using the light microscope to capture the specimens'
colored images. The proposed algorithm is used built-in functions and plugging of
ImageJ v. 149v [http://imagej.nih.gov] as well as commands and macro language
functions. From this stage, digital colored images (RGB) were collected. In the
second stage, these images were processed to eliminate defects produced during
image acquisition such as noise and uneven illumination as well as to enhance the
contrast to emphasize the cells features of interest in the image for subsequent
analysis. In the third stage, the enhanced contrast image was then used to produce
standard algorithm as each image has different mean value, while the maximum
mean is 255 pixels with the highest of 8-bit image. To accomplish this, the following steps were done: (1) divide 255 to the mean gray value of image that results
in some coefficient, (2) multiply the image to these coefficients to increase all values
of pixels on image. In the fourth stage, further enhancement was performed using a
logarithmic transformation to remove the unwanted darkness from the image.
Applying the logarithmic transform will increase the brightness with preserving the
darkest objects (cells) with their darkness. These remaining dark objects are indicating dead cells. Conversely, using the exponential transform will decrease the

brightness, which is not the required case; therefore the logarithmic transformation was used. The fifth stage is for image binarizing, where the images are to be converted into binary before performing the object analysis, which leads to binary objects with rough boundaries. This step is required as the analysis particles algorithm [http://imagej.nih.gov] works only with this type of images. This stage is followed by optimization image step using median filter with radius = 20 used to smooth the boundaries for correct identification as it reduces the noise and filling holes. The output of this step is an area and other unwanted particles. Thus, threshold values were obtained experimentally to extract the dead cells (objects of interest) without false detection parts of the cells cytoplasm as dead cells, where the dark black spots indicate dead cells. Finally, the last stage is concerned with the "Detecting dead cells function" as some other descriptors were used.

The used descriptors in the last stage are (i) the solidity that indicates the ratio between the area of the object and the convex hull area. The solidity is required to prevent detection parts of cell cytoplasm with shape like arc. Since, the investigated cells are not oblong but more round, thus a roundness ratio between area and major axis with threshold value >0.4 is to be included as one of determination rule into the detection dead cells function. (ii) The circularity that is the ratio between the area and perimeter of object is included to detect objects with more right boundaries and (iii) the roundness. Consequently, in this work if the object is characterized by area > 10,000 pixels^2, solidity > 0.5, roundness > 0.4, circularity > 0.1 it can be accepted as dead cell. This threshold values are needed to prevent accepting parts of cells cytoplasm like dead cell.

To take the decision that this object is actually represents a dead cell; the algorithm restores boundaries of object, which is analyzed, on enhanced image and measures values of red and blue. If the red value is greater than the blue one, the algorithm considers this object as a dead cell. As if red value is greater than the blue value indicates that the object has more red colors, where dead cells are brown, thus this color is based on red. Consequently, the proposed system algorithm to detect the dead cells can be introduced as follows.

Algorithm 1: Dead Cells Detection

Input: *the original RGB image*
Procedure:
    1. Read *the original RGB image*   **2. Enhance contrast** *(saturated pixels = 0.35%)*
    3. Determine *mean grey value, calculate coefficient as 255 / (mean grey value)*
    4. Multiply *image on coefficient*   **5. Perform** *logarithmic transformation*
    6. Convert *into binary image, use the median filter (with radius 20) and filling*
    7. Determine *the dead cells*
Output: *marked dead cells*

Figure 2 illustrates through example the proposed system algorithm steps for dead cells detection. As shown in Fig. 2a, a RGB image that required enhance contrasting using automated contrasting default value of saturated pixels is 0.35 % that is performed in Fig. 2b. The mean grey value (MGV) of the enhanced image is

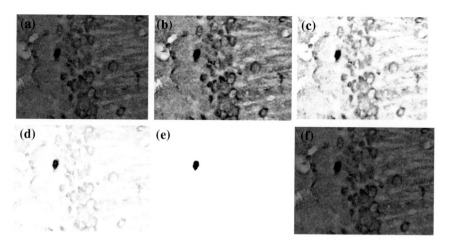

Fig. 2 Example illustrates the proposed system algorithm steps

calculated and determined the coefficients for multiplying. Theses coefficient = 255/MGV. Multiply the image under concern by the obtained coefficient results in the image shown in Fig. 2c. Afterwards, the logarithmic transformation is carried out in Fig. 2d, then Fig. 2e shows the binarized/optimized image that includes the results of using the median filter with radius = 20 pixels and filling the holes. After these manipulations, detecting dead cells function starts to analyze the objects on the binarized optimized image as obtained in Fig. 2f.

4   Result and Discussion

This study is performed on a dataset of albino rat hippocampus using caspase stain for light microscopic images captured in the Anatomy dept. laboratory, faculty of medicine, Tanta University, Egypt. The tested specimen images are obtained using a light microscope at magnification X1000, taken from a pool of 21 albino rats' images. To evaluate the performance of the proposed system, its detected dead cells are compared to manually marked dead cells by an expert anatomical physician. Figure 3a, c demonstrates free-of-dead cells specimen images as the cells remain blue without being affected by the Caspase stain. Figure 3b, d shows the results of the proposed system when the images in Fig. 3a, c were used. This output shows no marked dead cells exist. On the other hand, Fig. 3e, g shows images that contain dead cells affected by the Caspase stain and became brown. The proposed system output is shown in Fig. 3f, h; respectively, which contains marked red dead cells Dead cells on the binarized output images.

   Another example that compared the proposed system output to the manually marked dead cells specimens is shown in Fig. 4. In this figure, the red marks were

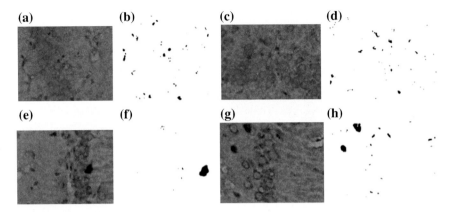

Fig. 3 The original stained images as analyzed using the proposed system: **a, c** original specimen images free-of-dead cells, **b, d** the binarized output of images in (**a, c**); respectively, **e, g** specimens contain dead cells, **f, h** the binarized output of the specimens (**e, g**), respectively

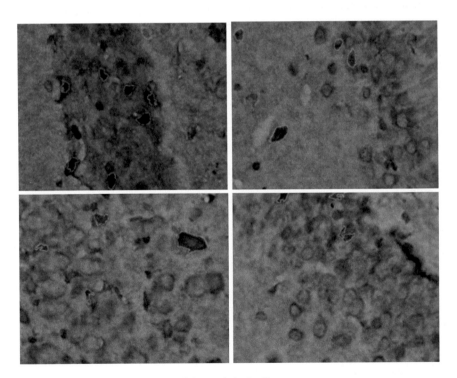

Fig. 4 Marked automated and manual detected dead cells

(a)   **(b)**   **(c)**

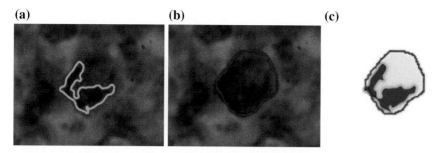

Fig. 5 True and false detected areas: **a** detected area by automated system with *green boundary* mark, **b** detected area manually with *red boundary*, **c** the true and false areas

done manually, while the green marks were done automatically using the proposed system that detected the dead cells. It is clear that, both manual and automatic detected cells marks are matching, which proves the performance of the proposed system.

Figure 5, illustrates the detected areas using both the automated proposed system and the manually detected area after zooming on the part of the specimen. In this figure, the green marked area represents the detected boundaries by the automated system, while the red boundaries represent the manually detected boundaries. In Fig. 5c, the colors, blue, yellow, black and white denote to the true positive, false negative, false positive and true negative, respectively. It can be seen from this figure that the false positive has small area which indicates the accurate detection performance of the proposed system. This was also proved as the area of the true negative area is accurately indicated in all the specimen regions that did not contain any dead cells.

To evaluate the performance of the automated proposed system, several standard performance metrics are applied. Specifically, True Positive (TP), False Positive (FP), True Negative (TN), and False Negative (FN) were used for the evaluation. Where, TP is the total area of dead cells that are detected correctly, the FP is total area of not dead cells which are detected as dead cells, TN is the total area of not dead cells that are detected by the proposed system as not dead cell, and the FN is defined as the total area of dead cells cases that are detected incorrectly.

These quantities (i.e. metrics) were then used to compute some metric measures such as: (i) the *sensitivity*: known as the probability that proposed system will give a dead cell mark from the used database, (ii) *specificity* is calculated as: TN/(TN + FP), (iii) *Positive Predictive Value (PPV)* is denoted as TP/(TP + FP), (iv) *Negative Predictive Value* (NPV) is calculated as TN/(TN + FN). Table 1, illustrates the mean and standard deviation (SD) of these performance metrics when using the automated proposed system that goes along with the results obtained in Fig. 5.

From Table 1, it can be noticed that the high mean specificity value of 99.36 ± 0.63 % indicates an ideal test scenario as well the proposed automated

Mean sensitivity analysis metrics	Mean ± SD (%)
Accuracy (ACC)	**98.89 ± 0.76**
Sensitivity (TPR)	72.34 ± 19.85
Specificity (SPC)	**99.36 ± 0.63**
Precision (PPV)	67.81 ± 21.71
Negative predictive value (NPV)	**99.52 ± 0.40**
False positive rate (FPR)	0.64 ± 0.63
False negative rate (FNR)	27.66 ± 19.85
False discovery rate (FDR)	32.19 ± 21.71
Balanced accuracy	85.85 ± 9.92

Table 1 Statistical measures of the performance (no. cases = 21)

system achieves high mean accuracy of 98.89 ± 0.76 %. These results prove the efficiency of the proposed system for dead cells (Apoptois) detection.

Consequently, the proposed system provides two advantages over traditional and manual methods of analysis are: (i) accurate scheme to extract information from image data, (ii) can be extended with huge number of microscopy images to facilitate the collection of large amounts of data for statistical analysis and providing dead cells markers.

5   Conclusion

In this current study, a fully automated system is proposedto detect apoptosis using image analysis methods. A specimens stained by Caspase stain of the albino rats hippocampus cultures images (21 images), taken by light microscope were used to evaluate the system performance. Comparing the marked detected dead cells with those marked manually indicates the high accuracy of the proposed system. Along with this comparison, the results prove efficient performance of the proposed system that gains high mean accuracy of 98.89 ± 0.76 % and mean specificity 99.36 ± 0.63 % besides good mean sensitivity level of 72.34 ± 19.85 %. Thus, the image processing and analysis could be helpful to automatically detect dead cells and avoid the time consuming and the error prone of the manual method.

Acknowledgment This paper has been elaborated in the framework of the project "New creative teams in priorities of scientific research", reg. no. CZ.1.07/2.3.00/30.0055, supported by Operational Programme Education for Competitiveness and co-financed by the European Social Fund and the state budget of the Czech Republic and supported by the IT4Innovations Centre of Excellence project (CZ.1.05/1.1.00/02.0070), funded by the European Regional Development Fund and the national budget of the Czech Republic via the Research and Development for Innovations Operational Programme and by Project SP2015/146 "Parallel processing of Big data 2" of the Student Grant System, VSB—Technical University of Ostrava.

References

1. Adamec, E., Yang, F., Cole, G., Nixon, R.: Multiple-label immunocytochemistry for the evaluation of nature of q cell death in experimental models of neurodegeneration. Brain Res. Protoc. **7**, 193–202 (2001)
2. Allen, D.A., Yaqoob, M.M., Harwood, S.M.: Mechanisms of high glucoseinduced apoptosis and its relationship to diabetic complications. J. Nutr. Biochem. **16**, 705–713 (2005)
3. Norbury, C.J., Hickson, I.D.: Cellular responses to DNA damage. Annu. Rev. Pharmacol. Toxicol. **41**, 367–401 (2001)
4. Levin, S., Bucci, T.J., Cohen, S.M., Fix, A.S., Hardisty, J.F., LeGrand, E.K., Maronpot, R.R., Trump, B.F.: The nomenclature of cell death: recommendations of an ad hoc Committee of the Society of Toxicologic Pathologists. Toxicol. Pathol. **27**, 484–490 (1999)
5. Lechuga-Sancho, A.M., Arroba, A.I., Frago, L.M., et al.: Activation of the intrinsic cell death pathway, increased apoptosis and modulation of astrocytes in the cerebellum of diabetic rats. Neurobiol. Dis. **23**, 290–299 (2006)
6. Fuchs, Y., Steller, H.: Programmed cell death in animal development and disease. Cell **147**(4), 742–758 (2011)
7. Hacker, G.: The morphology of apoptosis. Cell Tissue Res. **301**, 5–17 (2000)
8. Dey, N., Das, A.: Shape and size analysis of mineral grains from photomicrographs using Harris corner detection. Int. J. Adv. Eng. Sci. Technol. (IJAEST) **2**(2) (2012)
9. Dey, N., Nandi, B., Roy, A.B., Biswas, D., Das, A., Chaudhuri, S.S.: Analysis of blood cell smears using stationary wavelet transform and Harris corner detection. Published by Recent Advances in Computer Vision and Image Processing: Ethodologies and Applications, pp. 357–370 (2013)
10. Dey, N., Pal, M., Das, A.: A session based watermarking technique within the NROI of retinal fundus images for authencation using DWT, spread spectrum and Harris corner detection. Int. J. Mod. Eng. Res. **2**(3), 749–757 (2012)
11. Dey, N., Roy, A.B., Das, A.: Detection and measurement of bimalleolar fractures using Harris corner. In: ICACCI-2012, pp. 3–5. Chennai, India (2012)
12. Kerr, J.F., Wyllie, A.H., Currie, A.R.: Apoptosis: a basic biological phenomenon with wide-ranging implications in tissue kinetics. Br. J. Cancer **6**, 239–257 (1972)
13. Elmore, S.: Apoptosis: a review of programmed cell death. Toxicol. Pathol. **35**(4), 495–516 (2007)
14. Galluzzi, L., Maiuri, M.C., Vitale, I., Zischka, H., Castedo, M., Zitvogel, L., Kroemer, G.: Cell death modalities: classifications and pathophysiological implications. Cell Death Differ. **14**, 1237–1243 (2007)
15. Huh, S., Ker, D.F., Bise, R., Chen, M., Kanade, T.: Automated mitosis detection of stem cell populations in phasecontrast microscopy images. IEEE Trans. Med. Imaging **30**(3), 586–596 (2011)
16. Huh, S., Su, H., Kanade, T.: Apoptosis detection for adherent cell populations in time-lapse phase-contrast microscopy images. In: Medical Image Computing and Computer-Assisted Intervention–MICCAI 2012, pp. 331–339. Springer, Berlin, Heidelberg (2012)
17. Dunkers, J.P., Lee, Y.J., Chatterjee, K.: Single cell viability measurements in 3D scaffolds using in situ label free imaging by optical coherence microscopy. Biomaterials **33**(7), 2119–2126 (2012)
18. Cheng, N., Hsieh, T., Wang, Y., Lai, C., Chang, C., et al.: Cell death detection by quantitative three-dimensional single-cell tomography. Biomed. Opt. Express **3**(9), 2111–2120 (2012)
19. Scherf, N., Herberg, M., Thierbach, K., et al.: Imaging, quantification and visualization of spatio-temporal patterning in mESC colonies under different culture conditions. Bioinformatics **28**(18), i556–i561 (2012)
20. Adiga, U., Taylor, D., Bell, B., et al.: Automated analysis and classification of infected macrophages using bright-field amplitude contrast data. J. Biomol. Screen. **17**(3), 401–408 (2012)

21. Buggenthin, F., Marr, C., Schwarzfischer, M., et al.: An automatic method for robust and fast cell detection in bright field images from high-throughput microscopy. BMC Bioinform. **14**, 297 (2013)
22. Samanta, S., Dey, N., Das, P., Acharjee, S., Chaudhuri, S.S.: Multilevel threshold based gray scale image segmentation using cuckoo search. In: International Conference on Emerging Trends in Electrical, Communication and Information Technologies—ICECIT, pp. 27–34 (2012)

Digital Opportunities for 1st Year University Students' Educational Support and Motivational Enhancement

Dace Ratniece, Sarma Cakula, Kristaps Kapenieks
and Viktors Zagorskis

Abstract Young people face a real and increasing difficulty in finding a decent job with each day. Three additional combining factors are worsening the youth employment crisis even further, causing challenges while transiting to decent jobs. Namely (i) numbers of discouraged young people, who are neither in education nor employment or training (NEETs) are increasing. In tertiary education, (ii) unemployment among university graduates, in general, are rising. Therefore, (iii) potential NEET group students, especially in the 1st year, who, apart from reduced study fees, require extra motivation and moral support from educators. The aim of the paper is to find out how to solve the dropout crisis applying digital opportunities for 1st-year universities students' by an educational support and motivation enhancement. Educational institutions should combine the traditional forms of study and e-learning. However, direct contact between the teacher and the student plays very important role in acquiring a quality education, and should not be left out.

Keywords Digital opportunities · 1st year students · NEET group · Course management system MOODLE · Quality of educations

D. Ratniece (✉)
Distance Education Study Centre of Riga Technical University, Latvia,
University of Liepaja, Liepaja, Latvia
e-mail: ratniece.dace@gmail.com

S. Cakula
Faculty of Engineering, Vidzeme University of Applied Sciences, Valmiera, Latvia
e-mail: sarma.cakula@va.lv

K. Kapenieks
Distance Education Study Centre, Riga Technical University, Riga, Latvia
e-mail: kristaps.kapenieks@gmail.com

V. Zagorskis
Riga Technical University, Riga, Latvia
e-mail: viktors.zagorskis@rtu.lv

© Springer International Publishing Switzerland 2016
T. Gaber et al. (eds.), *The 1st International Conference on Advanced Intelligent System and Informatics (AISI2015), November 28–30, 2015, Beni Suef, Egypt,*
Advances in Intelligent Systems and Computing 407, DOI 10.1007/978-3-319-26690-9_7

69

1 Introduction

In the academic year of 2012/2013, 94,474 students studied at the Latvian higher education institutions. Compared to previous years the number of students was decreased by 6 %. The number of students in Latvia's universities and colleges continued to drop to 89,671 students in the 2013/2014 academic year.

Over the last five years, the number of students has decreased due to demographic factors, the economic crisis, and the general economic immigration. A large number of education reforms have been halted, including the higher education reform, that have been stipulated reduction of the number of universities, consolidation of institutions, and modified education programs. By the year of 2020 Latvian National Action Plan is aimed to ensure that 34 %–36 % of the population of Latvia higher education have to be acquired in the age group of 30–34. It is stated that Ministry of Education and Science handles the implementation of the following key policies and measures to enhance the access to higher education among the population of Latvia:

- Modernization of the study material and technical resources;
- Increasing of resource usage efficiency;
- Equal access opportunities to higher education resources;
- Improvement of quality of Research and Studies;
- Attraction of foreign students [1].

European Commission's report "Opening up Education: innovative teaching and learning for everyone through the use of new technologies and open educational resources" /COM/2013/0654/ highlights that nowadays students need more personalized contact with teachers. Greater cooperation and better links between formal education[1] and non-formal education[2] can be implemented by using digital technologies. Use of new technologies and open educational resources can also help to reduce the costs of educational institutions and students. Therefore, the following groups of individuals should be granted access to massive open on–line courses (MOOC) [3]:

- Those students, who discontinued their university studies due to financial reasons and started a full-time job;
- Study interrupted, who belong to NEET[3] group [4, 5];
- Potential future NEET group students [4, 6].

---

[1]Formal learning is always organised and structured, and has learning objectives. From the learner's standpoint, it is always intentional: i.e. the learner's explicit objective is to gain knowledge, skills and/or competences [2].

[2]Non–formal learning gives some flexibility between formal and informal learning, which must be strictly defined to be operational, by being mutually exclusive, and avoid overlap. Informal learning is never organised, has no set objective in terms of learning outcomes and is never intentional from the learner's standpoint. Often it is referred to as learning by experience or just as experience [2].

[3]NEET—Not in Employment, Education nor Training. A person who is not employed and is not involved in any educational process.

The Universities worldwide have been stated: "...blending e-learning together with habitual face-to-face learning could foster an active and deep approach to learning that enhance the intended learning outcomes." [5, 6].

Other researchers express the strong necessity for a flexible teaching and learning approach and possibility for offering the students a variety of different study materials in modern formats in order to meet their requirements across the university and campuses including the possibility learn from home applying e-learning platform MOODLE usage [7, 8].

Aims of the research.

- To identify the dropout crisis factors for universities 1st-year students caused by quality of the educational support and individual motivation;
- To define recommendations to reduce students' dropout by applying digital opportunities in educational support;
- To determine, whether digital opportunities impact the motivation enhancement of individual.

Questions of the research.

- The use of digital technologies in itself does not constitute an enhancement of the quality of teaching and learning, but it is a potential enabler for such enhancement;
- Evaluation of students' work using online course management system motivates them to study.
- Use of traditional forms and e-learning opportunities is necessary in similar proportions.

Research methods. As the theoretical framework, the following methods were applied:

- A survey with the assessment of the course;
- The risk-taking assessment by Schubert's method;
- Diagnostics of the person on motivation by T. Elersa methods;
- Failure avoidance motivation method;
- Survey about optimal proportion between traditional and e-learning studies;
- Feedback about usability of on–line learning platform.

2 Testbed

Web–based e–learning platforms allow educators to construct effective on-line learning study courses by uploading of various categories of study materials. E-learning platform allows usage of a wide range on-line learning tools as forums, discussion forums, e-mail messaging, as well as combining face-to-face and on-line approaches.

The purpose of these technologies is: to deliver study materials to a student, improve students' skills, assess skills and knowledge, and achieve better learning outcomes [7–13]. Fast and immediate feedback is possible.

In e–learning platforms produce data logging. Logged data can be used for later analysis. There are two types of data:

- Data produced by students or teachers and represent the content of the learning course;
- Data made by the system based on students activities like in system-spent time, kept sessions, the number of clicks on items of the content, etc.

At Riga Technical University (RTU), e-learning platform MOODLE had been maintained. In the period from October until December of academic year 2013/2014th and academic year 2014/2015th the course "Entrepreneurship (Distance Learning e-course)" to 1st-year students was provided. The course was conducted by RTU Professor A. Kapenieks. Author, as the Assistant to the Professor, supplemented the lecture content. Authors study, entitled "Use of Social microblogging to motivate young people (NEETs) to participate in distance education", was presented.

3 Getting of the Raw Data

The research was carried out during the lectures and the final exam of a course "Entrepreneurship (Distance Learning e-course)" with the 1st year students (respondents) participating on a voluntary basis.

Methods for diagnostics of the degree of risk preparedness, motivation to success and motivation to avoiding failures have been tested for all students three times at course: at the beginning, in the middle of course and at the end.

Homeworks. The course had two homework assignments:

- "Business Ideas searches on the Internet";
- "Your business idea".

The author has evaluated all homeworks uploaded in the e-learning environment. Reviews and comments were added, with the aim to encourage and motivate students to prepare their business plans in time and good in quality. Each comment was prepared according to the results of the content analysis. The feedback comments to students were written in a positive, supportive and motivational manner, personally addressing each student.

The assessment of the students homeworks was done by author concerning seven criteria:

1. actuality or viability of idea;
2. technological solution or how to enforce;

3. marketing—promotion of goods or services in the market;
4. competition;
5. financial security (e.g., planned revenues, expenses, financial support for the company's start-up and ongoing development (bank loan, other resources etc.);
6. the ability of a company to realize the idea;
7. the potential risks [6].

Questionnaries. The feedback from students is reflected in the questionnaires that were submitted at the end of the semester. Respondents were able to provide an objective criticism in regards to what situations required a direct contact with the teacher, and when e-learning was the most efficient option. The questionnaire had five key questions:

- Rating of the knowledge acquired during the course: low (1–3 points), moderate (4–7 points) or high (8–10 points), and a short commentary.
- Assessment of the effectiveness of the learning methods practiced on a scale from 1(the lowest) to 10 (the highest), and a short commentary.
- What the desired balance between E-learning and traditional learning would be, indicating the relationship between traditional forms of study and e-learning forms, as well as adding a comment.
- The three main competences of the teachers, and a short commentary.
- Details about the student: gender, age, the name of the educational institution they graduated from.

4 Results of Feedback

Most answers indicate an overall successful learning experience of the students. In the present study, the author have been evaluated the impact of managerial efforts for facilitating MOODLE implementation at RTU organization preparing to develop the new pattern for assessing the completeness of on–line study materials. Students highly appreciate the ability to use the course management system.

Respondents indicated that both, e-learning and traditional forms of the study had to be kept in balance. Particularly, e-learning provides a great advantage to learn anywhere, anytime. A successful guidance through the study process, however, is just as important, and can only be ensured when a teacher is present.

Respondents were also asked to state the core competence of the teacher:

- responsive, intelligent, able to establish a good contact with students; sociable, friendly yet demanding when it comes to the quality of students' performance; understanding, able to listen and motivate their students;
- possesses the ability to initiate discussions, to prepare a training plan, has a comprehensive understanding of different study subjects, ability to keep students motivated and interested in the particular subject of study.

In some comments, the positive atmosphere in the lectures, as well as the competence and responsiveness of teachers were emphasized. They rate high the ability to access teaching components via the Internet, therefore promoting the learning outcomes.

According to the viewpoint of the students, important are teachers experience, their level of expertise, and professionalism. These are positive traits of character, suggesting the need to develop an emotionally positive communication between the students and the teacher.

Author identify as very important the information provided by respondents regarding the educational institutions that they have been graduated. Based on the data collected, many of the students who currently study electronics and telecommunications sciences in RTU have been graduated from the same educational institutions.

The number of students has been decreased over the period of one semester: only two-thirds of all students turned in the second assignment. The author has identified that there exist some certain "*risk factor*" level for student to be dropped out of the course that correlate with a student's level of preparedness, readiness, and eagerness.

The "*risk factor*" is one of many that motivates students to work.

In 2013/2014 academic year 111 questionnaires were issued. Only 107 students had finished the course. In 2014/2015 academic year, 77 questionnaires were filled in by respondents.

5 Numerical Results of Student Activities

The measurement of students' activity was provided each week. The analysis of data shows that at the beginning of the course students activity,at a confidence level of 95 %, is the same as during period starting from the week 7 and continuing to the end of the course. There is a reducing of activity in the first part of the course between week 2 and week 7.

Table 1 shows results of "*t-test*" comparing the activity of the students during weeks 7–8 and 9–10. The null hypothesis H_0 and the alternative hypothesis H_a were stated. The null hypothesis (H_0) is about of statistical equality of independent activities during the weeks 7–8 and weeks 9–10. The *t-test* result (-0.146) do not exceed critical values of t value. Therefore, H_0 is not rejected, so we can conclude that both samples are not significantly different from each other and can be used for further analysis.

Additional *Mann–Whitney* "*nonparametric test*" of the Equality of students activity distributions was provided. In this case, H_0 stated that statistical distributions of students activity in weeks 7–8 and 11–12 are equal, assuming that activities are unrelated. Although, test results reveal that H_0 is rejected if the number of the weeks that students spent to study is less than seven.

Table 1 Test about independence of sampled activities comparing weeks 7–8 and 9–10

| | Levene's Test | | t-test for equality of means | | | | | | |
	F	Sig.	t	df	Sig. two-tailed	Mean diff.	Std. error diff.	Lower	Upper
								Confidence interval (95 %)	
Equal variances assumed (H_0)	0.070	0.792	−0.146	314	0.884	−0.487	3.346	−7.070	6.096
Equal variances not assumed (H_a)	–	–	−0.146	297.532	0.884	−0.487	3.346	−7.072	6.097

Table 2 Results of evaluation of the effectiveness of the study process by students

Form of study	Rated low			Average rating				Rated high			Mean
	1	2	3	4	5	6	7	8	9	10	
Lectures	0	0	0	3	6	16	34	29	11	6	7.30
Discussions	0	0	2	3	5	13	18	35	21	8	7.73
Assignments	0	0	1	0	7	11	19	35	15	17	7.90
Staff comments	0	0	1	2	7	12	15	33	17	18	7.88
e-portfolio utility	9	5	6	3	10	16	26	16	10	4	7.47

Course "Entrepreneurship (Distance learning course)" (2013/2014)

Table 3 Results of evaluation of the effectiveness of the study process by students

Form of study	Rated low			Average rating				Rated high			Mean
	1	2	3	4	5	6	7	8	9	10	
Lectures	1	1	2	1	1	10	25	18	12	8	7.80
Discussions	0	1	1	1	1	3	13	24	18	17	8.40
Assignments	0	0	1	3	4	5	15	20	15	16	8.01
Staff comments	1	0	1	2	4	4	9	14	12	31	8.54
e-portfolio utility	1	2	0	0	0	4	17	24	12	19	8.39

Course "Entrepreneurship (Distance learning course)" (2014/2015)

Additional correlation was observed between the level of readiness of the students' and course quality evaluation by themselves. The correlation was measured at the beginning (+0.734), at the middle (+0.633), and the end (+0.654) of the course. The correlation coefficient is slightly decreasing in the middle of the course, and characterizes that students in the beginning, as well as in the end of the course are higher motivated than in the middle phase of the learning.

The evaluation of the course by first group students' illustrate that "*Rated High*" weighted mean has reached grading of 7.42 mapped to the scale 0–10 (Table 2). Second group evaluation outcome—weighted mean has reached 8.08 (Table 3). Notably that all the students from the second group have been finished the course.

A significant correlation between the number of clicks produced by a student and final grade was observed (Fig. 1). The correlation value between students' activity and final grade was computed and is +0.506. That result shows that more active students can get better learning outcomes.

As it is depicted (Fig. 2), the number of clicks made by a student is related with positive course graduating outcome.

Fig. 1 Correlation between numbers of clicks produced by a student and final grade that was received at the end of a semester. Each *circle* denotes one student data. Values equal to 0 on vertical axis depicts students that had been participated in the course but were dropped out

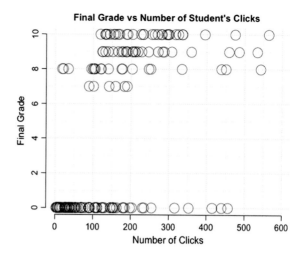

Fig. 2 Three functional relationships. How the number of clicks produced by the students impacts overall statistics of students activities. Students are divided into three groups. First group—successfully finished the course, second group—took part in activities, but had not finished, and third group—have been dropped out of the course without indication to be active

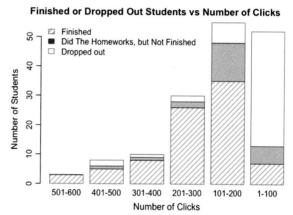

6 Conclusions

- According to learning and teaching quality assessment made by students, it can be concluded that using of traditional forms and e-learning opportunities is necessary for similar proportions. Higher education institutions should take into account the students' feedback. They could help to detect problems in teaching and learning environment, as well as faster and more efficiently to improve the environment.
- Proper motivation should be considered as extremely important leverage fostering e-courses preparing and raising faculty (University) interest for creating and maintaining study in course management system MOODLE.
- In addition to increasing access to education, greater use of new technologies and open educational resources can also help to reduce the costs of educational

institutions and students, particularly for disadvantaged groups, including existing and potential NEET group of people.

- The determination of risk motivation degree is recommended for Higher Educational institutions' first-year students. That is necessary to be able to predict the pedagogical support, balancing traditional and e-learning form of quantity.
- The author learned from RTU experience that those students who regularly uses e-learning resources and traditional teaching methods are slightly better knowledge compared with students who use e-resources very rarely.
- Educational institutions should combine the traditional forms of study and e-learning, as e-learning provides the opportunity to learn anywhere, anytime. However, a direct contact between the teacher and the student plays a very important role in acquiring a quality education, and should by no means be left out.

References

1. Progress Report on the Implementation of the National Reform Programme of Latvia within the "Europe 2020" Strategy. http://ec.europa.eu/europe2020/pdf/nd/nrp2013_latvia_en.pdf (2013)
2. Recognition of Non-formal and Informal Learning—Home. http://www.oecd.org/education/skills-beyond-school/index.htm
3. Kapenieks, A., Zuga, B., Stale, G., Jirgensons, M.: E-ecosystem drivene-learning versus technology driven e-learning. In: Proceedings of the 4thInternational Conference on Computer Supported Education, 16–18April 2012, pp. 436–439. CSEDU, Porto (2012)
4. Ratniece, D.: Use of social microblogging to motivate young people (NEETs) to participate in distance education through www.eBig3.eu. In: The Third International Conference on Data Analytics, 24–28 August 2014, pp. 7–11, Rome, Italy (2014)
5. Eurofound Yearbook 2012: Living and Working in Europe, Luxembourg:Publications Office of the European Union (2013)
6. Ratniece, D., Cakula, S.: Digital opportunities for student's motivational enhancement. In: Procedia Computer Science, 2015, Vol.: International Conference on Communication, Management and Information Technology, pp. 22–22. Prague (2015)
7. Salmon, G.: E-moderating. The Key to Teaching and Learning Online, 2nd edn. Abingdon, RoutledgeFalmer (2004)
8. Kibble, J., Kingsbury, J., Ramirez, B., Schlegel, W., Sokolove, P.: Effective use of course management systems to enhance student learning: experimental biology. AdvPhysiolEduc **31**, 377–379 (2007)
9. Taradi, S., Taradi, M., Radic, K., Pokrajac, N.: Blending problem-based learning with web technology positively impacts student learning outcomes in acid-base physiology. AdvPhysiolEduc **29**, 35–39 (2005)
10. Cole, J., Foster, H.: Using Moodle: Teaching with the Popular OpenSource Course Management System. O'Reilly Media, Sebastopol(2007)
11. Vasiljeva, T., Kremer L.: E-learning system MOODLE at higher school—implementation results. In: Proceedings of the 2nd Electronic International Interdisciplinary Conference EIIC, Zilina, Slovak Republic (2013)
12. Ynoue, Y.: Cases on Online and Blended Learning Technologies inHigher Education–Concepts and Practices. EditorialInformation Science Reference, New York (2010)
13. Dantas, A., Kemm, R.A.: Blended approach to active learning in a physiology laboratory-based subject facilitated by an E-learning component. AdvPhysiolEduc **32**, 65–75 (2008)

Building Numbers with Rods: Lesson Plans with Cuisenaire Method

Krokos Ioannis, Vasilis Tsilivis and Kosmas Mantsis

Abstract This paper reports on the pedagogical use of Cuisenaire method in mathematical teaching for students on preschool and primary school. Cuisenaire is being a teaching tool for more than 60 years, so our goal is to reintroduce the method and reinforce it with the use of ICT, guiding teachers and parents together to teach children. In the first part of the paper, we describe shortly the method. In the second part, we present exemplar lesson plans for preschool students and students in the first years of primary school. In the last part, we describe in detail how the method will be implemented in the field of ICT. We give the analysis and the functions of educational platform, which will be fully equipped with online interactive applications, complete educational material, video tutorials and lesson plans, as well as online training webinars for teachers and parents.

Keywords Structuring the numbers · Rods · Primary school · Mathematics

1 Introduction

This paper provides three lesson plans for applying Cuisenaire Method in a preschool level. The philosophy to end up in mathematical conclusions and build mathematical thinking through passive observations, comes alive through the Silent Way of Cuisenaire method. Cuisenaire Rods were created by Georges Cuisenaire in the early 1950s, a Belgian primary school teacher. His knowledge as a musician, helped him develop the idea of using colors for numbers, so as to use the specific

K. Ioannis (✉) · V. Tsilivis · K. Mantsis
Arnos Online Education, Solomou 29, 69121 Athens, Greece
e-mail: krokos@arnos.gr

V. Tsilivis
e-mail: vtsilivis@arnos.gr

K. Mantsis
e-mail: kosmas@arnos.gr

© Springer International Publishing Switzerland 2016   79
T. Gaber et al. (eds.), *The 1st International Conference on Advanced Intelligent System and Informatics (AISI2015), November 28–30, 2015, Beni Suef, Egypt*,
Advances in Intelligent Systems and Computing 407, DOI 10.1007/978-3-319-26690-9_8

colored rods of varying lengths to primarily help children understand arithmetic. By blending touch and sight, he created a synesthetic learning experience, where colors could create mixed shapes or become numbers and numbers, in turn, become colors. This amicable interconnection of the senses, acts as a tool that enhances learning and memory [1].

As we can easily understand, the premise behind this educational material is based on the creation of enriched learning environments through the use of teaching materials, tools and means. The use of teaching stuff and tools is considered to gain a prevalent role in the formation of the appropriate conditions, so that children's actions and perception skills can be studied with reference to mathematical concepts. The teaching stuff as well as their qualities on their own, do not lead to the elaboration of children's mathematical thought. In contrast, it is believed that children's physical actions and their interaction between them along with their reasoning (concerning these actions), contribute to shaping and reshaping of children's mental figures and functions. Teaching mathematics using this material, renders dealing with mathematics a pleasant, appealing as well as a life – like procedure, which facilities the student in particular into being led from this concrete stuff to the mathematical concepts represented. Thus, a bridge is being built in a stable way, gapping the child's work on this particular mathematical stuff with its abstractive thought, relating to symbols and numbers. The Cuisenaire rods are that kind of ways that puss the child through special actions to transform the specific to abstract, in other words to establish the mathematical way of thinking [2, 3] (Fig. 1).

Caleb Gattegno, an educator, theorist and a scientist who devoted his life to learning, met Georges Cuisenaire in 1953 and realized the power of rods and their ability to make algebra come alive, in the minds of the students. Later, seeing how learners were able to improve mathematical understanding, Gattegno utilized these rods for teaching foreign languages [4].

However, at this point, let us recall to our memory, the material composing Method Cuisenaire and individual features of. All the rods are divided into 65 families according to their colors. Namely, they are length models which children

Fig. 1 Cuisenaire method by its inspirer the Belgian Mathematician George Cuisenaire

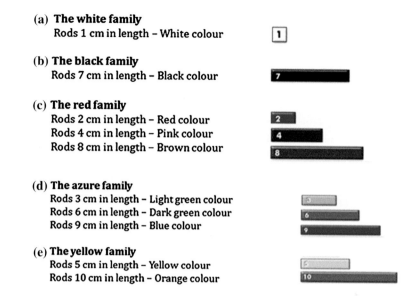

(a) **The white family**
    Rods 1 cm in length – White colour

(b) **The black family**
    Rods 7 cm in length – Black colour

(c) **The red family**
    Rods 2 cm in length – Red colour
    Rods 4 cm in length – Pink colour
    Rods 8 cm in length – Brown colour

(d) **The azure family**
    Rods 3 cm in length – Light green colour
    Rods 6 cm in length – Dark green colour
    Rods 9 cm in length – Blue colour

(e) **The yellow family**
    Rods 5 cm in length – Yellow colour
    Rods 10 cm in length – Orange colour

Fig. 2 The "color family" of Cuisenaire rods

handle for composing and comparing length models technological development is an indisputable fact, education can do nothing but include it in its future dimensions [5] (Fig. 2).

Aiming at further facilitation, the rods are accompanied with a track, a wooden rectangular framework with two grooves where students can put the rods while in the space between there is the numbers line (Fig. 3).

The child using the rods on the track has the chance of gradually structuring the numbers tracing various arithmetic combinations (Fig. 4).

Fig. 3 The rod track, used in Cuisenaire method

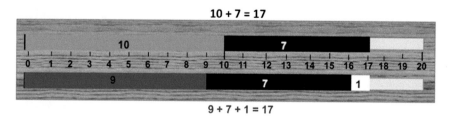

Fig. 4 Building number 17

2 Lesson Plans

In this section, we present some exemplary lesson plans in order to understand in depth the use and functionality of the method for early years and primary school age. The theoretical approach of *Comenius* and *Pestalozzi,* who recognized the necessity and the special value of teaching via visualization are to find application through the lessons plans, are described in this session [6].

It is worth to mention than for the educators the use of lesson plants and material for the support of the educational process, is not granted. For some, the use of them is improved especially affective in the construction of knowledge, in the development of mathematical thinking, in the improvement of performance and in having a positive attitude towards mathematics. But there are some others educators that don't support not even theoretically the use of educational planning and others that even though they understand that they help to make clear the abstract nature of mathematics they use them in a wrong way. So the results are not the desirable ones and they stop using them [7, 8].

For the accomplishment of whichever act and evaluation, it must be found an educational material that is appropriate for the act and can deliver the content with precision. Additionally for such a material to be an affective supporting way of teaching and learning of mathematics it has to be educationally planed. Even the best material is doomed to fail if the educator cannot handle it in the right way inside the classroom. Through the appropriate planning of introduction in the educational act, who will take in consideration the researching data that will occur for the possibilities of evaluating use of difficulties may cause the best way to achieve effective education through its use. So let's familiarize ourselves with the method and the visualization of mathematical concepts, through lesson plans we have created especially for this purpose.

In the first lesson plan, we give instructions, in order to make the first contact with the Cuisenaire Method. This could take time. Dedicate, as much as you need, to help your pupils get familiar with the objects. The second plan suggests a solution in how to build the concept of numbers for your students. The last one shows a way of explaining addition with Cuisenaire Method.

We see that the cornerstone of the method consists of three parts: observation, activation of all senses and self-activity of the learner. Let us not forget however that the experiential teaching, turned in the manner mentioned in the previous section, can lead to interesting results if run through the frame of New Technologies and Communications Networks. Already at a very young age, children are familiar to interact, play and perceive through online applications and programming environments. Therefore, it becomes urgent and necessary, our direct actuation in that direction, so that children become familiar with Cuisenaire method, through both of said material, and through the development of appropriate training for online applications, an act which are committed in their direct implement, either with paper and pencil, or using Web 1.0 or Web 2.0 tool.

2.1   Worksheet 1

The Subject: A rod game—Forming pictures
Grade: Kindergarten
Object:

- The child learns how to create pictures and objects, using the Cuisenaire rods.
- It can match a rod of a certain color with a certain rod sixe, which takes up a certain space and dimension on the cm graph paper.
- It can realize the way in which combinations of rods of various colors and sizes can form objects and whole entities.
- It can familiarize itself with the concept and definition of the location of an object in a predetermined coordinates system.

Materials:

- Cuisenaire Rods for each child
- 1 cm graph paper
- Worksheet

Procedure
Free exploration: A set of Cuisenaire rods, a sheet of 1-cm graph paper and the respective worksheet, are given out to children.

Then, they are shown a picture of the object as it is depicted on the worksheet and are asked to form the same picture on the 1-cm graph paper using the Cuisenaire rods.

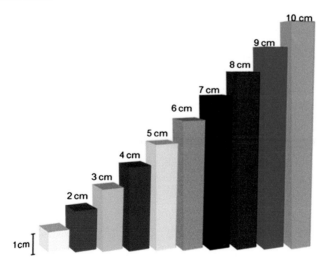

Worksheet 1

Name:………………………………………

Date:……………………………………….

Assignment: Using the 1-cm paper available, form each of the two depicted objects, using the Cuisenaire rods, as shown in the example below:

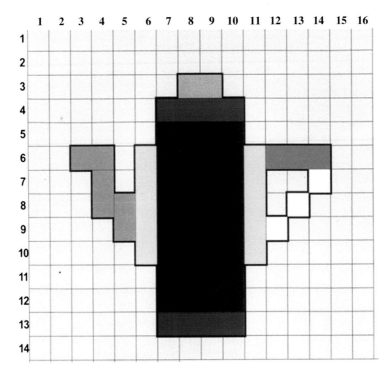

2.2   Worksheet 2

Worksheet

Name:………………………………………

Date:……………………………………….

Assignment: Using the Cuisenaire rods and the rod track, build numbers 16, 13 and 18 in two different ways, as shown in the following examples.

Example 1 Building number "17", following one of the following ways:
  Add number 7 and number 10 (*top of the rod track*)
  Add number 2, number 6 and number 9 (*bottom of the rod track*)

Example 2 Building number "12", following one of the following ways:
  Add number 2 and number 10 (*top of the rod track*)
  Add number 6 and number 6 (*bottom of the rod track*)

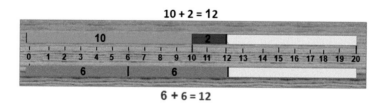

3 Application in the Field of ICT

The analysis of the method, as performed in the previous sections, makes especially appreciated the fact that the systematic occupation with activities, which utilize these plans, facilitates children into being accustomed to the relations between sizes, colors and numbers as symbols. It is very important for children to understand and develop, through a group of selected activities, the potential relations between the first ten numbers and the first one hundred numbers accordingly. These potential relations, which are stabilized and fully assimilated through systematic use, constitute the cornerstone of any arithmetic operation. That is the reason why the second lesson plan focuses on the formation of double digit numbers by adding suitable single digit ones. Through these plans, the Cuisenaire method, as we mentioned before, enables every student to work independently and in the group on meaningful mathematics contents, while the teacher provides individual attention to other students [9].

It is a proven educational method of mathematical perception's gradual construction, which, if learned and applied properly by teachers, promises to lead to innovative results and open new horizons in mathematical thinking. Our philosophy is that these horizons should be governed not by memorizing and sterile methodology, but by observation, perception and activation of all senses.

As a provider of educational services to parents, teachers and students, ARNOS Online Education Center is committed to enhance the understanding of the method through the field of ICT's, by offering both technological solutions to parents and teachers and also creating online educational applications for children of preschool and early school age.

More specifically, one of the main axes of Cuisenaire method's application in the field of new technologies, will be the design and operation of an educational platform, accessible from parents, students and teachers, through which, online webinars taking place at a specified time, will allow users to connect in real time and watch live the implementation of the method. Participants will have the opportunity to interact not only among themselves but also with the presenter, a fact that enables parents, teachers and students to cooperate and take an active role in the—understanding of the method- procedure. The didactic approach of Cuisenaire method is aimed to become a chain with the teacher as its first link, who is going to acquire, through the platform, the opportunity to become its partaker and initiator. By giving alternative approaches and solutions in thoughts and ideas, a complete training material consisting of video tutorials and lesson plans, will be posted on our platform and being continuously enriched, will provide complete guidance to the teacher for the degradation and the excellent teaching of this method.

Our immediate priority, as a provider of educational services, is the implementation of an online application that allows children creative and systematic involvement with the Cuisenaire rods. Through this application, the surface of 1-cm squared paper will display on every child's computer screen, as well as a set of Cuisenaire rods. By right-clicking on the specific option of the application, the child will be able to choose a Cuisenaire rod and then drag it onto the squared background. In a similar way, more rods can be added and aligned, as desired. This application also comes with a set of interactive assessment tests, formulated in the corresponding style with the above, so that each child has the opportunity of online training on specific math activities. At the end of each test, each user of the application, will be shown the option of direct storage and printing of results, accompanied by detailed explanation of the points where the child should give particular emphasis for the next time. The educational platform will be available in English for teachers, parents and students in late October 2015.

Furthermore, through the "Erasmus+KA2" program, we have already submitted proposals for cooperation with five nursery schools from five different European countries, in order to apply directly both our lesson plans and online application. The program which is going to have a total planned duration of 2 years will provide us with complete statistics and data collection regarding with the performance of children, as well as a comparative study of the each child's performance before and after the application of the method.

4   Conclusion

The systematic occupation with the aforementioned application facilitates children's observation, self—initiated action, creativity, verification and comprehension skills thus leading them to mathematical thought. The systematic occupation with activities which utilize this stuff facilitates children into being accustomed to the relations between sizes, colors and numbers as symbols. It is very important for children to understand and develop, through a group of selected activities, the potential relations between the first ten numbers and the first one hundred numbers accordingly.

Therefore, the Cuisenaire material will facilitate children into doing plenty of preparatory exercises, so that they can fully understand the meaning and value of each number as well as the meaning of all four arithmetic operations before starting to occupy with mental calculations and the symbols representing numbers. The use of teaching material and tools is considered to gain a prevalent role in the formation of the appropriate conditions, so that children's actions and perception skills can be studied with reference to mathematical concepts. It enables every student to work independently and in the group on meaningful mathematics contents, while the teacher provides individual attention to other students. For the most part, parents and teachers have the opportunity to enjoy and discover new things, which previously considered settled and stereotypes, but also children will have the opportunity for structured teaching outside school.

References

1. Heyrman, H.: Extending the synesthetic code: connecting synesthesia, memory, and art. http://www.doctorhugo.org/synaesthesia/art2/index.html (2007, March). Accessed 8th July 2012
2. Mullen, J.: Cuisenaire rods in the language classroom. http://john.mullen.pagesperso-orange.fr/cuisenaire.htm (1996, Dec). Accessed 9th July 2012
3. Gregg, S.: How I teach using Cuisenaire rods. http://mathagogy.com. Accessed 22 April 2014
4. Gattegno, C.: Teaching Reading and Writing in Color; A Scientific Study of the Problems of Reading. Educational Solutions, New York (1968)
5. Learning Resources (Firm), Learning Resources, Incorporated, Winstead, R., Dresen, K.: Using Cuisenaire rods: addition & subtraction (2001)
6. Streefland, L.: Fractions in Realistic Mathematics Education: A Paradigm of Developmental Research, pp. 63, 64, 80, 81. Kluwer Academic Publishers, Dordrecht (1991)

7. David, B.D. (ed.): Working on Awareness for Teaching and for Research on Teaching (transcript of a seminar by Caleb Gattegno). Abon Language School, Bristol (1989)
8. Gee, P.J.: Part II: Theories and Mechanisms: Serious Games for Learning. Routledge Taylor & Francis Group, New York and London (2009)
9. Van de Walle, J.A.: Elementary and Middle School Mathematics: Teaching Developmentally, 6th edn. Pearson Education Inc, Boston (2007)

Part II
Multimedia Computing
and Social Networks

Telepresence Robot Using Microsoft Kinect Sensor and Video Glasses

Mahmoud Afifi, Mostafa Korashy, Ali H. Ahmed, Zenab Hafez
and Marwa Nasser

Abstract Developing telepresence robots is one of the most important trends in the robotics research area, where the user acts as he/she is located in a remote location. In 2010, telepresence robots became a noticeable trend after the robot "QB" that introduced by Silicon Valley start-up Anybots (Robotics trends for 2012. IEEE Robot. Autom. Mag. 19(1):119–123, 2012). Although, the availability of "QB" as a commercial telepresence robot, its cost made it unavailable for most users. In this work, a low-cost telepresence robot is presented using iRobot-Create, Microsoft Kinect sensor, and video glasses. The proposed system makes the user feels like he/she is located in a different location and acting as in the normal life (walking, stop, rotating his/her head). The user takes feedback via a streaming video from the remote location to a pair of video glasses worn by him. The remote unit consists of three components: a single iRobot-Create, a laptop, and two web-cams. In the user side, the user's movements are recognized using Microsoft Kinect sensor. We use the RGB camera in Microsoft Kinect sensor for streaming the video of the user to the remote side. So, People in the remote side see the user, as he/she is located with them. The results of the proposed system show that the user is integrated in another environment using low-cost hardware components.

M. Afifi (✉) · M. Korashy
ACM Student Research Laboratory, Assiut University, Assiut, Egypt
e-mail: mahmouda1@acm.org; m.afifi@aun.edu.eg

M. Korashy
e-mail: m.korashy@acm.org

A.H. Ahmed · Z. Hafez · M. Nasser
IT Department, Assiut University, Assiut, Egypt
e-mail: ali.hussein@fci.au.edu.eg

Z. Hafez
e-mail: zenab_hafez@hotmail.co.uk

M. Nasser
e-mail: eng.marwa.nasser@compit.aun.edu.eg

© Springer International Publishing Switzerland 2016   91
T. Gaber et al. (eds.), *The 1st International Conference on Advanced Intelligent System and Informatics (AISI2015), November 28–30, 2015, Beni Suef, Egypt,*
Advances in Intelligent Systems and Computing 407, DOI 10.1007/978-3-319-26690-9_9

1 Introduction

Telepresence refers to giving the user a sense as he/she is located in another location [2]. In 1980, Marvin Minsky used the "telepresence" term for referring to teleoperation systems that manipulate remote physical objects. In 1992, Sheridan and Furness adapted the "telepresence" term to be "presence" for referring to machines that are operated by remote controls, and virtual reality systems [2].

Robots are used for achieving telepresence by making the user controls a robot in a remote location as he/she is located in this location. In 2010, telepresence robots became a noticeable trend after the telepresence robot "QB" that was introduced by Silicon Valley start-up Anybots. In 2011, more commercial telepresence robots were introduced, such as "Vgo" and "Jazz" robots [1].

Telepresence robots have many applications in different fields. One of those applications is helping elderly people in their home [3–5]. Another application of telepresence robots is embodied video conferencing systems [6]. Traditional video conferencing systems are restricted in a meeting room that has a static camera that streams to another side of the meeting. This constraints of the traditional video conference leads to use robots for remote connectivity in video conference [7], where the remote member can walk through the meeting room via a robot that streams the view of the meeting room to this remote member. Many other applications of telepresence robots were presented, such as explosives detection, distance education, digital tutors, and home security.

Many commercial telepresence robots were introduced with a variety of differences in drive (wheels and casters), speed, height, weight, battery life, navigation control methods, camera features, and cost. However, most of those telepresence robots are not available to many people because of their cost. In this work, we present a low-cost telepresence robot using Microsoft Kinect sensor and video glasses. The proposed system consists of two main units, the remote unit and the user unit. In the remote side (the remote unit), the robot is controlled using the recognized movements of the user that are captured and recognized in the user side (the user unit). The rest of the paper is organized as follows: In Sect. 2, the details of the proposed system are presented. In Sect. 3, the experiment results and comparisons are presented. Finally, we conclude the paper in Sect. 4.

2 The Proposed System

The proposed system is designed for making the user feels as he/she is located in a remote location. In addition, persons in the remote location feel that the user as if he is located with them. The proposed system is divided into two parts: **User side** and **Remote side**. In the **user side**, user acts as he/she acts in the daily life. If he/she wants to move from the current location to another one, he/she walks. If he/she want to change the current viewpoint, he/she rotates his/her head, and so on.

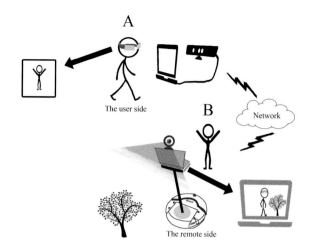

The user side consists of a personal computer, Microsoft Kinect sensor, and video glasses. The user's body and head movements is captured via Kinect and analyzed to control the robot and the controllable camera consequently in the remote side. In the **remote side**, a single laptop and a controllable web-cam is attached to the robot. The controllable web-cam is used for capturing the scene around the robot and streaming it to the video glasses worn by the user. Figure 1 shows an overview of the proposed system, where the user "A" communicates with the user "B" via the proposed system. The user "A" sees the remote scene that is captured using the controllable web-cam via the video glasses. He/she rotates his/her head for changing his/her viewpoint. When the user walks, the robot moves according to his/her movements and direction.

2.1 Human–Robot Interaction Using Microsoft Kinect Sensor

To provide the user with a feeling of interacting with a realistic environment, not just a communicating through a video chatting, we allow the user to act in natural movements through his/her daily life. The user walks when he/she wants to move from his/her current location to another one, rotate his/her body to change his/her direction of walking to another direction, and rotate his/her head for looking from another viewpoint. For achieving that, we use Microsoft Kinect sensor as a low-cost depth sensor to get the depth information of the current environment of the user. From the depth information, we extract the user from his/her current environment to be integrated with another location, as described in Sect. 2.2. In addition, Microsoft Kinect sensor is used for tracking the skeletal movements of the user to recognize what the user means of those movements.

(a)   **(b)**

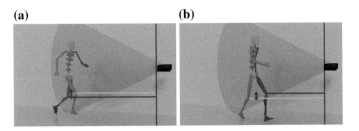

Fig. 2 Walking recognition. **a** Right leg moves forward. **b** Left leg moves forward

Walk Recognition

Recognizing the user's walking is based on the following observation, when a person walks his/her knees are not in the same distance from Microsoft Kinect sensor, as shown in Fig. 2. In addition, the user's knees are not in the same vertical distance. For each captured frame via Microsoft Kinect sensor, the user's walking is recognized using the following equation

$$Walking = \begin{cases} true & \text{if } \ |KneeL_Z - KneeR_Z| > T_Z \text{ AND } |KneeL_Y - KneeR_Y| > T_Y \\ false & otherwise, \end{cases} \tag{1}$$

where $KneeL_Z$ and $KneeR_Z$ are Z values, $KneeL_Y$ and $KneeR_Y$ are Y values of vectors contain the X, Y, and Z of the left knee and right knee according to Microsoft Kinect sensor, respectively. T_Z and T_Y are predefined thresholds for avoiding the error of depth measurements of Microsoft Kinect sensor which is directly proportional to increasing distance to Microsoft Kinect sensor [8].

Direction Recognition

Recognizing the user's orientations is based on trigonometry. The primary walking direction of the user (forward or backward) is recognized using the following equation

$$Forward = \begin{cases} true & \text{if } \ |HandR_Y - Head_Y| < T \\ false & otherwise, \end{cases} \tag{2}$$

where $HandR_Y$ and $Head_Y$ are Y values of vectors contain the X, Y, and Z of the right hand and head joints in the Microsoft Kinect skeleton, respectively, and T is a predefined threshold. If the user wants the robot to walk backward, he/she raises his/her right hand.

In the case of walking whether forward or backward, we recognize the hip rotation angle of walking around Y axis. The rotation matrix around Y axis is given by the following equation [9]

$$\begin{bmatrix} x' \\ y' \\ z' \\ 1 \end{bmatrix} = \begin{bmatrix} cos\theta & 0 & sin\theta & 0 \\ 0 & 1 & 0 & 0 \\ -sin\theta & 0 & cos\theta & 0 \\ 0 & 0 & 0 & 1 \end{bmatrix} \begin{bmatrix} x \\ y \\ z \\ 1 \end{bmatrix}. \tag{3}$$

Where x,y, and z are the joint coordinates before rotating by θ around Y axis, and x',y', and z' are the joint coordinates after the rotation relative to the origin of a reference system. The reference system of hip rotations that affects on the rotation angle of walking is the Hip Joint Center (HJC) [10]. We assume that, the HJC is calculated approximately using the following equation

$$HJC = \left| \frac{PHip_R - PHip_L}{2} \right|, \tag{4}$$

where $PHip_R$ and $PHip_L$ are the right hip joint and the left hip joint of the previous received frame before the rotation, respectively.

The angle θ of rotation of the body around Y axis is calculated using the following equations

$$\Theta = tan^{-1} \left(\frac{| CNearHip_Z - PNearHip_Z |}{| CNearHip_X - PNearHip_X |} \right), \tag{5}$$

$$D = \sqrt{\Delta NearHip_X^2 + \Delta NearHip_Y^2 + \Delta NearHip_Z^2}, \tag{6}$$

$$\Delta NearHip = CNearHip - PNearHip, \tag{7}$$

$$C = | HJC_X - PNearHip_X |, \tag{8}$$

$$I = \sqrt{D^2 + C^2 - 2 \times D \times C \times cos\Theta}, \tag{9}$$

$$\theta = sin^{-1} \left(D \times \frac{sin\Theta}{I} \right). \tag{10}$$

where $CNearHip_Z$ and $PNearHip_Z$ are Z values of the current near hip joint and the previous near hip joint, respectively, and $CNearHip_X$ and $PNearHip_X$ are X values of the current near hip joint and the previous near hip joint, respectively. Figure 3 shows how to calculate the angle θ of the body rotation around Y axis. The current near hip joint determines the rotation direction, i.e. if the current near hip joint is the right hip joint, the rotation direction is counter clockwise, and vice versa, as shown in Fig. 4. If the depth distance between the left and right hip joints is less than a predefined

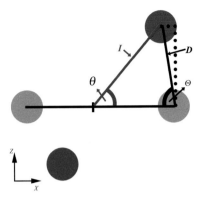

Fig. 3 The *top view* of the hip joints for estimating the angle of the body rotation around Y axis. The *pale colored circles* refer to the previous right and left hip joints. The *full colored circles* refer to the current right and left hip joints after rotating around Y axis by θ

(a)   **(b)**

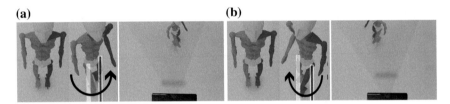

Fig. 4 The relation between the direction of the user's orientation, and the left and the right hip joints. **a** The user rotates along Y axis in counter clockwise direction. **b** The user rotates along Y axis in clockwise direction

threshold, there is no rotation. During the user walks, his/her body moves a little bit, but these movements are applied to all hip joints. So, they are fading each other.

Head Orientation Recognition

The head's orientation of the user is recognized for controlling the controllable webcam that is attached to the remote side. If the user rotates his/her head along Y axis, the camera rotates to the same direction for giving the user a sense of acting in the remote location. Recognizing the rotation of the user's head along Y axis is performed using the direction of the normal vector N that is generated using the cross product in the following equation

$$N = SRMatrixQuaternion_X \times HRMatrixQuaternion_X \qquad (11)$$

where $SRMatrixQuaternion_X$ is the X vector in the rotation matrix of the shoulder center joint, and $HRMatrixQuaternion_X$ is the X vector in the rotation matrix of the head joint. If the direction of N is upward, so the user rotates his head in the right direction. If the user rotates his head in the left direction, the direction of N is downward. If N is zero, so there is no rotation of the head around Y axis. For avoiding the

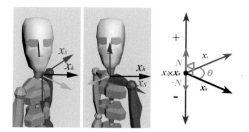

Fig. 5 The relation between the X vector in the rotation matrix of the shoulder center joint and the head joint

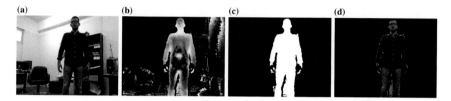

Fig. 6 Separating the user from his/her original location. **a** The color frame. **b** The depth frame. **c** The generated binary mask M. **d** After separating the user from the original scene

error of the depth sensor, we assume that there is no rotation of the user's head if the difference between the two X vectors in the head and shoulder center joints is less than a predefined threshold. Figure 5 shows the relation between the X vector in the rotation matrix of the shoulder center joint and the X vector in the rotation matrix of the head joint.

2.2 Compositing User's Video and Robot Scene's Video

Depth frames that are generated by Microsoft Kinect sensor describe depth data of the current scene of the user. We assume that the user is the nearest object to Microsoft Kinect sensor. From the previous assumption, we use depth data for extracting the user from the received frame via Microsoft Kinect sensor to be composited with the received frame via the web-cam that is attached to the robot, see Fig. 6. The web-cam streams the back scene of the remote location. We use alpha blending technique [11] for compositing the both frames together. We use a simple threshold operation for generating the binary mask M that describes the nearest pixels to Microsoft Kinect sensor (the user). After that, we composite the received frame via Microsoft Kinect sensor K with the received frame via the web-cam W using the following equation for each channel in the true color frames

$$V_{(x,y)} = K_{(x,y)} \times M_{(x,y)} + W_{(x,y)} \times M'_{(x,y)}, \tag{12}$$

where $V_{(x,y)}$, $K_{(x,y)}$, $W_{(x,y)}$, $M_{(x,y)}$, and $M'_{(x,y)}$ are the intensity of the pixel at location (x, y) in the composited frame, the received frame via Microsoft Kinect sensor K, the received frame via the web-cam W, the binary mask M, and the inverse of the binary mask M', respectively. After generating the composited frame V, the frame is displayed in the laptop screen that is attached to the robot for persuading persons in the remote location that the user stands with them in the same place.

2.3 Telepresence Using Video Glasses

For making the illusion of placing the user in a remote location, the video that is received from the remote side is streamed to a pair of video glasses that is worn by the user. The user sees as he/she is located in the remote location. As described, the controllable web-cam in the remote side responds to the user's head orientations. So, the user largely feels like he/she is located in the remote locations.

3 Experiment Results

The proposed system is considered a low-cost system that consists of iRobot-Create as a main unit on the remote side of the system. There are two web-cams that are attached to the robot. Logitech QuickCam Orbit AF[1] streams the front view of the remote location and a regular web-cam streams the back view. There is a laptop that is attached to the robot for communicating the remote unit with the user side. The embedded microphone in the laptop is used for the audio communication between the user and persons who are located in the remote location. In the user side, the user controls the robot using his/her movements that are captured and recognized using Microsoft Kinect sensor. The front view of the remote location is streamed to the user via Innovatik video glasses,[2] see Fig. 7. Figure 8 shows the recognition process of the user's movements, where the user controls the robot using his/her natural movements. Figure 9 shows the success rate of the recognition process of the user's movements using random movements of several users. We test a variety of movements (750 movements) of 12 different persons (25 % female and 75 % male).

As described, the proposed system is considered a low-cost system compared with other systems. Table 1 shows a comparison between the proposed system and other systems [12].[3] The proposed system is considered a low-cost systems compared with other systems, that makes it contributors in the availability of telepresence systems for a large-scale of the users.

---

[1]http://support.logitech.com/product/quickcam-sphere-af.

[2]http://www.innovatek.cn/products/video20glasses/v-490.htm.

[3]http://telepresencerobots.com/comparison.

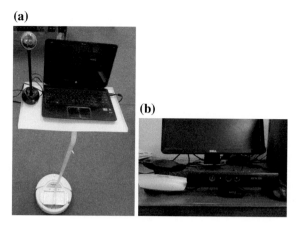

Fig. 7 The proposed system. **a** The remote side. **b** The user side

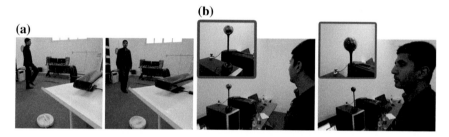

Fig. 8 The recognition process of the user's movements using Microsoft Kinect sensor

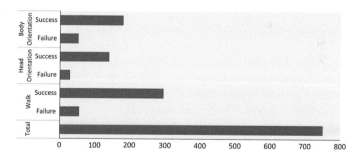

Fig. 9 The success rate of recognizing the user's movements

Figure 10 shows the final result of the proposed system. The user moves through the remote scene using the robot and interacts with persons who are located in the remote location via the movements of his/her head and body, and his/her voice.

Table 1 A comparison among several robotic telepresence systems

Robotic telepresence systems	Adjustable height	Cost
QB	Yes	Very high
VGo	No	High
Jazz	No	High
TeleMe	No	High
Beam	No	High
The double	Yes	Medium
The proposed system	No	Low

The cost column shows the price of the system, where "Very High" refers to more than 8,000$, "High" refers to more than 5,000$, "Medium" refers to more than 2,000$, and "Low" refers to less than 1,500$. The Price may change according to the time

(a)   **(b)**

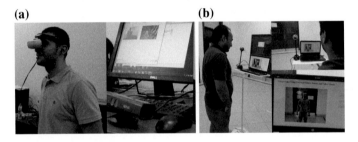

Fig. 10 The result of the proposed system, where the user in **a** communicates with the remote location in **b**

Limitations

1—In the user side, the proposed system requires only one user for confusion avoidance.

2—In human–robot interaction process, the user must perform his/her motions in frontal view of Microsoft Kinect Sensor.

3—The user must perform his/her walking movements in a specific walking technique (raise the knee to a proper level to be detected).

4—The robot's responses are slower than the natural movements of the human, so the user must perform slow rotations.

4 Conclusions

In this paper, a low-cost telepresence system has been presented. The system consists of iRobot-Create that is located in a remote location. Two web-cams attached to the robot for streaming the front scene and the back scene. The front scene is streamed via video glasses to the user who is located in another location. The back scene is used for compositing the user with the background of the remote scene. The user controls the robot using his/her natural movements. The user walks, rotates his/her head, changes his/her walking direction, and the robots responds to those commands. Persons in the remote location see the user who is integrated with the background scene of the remote location via the attached laptop to the robot. The proposed system is considered a low-cost system compared with the state-of-art systems.

Acknowledgments Many thanks to Dr. Khaled F. Hussain, CS Dept., Assiut University, and Dr. Nagwa M. Omar, IT Dept., Assiut University, for supporting us with the required equipment.

References

1. Guizzo, E., Deyle, T.: Robotics trends for 2012. IEEE Robot. Autom. Mag. **19**(1), 119–123 (2012)
2. Steuer, J.: Defining virtual reality: dimensions determining telepresence. J. Commun. **42**(4), 73–93 (1992)
3. Tsai, T.-C., Hsu, Y.-L., Ma, A.-I., King, T., Chang-Huei, W.: Developing a telepresence robot for interpersonal communication with the elderly in a home environment. Telemedicine and e-Health **13**(4), 407–424 (2007)
4. Michaud, F., Boissy, P., Labonte, D., Corriveau, H., Grant, A., Lauria, M., Cloutier, R., Roux, M.-A., Iannuzzi, D., Royer, M.-P.: Telepresence robot for home care assistance. In: AAAI Spring Symposium: Multidisciplinary Collaboration for Socially Assistive Robotics, pp. 50–55. California, USA (2007)
5. Lowet, D., Isken, M., Lee, W.P., van Heesch, F., Eertink, E.H.: Robotic telepresence for 24/07 remote assistance to elderly at home. In: Proceedings of Social Robotic Telepresence Workshop on IEEE International Symposium on Robot and Human Interactive Communication. Paris, France (2012)
6. Desai, M., Tsui, K.M., Yanco, H.A., Uhlik, C.: Essential features of telepresence robots. In: 2011 IEEE Conference on Technologies for Practical Robot Applications (TePRA), pp. 15–20. IEEE (2011)
7. Kanigoro, B., Budiharto, W., Moniaga, J.V., Shodiq, M.: Web based conference system for intelligence telepresence robot: a framework. J. Comput. Sci. **10**(1), 10 (2013)
8. Khoshelham, K., Elberink, S.O.: Accuracy and resolution of kinect depth data for indoor mapping applications. Sensors **12**(2), 1437–1454 (2012)
9. Vince, J.: Mathematics for Computer Graphics. Springer Science & Business Media (2013)
10. Kadaba, M.P., Ramakrishnan, H.K., Wootten, M.E.: Measurement of lower extremity kinematics during level walking. J. Orthop. Res. **8**(3), 383–392 (1990)
11. Wright, S.: Digital Compositing for Film and Video. Taylor & Francis (2013)
12. Kristoffersson, A., Coradeschi, S., Loutfi, A.: A review of mobile robotic telepresence. Adv. Hum.-Comput. Interact. **2013**, 3 (2013)

Video Flash Matting: Video Foreground Object Extraction Using an Intermittent Flash

Mahmoud Afifi

Abstract Video foreground object extraction has many useful applications, such as changing background, duplicating foreground objects in the scene, and in virtual reality. The main task is to create an accurate alpha matte that specifies the foreground objects to be extracted. In this work, a new approach for video foreground object extraction is presented; this approach is called "Video Flash Matting". An intermittent flash is used for creating bright frames that are used for creating the alpha matte of the foreground objects through the video. Bright frames contain lit foreground objects that are illuminated using the intermittent flash; however, the background is still under the same illumination without flash. According to that, foreground objects are extracted from the rest of the video using those bright frames. The results of the proposed approach show that Video Flash Matting extracts foreground objects using good alpha mattes without requiring trimaps. Video Flash Matting handles free camera and dynamic background conditions.

1 Introduction

Video foreground object extraction has many useful applications, such as changing background, duplicating foreground objects in the scene, video compression, and in virtual reality [1, 2]. The main task is to create an accurate alpha matte that specifies the foreground objects to be extracted [2]. Traditional techniques use either green-screens or depth sensor for extracting objects from videos. However, these techniques require additional hardware equipment or a specific environment for recording the

M. Afifi (✉)
ACM Student Research Laboratory, Assiut University, Assiut, Egypt
e-mail: mahmoudafifi1@acm.org; m.afifi@aun.edu.eg

M. Afifi
IT Department, Assiut University, Assiut, Egypt

© Springer International Publishing Switzerland 2016
T. Gaber et al. (eds.), *The 1st International Conference on Advanced Intelligent System and Informatics (AISI2015), November 28–30, 2015, Beni Suef, Egypt,*
Advances in Intelligent Systems and Computing 407, DOI 10.1007/978-3-319-26690-9_10

videos [2, 3]. Green-screens are used in the cinema industry for changing the background of the actors instead of the background of the studio which requires special settings [2]. Depth sensors, such as Microsoft Kinect and ASUS Xtion generate mid-quality alpha mattes in many cases because of their resolution of the depth information that decreases with the camera distance and restricted with specific resolutions [4, 5]. Time-of-Flight (TOF) cameras are used for extracting foreground objects, such as L. Wang et al. [3] who used a TOF camera for separating foreground objects in real-time.

On the other hand, many computer vision techniques have been presented for extracting foreground objects from a given video. N. R. Howe et al. [6] used a minimum graph cut technique for extracting foreground objects from videos. H. Myeong et al. [7] presented a technique for extracting motion-blurred objects from a video with the assumption of the appearance of only one blurred object in the video. M. Zhao et al. [1] improved the missing data of the RGBD cameras for extracting foreground objects at real-time using GPU hardware. There are many alpha matting techniques have been presented for extracting foreground objects from still images using accurate alpha mattes, such as Y. Zheng et al. [8] who presented two alpha matting techniques: a local learning based technique and a global learning based technique which learn a more general alpha color model, either linear or nonlinear using the neighboring pixels and the nearby labeled pixels of each pixel that being estimated. E. Shahrian et al. [9] used textures to complement colors for generating more accurate mattes. By using a sampling strategy for generating a comprehensive set of known samples from all color distributions in known regions, improved alpha mattes are generated [10].

In this work, a new approach for video foreground object extraction is presented; this approach is called "Video Flash Matting". An intermittent flash is used for creating bright frames that are used for creating the alpha matte of the foreground objects through the video. Video Flash Matting is based on Flash Cut technique [11] which has been presented for extracting foreground objects from still images using a pair of flash and no-flash images. In this work, improvements are presented for adapting the usage of the flash lighting for foreground objects extraction from videos. A set of bright frames is extracted from the original video. Flash Cut is based on the following observation, when frames are captured with and without flash, bright frames contain lit foreground objects that are illuminated using the intermittent flash; however, the background is still under the same illumination without the flash lighting. According to this observation, foreground objects are extracted from the rest of the video using those bright frames. Video Flash Matting handles free camera and dynamic background conditions.

The rest of the document is organized as follows: the details of Video Flash Matting are presented in Sect. 2. Next, the experiments and discussions are presented in Sect. 3. In Sect. 4, the conclusions and the future work are presented.

2 Video Flash Matting

The idea is to extract the foreground objects using a pair of frames. One of them is illuminated with flash light and the other one is illuminated with no-flash light. Bright frames are the frames that illuminated using the intermittent flash. For working with videos, the bright frames must be spread over the video with a high frequency. The video is split into a set of segments. Each segment starts with a bright frame. By aligning each no-flash frame (normal frame) with the closest similar flash frame (bright frame), foreground objects are extracted from the video. Video Flash Matting extracts foreground objects without using green-screens, depth sensors, or trimaps, see Fig. 1. For detecting the bright frames from the given video, the histogram of each frame is analyzed with the assumption of the appearance of the first bright frame at the beginning of the video, see Fig. 2. For each frame in the video, the bright frames are detected using the following equation

$$I \in B_F, \text{ if } | Max_c - Max_p | > T, \text{ with } Max \in [\frac{n}{2}, n], \qquad (1)$$

where I is the current frame, B_F is the set of the bright frames, Max_c and Max_p are the histogram values at the last half of the histogram domain $[0, n]$, and T is a threshold that is calculated experimentally. The bright frames are dropped from the

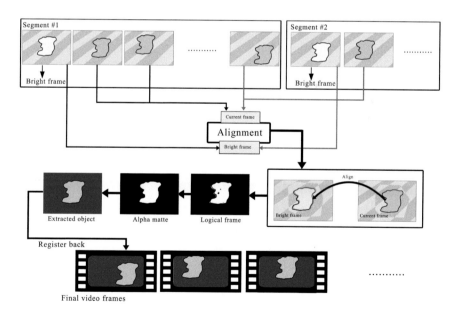

Fig. 1 Video Flash Matting approach is based on bright frames that are used in each segment of the video. Each frame is aligned with the closest bright frame and the alpha matte is extracted. The extracted object frame is aligned back to generate the final sequence of frames without the background

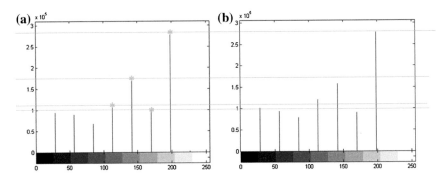

Fig. 2 Bright frame detection. (**a**) Histogram of a bright frame. (**b**) Histogram of a normal frame

final output video for generating a video using a sober illumination. The foreground object is tracked for determining the region of it in the given video using the planar tracking technique.[1]

Each bright frame starts a new segment in the given video. For each segment, each frame is aligned with the closest similar bright frame. For aligning each frame with the closest similar bright frame, Scale Invariant Feature Transform (SIFT) technique [12] is used for extracting features of each frame. Then, estimating the geometric transformation from the matching point pairs is performed using the following equation

$$A\chi = b, \tag{2}$$

$$A = \begin{pmatrix} x_I^{(1)} & -y_I^{(1)} & 1 & 0 \\ y_I^{(1)} & x_I^{(1)} & 0 & 1 \\ \vdots & & & \\ y_I^{(N)} & x_I^{(N)} & 0 & 1 \end{pmatrix}, \tag{3}$$

$$\chi = \begin{pmatrix} c \\ s \\ t_x \\ t_y \end{pmatrix}, \text{ where, } c = cos\theta, s = sin\theta, \tag{4}$$

$$b = \begin{pmatrix} x_B^{(1)} \\ y_B^{(1)} \\ \vdots \\ y_B^{(N)} \end{pmatrix}. \tag{5}$$

By solving the system of the linear equations, χ is found. Thus, the transformation matrix TM is generated for aligning the current frame I with the bright frame B,

---

[1]https://www.imagineersystems.com.

where TM is given by the following equation

$$TM = \begin{pmatrix} c & -s & t_x \\ s & c & t_y \\ 0 & 0 & 1 \end{pmatrix}, \tag{6}$$

For each aligned frame, the subtraction of the luma signal of the bright frame and the luma signal of the aligned frame is calculated using the following equation

$$L = Binarize(|\ Y_B - Y_A^*\ |), \text{ where } A^* \in S,\ B \in S \cup S_N. \tag{7}$$

where, L is the logical frame of subtracting the luma version Y_B of the bright frame B from the luma version Y_A^* of the aligned frame A^*, S is the current segment, and S_N is the next segment.

The logical frame is filtered for generating the binary mask of the foreground object. Fist, a pillbox filter within a square matrix is used for blurring the logical frame. Then, 8-connected component labeling is performed for removing small objects from the logical frame. Filtering the generated logical frame using an erosion followed by a dilation is performed as shown in the following equation

$$L_F = (L \ominus R) \oplus R, \tag{8}$$

where, L_F is the filtered logical frame, R is a morphological rectangle-shaped structuring element. Finally, the binary mask is generated after filling holes in the logical frame L_F using the morphological reconstruction [13]. For each aligned frame, the foreground object is extracted using the following equation

$$O_{(x,y)} = A_{(x,y)}^* B_{(x,y)}^*, \tag{9}$$

where, O is the frame that contains the foreground object, and B^* is the created binary mask of A^*. Then, the frame O is registered back using the geometric transformation as shown in the following equations

$$[x_F, y_F, 1]^T = TM^{-1}[x_O, y_O, 1]^T, \tag{10}$$

where, F is the final frame of Video Flash Matting, and TM^{-1} is the inverse of the transformation matrix of the registration process, see Algorithm 1.

3 Experiments and Discussions

Video Flash Matting approach is based on an intermittent flash. Using a higher frequency flash rate, leads to more accurate results. However, due to inability of

Data: A video V_I with an intermittent flash
Result: A video V_O of the foreground object
$n \leftarrow$ number of frames in V_I;
$i \leftarrow 0$;
$B_F \leftarrow$ ExtractBrightFrames(V_I);;
while $i < n$ **do**
    $I \leftarrow$ frame number (i) in V_I;
    if *IsBrightFrame(I)==false* **then**
        $[A^*, TM]$=AlignWithSimilarBrightFrame(I);
        L=Binarize(A^*, B);
        B^*=filter(L);
        $O=A^*B^*$;
        F=RegisterBack(O, TM^-1);
        $V_O \leftarrow F$;
    else
        Continue;
    end
end

Algorithm 1: Video Flash Matte

obtaining an intermittent flash with a high frequency, more than 25 flash/time, the flash of a smartphone is used. A mobile app was used for pulsing the flash of the smartphone. Samsung Galaxy S5 was used for this purpose. Canon Power Shot SX40 HS (24 frame/second) is used for recording the videos. Thus, the experimental results suffer from some problems with using a low frequency flash with a medium frame rate video camera. Those problems were solved by using a non-deformable foreground objects and dropping the multiple flash frames from the recorded video to be one bright frame per segment.

Figure 3 shows the steps of creating the binary mask of the foreground objects using Video Flash Matting. Figure 4 shows the final results of the proposed approach using a free camera and dynamic backgrounds. The proposed approach generates good alpha mattes that are used for extracting the foreground objects to change the original background with another real or virtual one. Figure 5 shows an example of integrating the extracted object with a real footage and a virtual scene that was modeled by the author.

A green screen in Multimedia Lab[2] was used for evaluating the accuracy of the generated alpha matte. A chroma-keying effect[3] was used for extracting the foreground objects frames. Those frames were used as the ground truth in the evaluation process.

Six videos, with approximately the same length and resolution (roughly 700 frames with 1280×720 pixels), were used in the evaluation. For each video, the Sum of Absolute Differences (SAD) and the Mean Squared Error (MSE) were calculated for evaluating the accuracy of Video Flash Matting approach compared with the ground truth, see Fig. 6. SAD is calculated using the following equation

[2] http://www.aun.edu.eg/multimedia/index.htm.
[3] Keylight effect | Adobe After Effects.

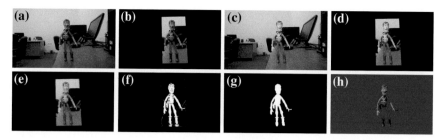

Fig. 3 Steps of creating the binary mask of the foreground objects. **a** The original frame without flash. **b** The tracked foreground region in the original frame. **c** The closet similar bright frame. **d** The tracked foreground region in the bright frame. **e** The alignment process. **f** The logical frame of subtracting the aligned frame from the bright frame. **g** The final binary mask. **h** The extracted foreground object using Video Flash Matting approach

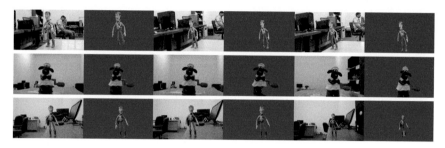

Fig. 4 The results of Video Flash Matting. Each row shows random frames of the original videos and the extracted foreground object

Fig. 5 Integrating the foreground object with a real background in **a** and a virtual one in **b**

$$SAD = \sum_{i=1}^{n} |\bar{Y}_i - Y_i|, \tag{11}$$

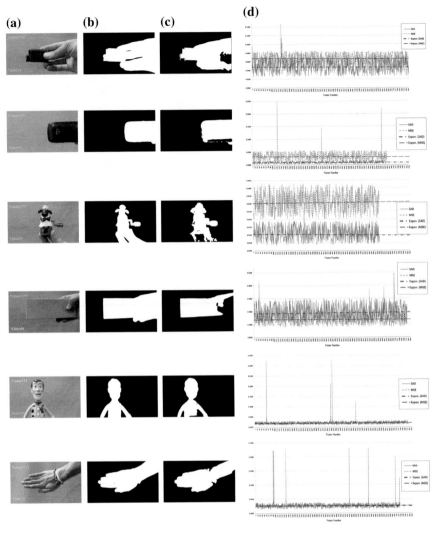

Fig. 6 Samples of videos that were used in the evaluation process. **a** The original frame. **b** The ground truth. **c** The alpha matte of Video Flash Matting approach. **d** Evaluating the results of the proposed approach using SAD (all values are coefficients of the order 10^6) and MSE

where n is the number of pixels in each frame, $\bar{Y}_i$ and Y_i represent the binary values of pixels number i in the ground truth alpha frame and the corresponding pixel in the generated alpha frame of the proposed approach, respectively.

MSE is calculated using the following equation

$$MSE = \frac{1}{n} \sum_{i=1}^{n} (\bar{Y}_i - Y_i)^2. \qquad (12)$$

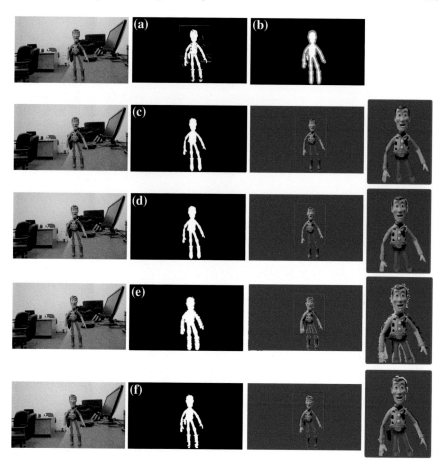

Fig. 7 Comparisons between the generated alpha matte of the proposed approach and other technique. The first column is the original frame. **a** The logical frame of the difference between the closest similar frame and the aligned frame. **b** The trimap image that is used by the other techniques. **c** The alpha matte of improving image matting technique [10]. **d** The alpha matte of weighted color and texture sample selection technique [9]. **e** The alpha matte of learning based digital matting technique [8]. **f** The alpha matte of the proposed approach without trimaps

Figure 7 shows a comparison between the alpha matte of the proposed approach and other techniques. As shown, Video Flash Matting generates an acceptable alpha matte compared with other techniques which require a trimap as a guidance, for SAD and MSE see Fig. 8.

Although the results appear to be valid, miss-alignment of frames with bright frames leads to undesirable matting as shown in Fig. 4, and in the divergence in MSE and SAD values as shown in Fig. 6.

The limitations of the proposed approach are:

1. Outdoor scenes require a strong flash, which is not easy to accessible.

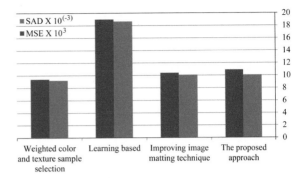

Fig. 8 SAD and MSE of the generated alpha matte of the proposed approach, improving image matting technique [10], learning based digital matting technique [8], and weighted color and texture sample selection technique [9]

2. In the case of foreground objects with high shininess or transparent foreground objects, foreground objects are hard to be extracted.
3. The need for a flash with a high frequency to achieve the extraction of deformable dynamic objects with a good quality.

4 Conclusions and Future Work

In this work, Video Flash Matting approach has been presented. The proposed approach is based on using an intermittent flash which is used for creating bright frames. Those frames are used in the alignment process for foreground objects extraction. For each frame, the closest similar bright frame is determined. Then, the subtracting operation is performed for extracting the foreground objects using an additional filtering operation. The results of the proposed approach show that Video Flash Matting extracts foreground objects without using green-screens, depth sensors, or trimaps. Video Flash Matting handles the camera movements and the dynamic background.

In the future work, Video Flash Matting can be improved using a high frequency flash and a video camera with a high frame rate for handling the extraction of deformable dynamic foreground objects. In addition, comparisons with other video matting techniques will be preformed.

References

1. Zhao, M., Chi-Wing, F., Cai, J., Cham, T.-J.: Real-time and temporal-coherent foreground extraction with commodity rgbd camera. IEEE J. Sel. Topics Signal Process. **9**(3), 449–461 (2015)
2. Afifi, M., Hussain, K.F.: What is the truth?: a survey of video compositing techniques. Accept. Int. J. Image, Graph. Signal Process. (IJIGSP) (2015)

3. Wang, L., Zhang, C., Yang, R., Zhang, C.: Tofcut: Towards robust real-time foreground extraction using a time-of-flight camera. In: Proceedings of 3DPVT (2010)
4. Wilson, A.D.: Using a depth camera as a touch sensor. In: ACM International Conference on Interactive Tabletops and Surfaces, pp. 69–72. ACM (2010)
5. Gonzalez-Jorge, H., Riveiro, B., Vazquez-Fernandez, E., Martínez-Sánchez, J., Arias, P.: Metrological evaluation of microsoft kinect and asus xtion sensors. Measurement **46**(6), 1800–1806 (2013)
6. Howe, N.R., Deschamps, A.: Better foreground segmentation through graph cuts. arXiv:cs/0401017 (2004)
7. Myeong, H., Lin, S., Lee, K.M.: Alpha matting of motion-blurred objects in bracket sequence images. In: Computer Vision—ECCV 2014, pp. 125–139. Springer (2014)
8. Zheng, Y., Kambhamettu, C.: Learning based digital matting. In: Proceedings of the 20th IEEE International Conference on Computer Vision, September–October (2009)
9. Shahrian, E., Rajan, D.: Weighted color and texture sample selection for image matting. In: 2012 IEEE Conference on Computer Vision and Pattern Recognition (CVPR), pp. 718–725. IEEE (2012)
10. Shahrian, E., Rajan, D., Price, B., Cohen, S.: Improving image matting using comprehensive sampling sets. In: 2013 IEEE Conference on Computer Vision and Pattern Recognition (CVPR), pp. 636–643. IEEE (2013)
11. Sun, J., Sun, J., Kang, S.B., Xu, Z.-B., Tang, X., Shum, H.-Y.: Flash cut: Foreground extraction with flash and no-flash image pairs. In: IEEE Conference on Computer Vision and Pattern Recognition, 2007, CVPR'07, pp. 1–8. IEEE (2007)
12. Lowe, D.G.: Distinctive image features from scale-invariant keypoints. Int. J. Comput. Vis. **60**(2), 91–110 (2004)
13. Soille, P.: Morphological Image Analysis: Principles and Applications. Springer Science & Business Media (2013)

Enhanced Region Growing Segmentation for CT Liver Images

Abdalla Mostafa, Mohamed Abd Elfattah, Ahmed Fouad, Aboul Ella Hassanien and Hesham Hefny

Abstract This paper intends to enhance the image for the next usage of region growing technique for segmenting the region of liver away from other organs. The approach depends on a preprocessing phase to enhance the appearance of the boundaries of the liver. This is performed using contrast stretching and some morphological operations to prepare the image for next segmentation phase. The approach starts with combining Otsu's global thresholding with dilation and erosion to remove image annotation and machine's bed. The second step of image preparation is to connect ribs, and apply filters to enhance image and deepen liver boundaries. The combined filters are contrast stretching and texture filters. The last step is to use a simple region growing technique, which has low computational cost, but ignored for its low accuracy. The proposed approach is appropriate for many images, where liver could not be separated before, because of the similarity of the intensity with other close organs. A set of 44 images taken in pre-contrast phase, were used to test the approach. Validating the approach has been done using similarity index. The experimental results, show that the overall accuracy offered by the proposed approach results in 91.3 % accuracy.

Keywords Region growing · Morphological operations · Filtering · Segmentation

A. Mostafa (✉) · H. Hefny
Institute of Statistical Studies and Researches, Cairo University, Giza, Egypt
e-mail: abdalla_mosta75@yahoo.com

M.A. Elfattah
Faculty of Computers and Information, Mansoura University, Mansoura, Egypt

A. Fouad
Faculty of Computers and Informatics, Suez Canal University, Ismailia, Egypt

A.E. Hassanien
Faculty of Computers and Information, Cairo University, Giza, Egypt

A. Mostafa · M.A. Elfattah · A. Fouad · A.E. Hassanien
Scientific Research Group in Egypt (SRGE), Giza, Egypt
URL: http://www.egyptscience.net

© Springer International Publishing Switzerland 2016
T. Gaber et al. (eds.), *The 1st International Conference on Advanced Intelligent System and Informatics (AISI2015), November 28–30, 2015, Beni Suef, Egypt,*
Advances in Intelligent Systems and Computing 407, DOI 10.1007/978-3-319-26690-9_11

115

1 Introduction

At the time being, A lot of efforts have been done by researchers to support radiologists to make their decision. All the efforts are trying to offer them an effective CAD system. CAD systems should be composed of some main phases. It starts with preprocessing phase where filtering is used to points out features and get rid of noise and distortion [11]. The preprocessing passes the image to another phase which is segmentation. Liver is segmented and separated from other organs, using a simple known region growing segmentation method. Liver separation requires a smoothing filter to conceal the unnecessary details such as arteries, veins and lesions. Also some morphological operations could be used to deep the boundaries of the liver. Segmentation process separates the anatomical structure of liver from the abdomen, depending on the intensity values in the image. Images with JPG format are implemented in this paper to segment liver and extract regions of interest (ROIs). It has intensity values range between 0 and 512 [2]. In CT images, there is a similarity between liver intensity and some other organs intensities, like spleen, skin and muscles. So, the extraction of liver area is appropriate for next steps of segmentation of ROIs and features extraction. The proposed approach in this paper smooths image and uses some morphological operations to separate or segment the whole liver using region growing method. Finally, The segmented image is tested using similarity index measure. Liver is an important gland in the body, that has many important functions. Liver is situated under the diaphragm in the upper abdominal cavity, and is held in place by several ligaments. It is a reddish-brown colour organ, but it may have different colours according to different lesions. Colours could be dark blue for cyst, dark brown for cirrhosis, yellow for fatty, green for biliary cirrhosis. These colours have different intensity values in CT grey images. This is one of the difficulties in liver segmentation. Also, falciform ligament separates the liver lobes as a boundary, which could be a kind of difficulty in segmenting the liver from other regions. Other difficulty, could be blood vessels as portal vein and hepatic artery [12]. A CT (computed tomography) is a special type of X-ray machines. Compared to X-ray machine which sends a single X-ray through the body, CT machine sends several beams simultaneously to the body from different angles. The X-ray that passes the body are detected and their strength are measured. The beams of X-ray that passes through tissues will be stronger than that passes through denser tissues. So, computer can differentiate between soft tissues of liver and hard tissues of bones. Using these information of sets of cross-section, intensities are calculated and interpreted as a two dimensional image [4]. CT images have an advantage against x-ray, it is more detailed than x-ray. The remainder of this paper is ordered as follows. Section 2 demonstrates the morphological operations. Section 3 describes filtering techniques which is tested for the next segmentation phase. Section 4 gives a brief description of region growing segmentation method. Details of the proposed approach is shown in Sect. 5. Section 6 shows the experimental results and analysis. while, conclusions and future work are discussed in Sect. 7.

2 Morphological Operations

Mathematical morphology is a set theory approach, developed by J. Serra and G. Matheron. It provides an approach to processing of digital images that is based on geometrical shape. Two fundamental morphological operations erosion and dilation are based on Minkowski operations.

It has been used for extracting edges and detecting characteristic objects in mobile photogrammetry systems for making maps from images taken from a car, called mobile mapping systems (Tao, 1998). The author points into robustness of results during achievement of specific information from series of images as a main goal for future researching task. Morphology is used mainly to decrease an area of interest and extracting specific objects.

Images have been treated by two basic morphology functions working oppositely erosion and dilation or those composition opening and closing. Each case needs proper structural element depending on size of grains in picture noise.

Order of these two functions is important because of existence of thin dark or light lines in imaged scene. Industrial objects have usually dark lines which appear on the edge of joining parts of the device, thus in the beginning should be used erosion or opening in order to prevent those edges from disappearing.Morphological filters base on a simple idea of alternating two operations: opening and closing removing, respectively, small high-valued and low-valued objects from the image. These filters, called generally alternate filters differ from each other in the type of opening and closing operations applied. One of the morphological filters seems to be very effective. It is called Alternate Filter with Multiple Structuring Function (Element) the sequence of opening and closing operations with multiple structuring element. It removes the small objects from the image but preserves edges and linear objects, which is very important for the purpose of image filtering as the preprocessing for edge detection [9]. The basic operations in the area of mathematical morphology are described as follows:

- **Structure element**

  It is a shape, used to interact with a given image, It is typically used in morphological operations, such as dilation, erosion, opening, and closing.
- **Dilation**

  It is the first basic operators of mathematical morphology, the second one is erosion. It can be applied to binary and grey-scale images. This operator affects binary image by enlarging the boundaries of regions of foreground pixels (typically, white pixels). So, the number of pixels of the foreground grow in size. And it reduces the size the holes within those regions.
- **Erosion**

  It is the second basic operators of mathematical morphology. It can be applied to binary and grey-scale images. This operator affects binary image by eroding away the boundaries of regions of foreground pixels.So, the number of pixels of the foreground shrinks in size. And it increases the size the holes within those regions.

- **Opening and closing**
  They are two important operators of mathematical morphology. Both of them are derived from the basic operations of erosion and dilation. They could be applied to binary images, and also grey-level images.

  The **opening** is dilation followed by erosion. Opening basic effect resembles the effect of erosion in that it removes some of the pixels of the foreground (bright) from the edges of different regions. But in general, it is less destructive than erosion. The operation uses a structuring element to preserve foreground regions which have a shape that resembles this structuring element, or that regions that contain the whole structuring element. And it eliminates all rest regions of foreground pixels.

  Somehow, **closing** is similar to the operation of dilation. It is erosion followed by dilation. It shrinks the holes, while enlarging the boundaries of the foreground regions [5].

3 Filtering Techniques: Preliminaries

An important role is played by initially filtering the image. It increases the accuracy in computer aided detection systems. This section will give a brief introduction to five image filtering techniques. They proved efficiency in aiding segmentation process. They include contrast stretching, texture range, inverse transformation, average, and median filters. Inverse transformation and logarithm transformation filters and combinations of them. Contrast and texture range filters aid in determining the contours of liver, while smoothing hides the small details like veins and ligaments. Sharpening filters are excluded because they show small details which is not needed to segment the liver. Also some smoothing filters that do not give a promising results are excluded such as correlation transformation, histogram equalization, Laplacian, Gaussian, Laplacian of Gaussian, Prewitt, and Sobel [11]. An application of CT liver lesion imaging has been written and each filter have been applied to see their ability and accuracy to segment the whole liver images. This section reviews the used filtering techniques.

- **Contrast stretching filter** is a process of expanding the range of intensity levels in an image to span the full intensity level. Data is saturated at low and high intensities increasing the contrast of the image. The boundaries of low and high intensities of the input image is stretched to new boundaries in the output image. The used values are imadjust (im, [0.25 0.65], [0 .8]).
- **Texture range** is a statistical texture filter that calculates the local standard deviation of an image. Range filter is used to find edges within an image. In the range method, the colour value of each pixel is replaced with the difference of maximum and minimum of colour values of the pixels in a surrounding region [10].

- **Inverse transformation filter** uses the inverse transformation of contrast stretching approach with a condition of high boundary of the output image is less than the low boundary of the output image then the output image is reversed.
- **Average filter** smooths an image by reducing the variation in intensities. This is done by replacing the intensity level at a point by the average of the intensities in a neighbourhood of the point.
- **Median filter** is a non-linear filter that replaces each pixel with the median of the accumulated values of the mask. Median filter is used to reduce noise in an image. Moreover, it is used to reduce image resolution without losing the edges and resulting in a smoothed image.

4 Region Growing Segmentation

Segmentation methods are moving around two main categories, edge-based segmentation and region-based segmentation. While edge-based segmentation uses the gradient to determine edges using the discontinuity concept, region-based segmentation uses similarity difference to determine regions. Edge-based category is represented in many methods like Sobel, Prewitt, Roberts, zero crossings, Laplacian of Gaussian (LoG) and canny. Region-based includes methods like region growing, region split and merge [5], watershed [1] and level set [3]. Also thresholding methods can be used as Otsu [13] and Otsu improved by histogram [8]. Region growing has been chosen to be used in the process of segmenting liver from the abdomen image and testing the effect of filtering on segmentation process. When segmenting liver organ, some liver properties cause some limitations and problems in segmentation process. Observation provides some of these limitations such as (1) Liver is adjacent to the ribs that are covered by flesh cells and muscles that have similar intensities to liver. So liver segment might include this flesh and (2) The intensity values of stomach, kidneys and spleen are too close to liver intensity. If there is a slight band connecting between them, they might be included in liver segment. Sometimes almost the whole abdomen could be included in a wrong way. A number of filters and its combinations will be used to aid solving this problem.

5 CT Liver Segmentation Proposed Approach

The CT liver segmentation approach is comprised of some fundamental building phases:

- **Pre-processing phase**
  In the suggested approach, a pre-processing algorithm is performed before segmentation, to perform some operations to enhance contrast, remove noise, stress

boundaries of liver. It also emphasizes certain features and connects ribs around the liver.

- **Region growing phase**
  In the second phase, region growing segmentation method is used to separate the whole liver.

Figure 1 illustrates the architecture of the proposed liver image segmentation approach. These two phases are described in detail in the following section, along with the involved steps and the characteristics of each phase.

5.1 Preprocessing Phase: Applying Filters and Ribs Boundary Detection

In this paper, there is a main role for filters to play in segmentation process. Filters can enhance the sensitivity and accuracy of CAD systems for lesions detection. Smoothing concept is used to conceal the very little details in liver as veins, arteries and ligaments. That enables the region growing method to grow the initial seed points till extracting the whole liver. On the other hand, contrast concept combined with thresholding and morphological operations, could be used to emphasize the boundaries of all ROIs in the extracted liver. That includes the anomalies that have different intensities and features. The most relevant filters could be average, gabor, gaussian and contrast stretching filters.

Liver segmentation is the process of deriving liver from the abdomen image. The organ of liver has some properties, that may cause some limitations and problems while processing segmentation [12]. Observations can lead to some of these limitations:-

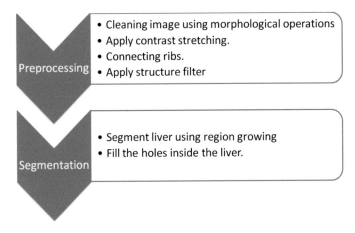

Fig. 1 Liver image segmentation approach: Phases

- Liver has different shapes, according to the abdomen slice, taken by the scanning machine.
- It also has different shapes according to the patient, especially for enlarged liver and exposure to surgery operations.
- Liver is adjacent to the ribs, that are covered by flesh cells, which have the same intensity of liver. So liver segment can include this flesh.
- Belly muscles, and flesh under the skin and outside the ribs, are difficult to separate from liver, because they have intensity values similar to liver, and connected by flesh over the ribs.
- The intensity values of stomach, kidney, and spleen are too close to liver intensity. If there is a slight band connecting them, they may be included in liver segment. Sometimes almost the whole abdomen, could be included in a wrong way.
- In some cases, dark infected spots near edges in liver are excluded because they have different intensity values.

Cleaning image is the first step that removes the annotation around the abdomen, the machine's bed, to ease the next operation of connecting ribs. The following algorithm (1) shows the steps that uses the morphological operations (open, close, erode and fill) to clean the image as follows:

Algorithm 1 Image cleaning

1: Using the graythreshold.
2: Use open operation to remove all objects that have pixels fewer than 5500 pixels, that removes all annotations.
3: Some small areas in the liver are removed, so strl (structure element) is used to close the binary image.
4: Fill all holes inside the abdomen.
5: Erode the edges of abdomen.

The step of opening includes cleaning image by removing machine's bed and the annotation of the image that includes patient information. It erodes the skin of the abdomen. It also removes the bed of the machine. Ribs boundary algorithm is used, to overcome the problem of muscles which are adjacent and connected to liver [11]. A thick line is drawn to connect the pixels of the ribs bones. The ribs are connected using black pixels. This will guarantee a better connection between the ribs. For the above mentioned intensity values of stomach, kidneys and spleen problem, where its values is similar to liver. This is manipulated by some other filters contrast stretching and structure filters. Figure 2 demonstrates visually, the results of such process using adjust contrast and structure filters.

(a)   **(b)**   **(c)**

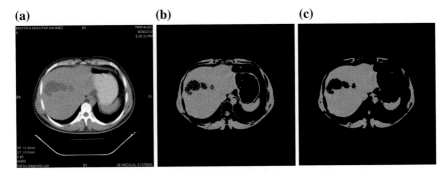

Fig. 2 Adjust contrast and texture filters to emphasize the boundaries, **a** Original image, **b** Adjusted image, **c** Texture filter ribs

5.2 Region Growing Phase

Segmentation methods are divided into two main branches, edge-based segmentation and region-based segmentation. While edge-based segmentation uses the gradient and discontinuity to detect edges, region-based segmentation methods uses similarity difference to get regions. Level set, Fast marching, and watershed segmentation methods are some of region-based segmentation methods [14]. Region growing is chosen to be used in this paper to implement the extraction of the whole liver. For automatic seed points, we can use statistical calculation of all available liver segmented images. Algorithm (2) shows how to calculate the seed points to be passed to region growing technique.

Algorithm 2 Proposed

1: Prepare all segmented images.
2: Convert them into binary images, which have 1 when the pixel value > 0.
3: Sum all binary images to an empty image.
4: Calculate the average of each pixel according to the number of images.
5: The pixels which have average value equals one are the required seed points.

Algorithm (3) demonstrates the steps of liver segmentation proposed approach.

Algorithm 3 Proposed

1: Clean the image, removing the annotation, machine bed and body flesh.
2: Apply contrast stretching filter.
3: Connect ribs.
4: Apply structure filter.
5: Segment the liver by using region growing segmentation method.
6: Fill all holes inside the liver.

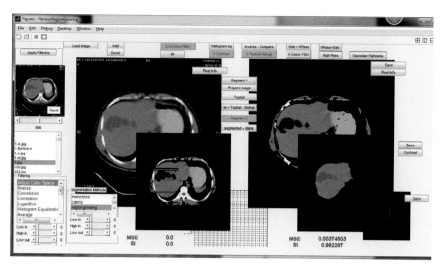

Fig. 3 Application of proposed approach

Figure 3 shows the implemented application to the proposed approach. It shows other different tested combinations of filters including gabor, inverse, histogram, high pass and gaussian. It also shows top-hat, bottom-hat and dilation. Open and close morphological operations are included when using region growing segmentation.

6 Experimental Results and Discussion

The experiments on the proposed approach have been applied using 44 real CT images of patients in the first phase of CT scan before patient injection of contrast materials. Region growing method is compared to the proposed approach. Figure 4 shows the result of the procedure Fig. 5 shows the result of connecting ribs process changing the ribs white colour into black. The resultant image is compared to an image manually segmented as shown in Fig. 6.

Evaluation is performed using SI where *SI* is the similarity index defined using the following equations:

$$SI(I_{auto}, I_{man}) = \frac{I_{auto} \bigcap I_{man}}{I_{auto} \bigcup I_{man}} \tag{1}$$

where I_{auto} is the binary automated segmented image, resulting from the phase of segmentation of the whole liver in the used approach and I_{man} is the binary manual segmented image by a radiology specialist.

According to Table 1, the average performance of liver images segmentation is improved using the proposed approach. Segmentation using region growing has average result of SI = 84 %. This result is improved using the proposed approach with

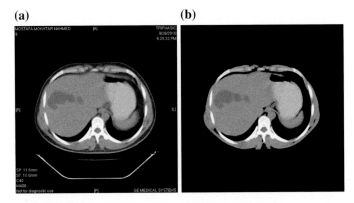

Fig. 4 Cleaning image annotation: **a** Original image, **b** Cleaned annotation

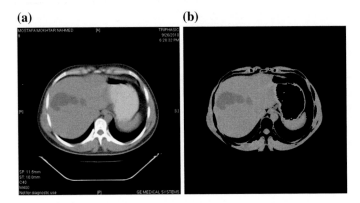

Fig. 5 Ribs boundary, **a** Original image, **b** Connected ribs

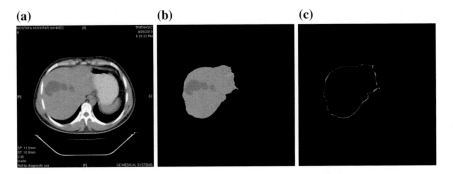

Fig. 6 Segmented image compared to the annotated image, **a** Original image, **b** Segmented image, **c** Annotated image

Table 1 Results of proposed approach compared to region growing

Img.	RG	Prop.	L. set	Img.	RG	Prop.	L. set	Img.	RG	Prop.	L. set	Img.	RG	Prop.	L. set
1	0.881	0.942	0.961	12	0.893	0.88	0.93	23	0.708	0.939	0.87	34	0.947	0.957	0.946
2	0.927	0.936	0.904	13	0.757	0.772	0.92	24	0.848	0.95	0.93	35	0.941	0.935	0.93
3	0.831	0.876	0.928	14	0.917	0.959	0.93	25	0.616	0.807	0.888	36	0.942	0.936	0.91
4	0.74	0.782	0.88	15	0.902	0.946	0.96	26	0.696	0.911	0.92	37	0.866	0.915	0.919
5	0.868	0.903	0.91	16	0.919	0.943	0.89	27	0.682	0.927	0.87	38	0.909	0.943	0.934
6	0.849	0.938	0.89	17	0.938	0.95	0.96	28	0.82	0.904	0.92	39	0.888	0.952	0.915
7	0.956	0.945	0.89	18	0.91	0.937	0.95	29	0.913	0.954	0.91	40	0.856	0.902	0.862
8	0.934	0.936	0.908	19	0.898	0.933	0.92	30	0.93	0.941	0.97	41	0.855	0.901	0.87
9	0.923	0.905	0.948	20	0.622	0.891	0.9	31	0.921	0.95	0.94	42	0.913	0.901	0.866
10	0.939	0.924	0.95	21	0.536	0.897	0.915	32	0.908	0.923	0.96	43	0.671	0.881	0.901
11	0.89	0.866	0.97	22	0.88	0.872	0.88	33	0.92	0.948	0.95	44	0.661	0.855	0.909
Res.	0.848	0.913	0.917												

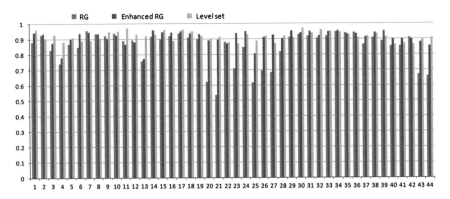

Fig. 7 Similarity index results for 44 segmented images

SI = 91.3 %. Compared to level set which results in 91.7 %, the proposed approach overcomes level set in simplicity and computational cost.

Figure 7 shows *SI* for forty four CT liver images, where each point in the figure represents the similarity of one segmented image, compared to its annotated image. Experimental results showed that the average accuracy for liver segmentation using proposed approach, investigated by SI was 91.3 %. But the figure shows that the proposed approach managed to segment 9 images that was not segmented before.

7 Conclusion and Future Work

The segmentation of liver is a complicated process, compromised of many steps for preprocessing and segmentation. The preprocessing is involved in finding the similar intensities of other organs like spleen, flesh and muscles. The algorithm of ribs connecting combined with contrast stretching, morphological operations and can help to handle this problem. The whole liver segmentation using region growing has a considerable average accuracy rate 91.3 % using *SI*. Optimizations techniques are a good trend that includes firefly algorithm, fish swarm, genetic algorithm, and bacterial foraging. Our future work is to use optimization techniques in segmentation and classification of the anomalies of liver.

References

1. Aayushi, A.K.: Image segmentation using watershed transform, Int. J. Soft Comput. Eng. (IJSCE), **4** (2014)
2. Bandkman, I.N.: Handbook of Medical Imaging Processing and Analysis. Academic Press, San Diego (2000)

3. Caselles, V., Kimmel, R., Sapiro, G.: Geodesic active contours. Int. J. Comput. Vis. **22**, 61–79 (1997)
4. Chen, M.Y.M., Pope, T.L., Ott, D.J.: Basic Radiology, 2nd edn. McGrow Hill, New York (2011)
5. Gonzales, R.C., Woods, R.E.: Image segmentation. Digital Image Processing, 3rd edn. Pearson Printice Hall, New jersy (2008)
6. Hassanien, A., Abraham, A., Peters, J.F., Schaefer, G., Henry, C.: Rough sets and near sets in medical imaging: a review. IEEE Trans. Inf. Technol. Biomed. **13**(6), 955–968 (2009)
7. Hsu, C.Y., Liu, C.Y., Chen, C.M.: Automatic segmentation of liver PET images. Comput. Med. Imaging Graph. **32**(7), 601–610 (2015)
8. Jun, Z., Jinglu, H.: Image segmentation based on 2D otsu method with histogram analysis. In: International Conference on Computer Science and Software Engineering (2008)
9. Kowalczyk, M., Koza, P., Kupidura, P., Marciniak, J.: Application of mathematical morphology operations for simplification and improvement of correlation of image in close-range photomography, The International Archives of the Photogrammetry, Remote Sensing and Spatial Information Sciences, vol. XXXVII, Part B5. Beijing (2008)
10. Mamatha, Y.N., Ananth, A.G.: Performance of texture based filters for the extraction of features from the sateliite imageries using confusion matrix and artificial neural networks. In: 4th International Conference on Computer Engineering and Technology (ICCET 2012) IPCSIT, vol. 40. Singapore (2012)
11. Mostafa, A., Hefny, H., Ghali, N.I., Hassanien, A., Schaefer, G.: Evaluating the effects of image filters in CT liver CAD system, IEEE-EMBS International Conference on Biomedical and Health Informatics (BHI2012). The Chinese University of Hong Kong, Hong Kong SAR, January 5–7 (2012)
12. Sherlock, S.: Sherlock's Diseases of the Liver and Biliary System, 12th edn. Wiley, New Jersey (2011)
13. Vala, H.J., Baxi, A.: A Review on Otsu Image Segmentation Algorithm. Int. J. Adv. Res. Comput. Eng. Technol. (IJARCET) **2**(2), 387 (2013)
14. Vanhamel, I., Pratikakis, I., Sahli, H.: Multiscale gradient watersheds of color images. IEEE Trans. Image Process. **12**(6), 617–626 (2003)

A Multi-Objective Genetic Algorithm for Community Detection in Multidimensional Social Network

**Moustafa Mahmoud Ahmed, Ahmed Ibrahem Hafez,
Mohamed M. Elwakil, Aboul Ella Hassanien
and Ehab Hassanien**

Abstract Multidimensionality in social networks is a great issue that came out into view as a result of that most social media sites such as Facebook, Twitter, and YouTube allow people to interact with each other through different social activities. The community detection in such multidimensional social networks has attracted a lot of attention in the recent years. When dealing with these networks the concept of community detection changes to be, the discovery of the shared group structure across all network dimensions such that members in the same group interact with each other more frequently than those outside the group. Most of the studies presented on the topic of community detection assume that there is only one kind of relation in the network. In this paper, we propose a multi-objective approach, named MOGA-MDNet, to discover communities in multidimensional networks, by applying genetic algorithms. The method aims to find community structure that simultaneously maximizes modularity, as an objective function, in all network dimensions. This method does not need any prior knowledge about number of communities. Experiments on synthetic and real life networks show the capability of the proposed algorithm to successfully detect the structure hidden within these networks.

M.M. Elwakil · A.E. Hassanien · E. Hassanien
Faculty of Computers and Information, Cairo University, Giza, Egypt
e-mail: m.elwakil@fci-cu.edu.eg

A.E. Hassanien
e-mail: aboitcairo@gmail.com

E. Hassanien
e-mail: e.ezat@fci-cu.edu.eg

M.M. Ahmed (✉) · A.I. Hafez
Faculty of Computers and Information, Minia University, Minya, Egypt
e-mail: moustafa_fci@yahoo.com

A.I. Hafez
e-mail: ah.hafez@gmail.com

M.M. Ahmed · A.I. Hafez · A.E. Hassanien
Scientific Research Group in Egypt (SRGE), Giza, Egypt
URL: http://www.egyptscience.net

© Springer International Publishing Switzerland 2016   129
T. Gaber et al. (eds.), *The 1st International Conference on Advanced Intelligent
System and Informatics (AISI2015), November 28–30, 2015, Beni Suef, Egypt*,
Advances in Intelligent Systems and Computing 407, DOI 10.1007/978-3-319-26690-9_12

Keywords Social network · Community detection · Complex Network · Multidimensional network · Multi-objective genetic algorithms

1 Introduction

Networks or graphs are a suitable model for representing interactions that take place between entities in the real world systems. The web, telecommunication network, biological networks, and social networks are examples of systems modeled naturally as networks, where nodes in the network represent entities and edges represent relationships between pairs of entities. A network is said to have community structure, if it can be divided into groups of densely connected nodes, with the nodes belonging to different communities being only sparsely connected. Therefore, communities are groups of entities that probably share some common properties and/or play similar roles within the system that is being represented. The identification of communities hidden within networks is an important task for a wide set of fields such as sociology, physics, and biology [1].

Community detection in social network has been widely studied in the literature over last few years [2, 3] (Results of a recent survey can be seen in [4, 5]). However, most of these studies have been discarded that, real world social networks are often multidimensional, i.e. many different types of connections may take place between pairs of entities at the same time, reflecting different types of interactions between them. For example, user in Facebook can be connected with his friends through friendship relationship; user can be a follower of other users; also user can like/share/comment on social content like photos and videos posted by other users. Each type of these interactions can be represented by single dimension. This imposes a great challenge to the conventional graph clustering algorithm.

In this paper we propose a multi-objective approach, named MOGA-MDNet, to discover communities in multi-dimensional networks by applying genetic algorithms. The method aims to find community structure that simultaneously maximizes modularity, as an objective function, in all network dimensions, so maximizing modularity in a single dimension can be considered an objective needed to be optimized. MOGA-MDNet exploits the benefits of these objectives and discovers the communities present in the network structure by selectively exploring the search space, without the need to know in advance the exact number of communities. This number is automatically determined by the optimal adjustment of objectives values. An interesting aspect of the multi-objective approach is that it aims to find the optimal community structure in the presence of trade-offs between all conflicting objectives, so it returns not a single partitioning of the network, but a set of solutions which are called Pareto-optimal solutions. Each of these solutions satisfies a different trade-off between all objectives. Experiments on synthetic and real life networks show the capability of the multi-objective genetic approach to correctly detect communities with results comparable to the state-of-the-art approaches.

The rest of the paper is organized as follows. In Sect. 2, we review some related work. In Sect. 3, the concept of multi-dimensional community is defined and the community detection problem is formalized as a multi-objective optimization problem. In Sect. 4, we described the genetic representation adopted and the variation operators used. In Sect. 5, the results of the method on synthetic and real life networks are presented. In Sect. 6, we give some concluding remarks.

2 Related Work

There are many studies on community detection, in different fields of research: computer science, physics, sociology, and others. Most of them define community detection as dividing network nodes into set of groups such that nodes in the same group interact with each other more frequently than those outside the group. The papers working with this definition rely on the concept of modularity, which is a function proposed by Newman [2] to measure the clustering quality by detecting the ratio between intra- and inter-community numbers of edges. One of the first community detection algorithms based in modularity is the Girvan-Newman algorithm, which introduces a divisive method that iteratively removes the edge with the greatest betweenness value [2].

Many optimization techniques have been applied to optimize the modularity as an objective such as greedy optimization [3], extremal optimization [6], simulated annealing [7], and genetic algorithms (GA) [8]. However, one aspect of such networks has been ignored so far: real networks are often multidimensional, i.e. two nodes may connect through many different types of connections, reflecting different types of relationships or interactions between entities.

Recently multidimensionality has been taken into account in a lot of works. In [9, 10] authors extended and defined a set of analytical measures which take into account the structure of multidimensional networks to open the way for new techniques to analyze and extract non trivial knowledge from this type of networks. In [11–13] authors presented many integration strategies in order to map multidimensional network into mono-dimensional network to be able to apply existing solutions to multidimensional network, but the most of these strategies don't consider the degree to which each node participate in each dimension, as in reality, social actors often participate with varied intensity in different dimensions of network, thus, within the same group, the interaction can be very sparse in one dimension but relatively more observable in another dimension. So using these strategies; dimensions with intensive interaction would force other dimensions to follow the same structure. In [14] authors introduce simple two-phase strategy called PMM to extend modularity maximization from one-dimensional networks to multi-dimensional networks, however this method ignores the fact that different relations might have different importance with respect to a certain query.

3 Community Detection in M-D Networks Problem

A network can be defined as a graph $G = (V, E)$, in which V is a set of vertices or nodes, and E is a set of ties that connect nodes. In the field of social networks, nodes represent persons or actors within the network, and ties represent the relationships or the interaction between those persons. To take into account the presence of more than one type of relationships, a multidimensional network is represented by a multiple graph; one graph for each type of relationships (dimensions), a D-dimensional network is represented as $G = \{G_1, G_2, \ldots, G_d\}$ where G_i represents the interactions between social actors in the ith dimension.

Community in multidimensional networks is a set of nodes densely connected including interactions in all network dimensions. The problem can be formulated as dividing network's nodes into k communities, where the number k is unknown, such that modularity is maximized for each dimension. Thus we treated this problem as a multi-objective optimization problem, where given the modularity as a quality measure we want to find community structure S that simultaneously maximize modularity in each dimension.

A multi-objective optimization problem $(\Omega; f_1, f_2, \ldots, f_d)$ is defined as

$$min f_i(S), i = 1, 2, \ldots, t \ s.t \ S \in \Omega \tag{1}$$

where $\Omega = \{S_1, S_2, \ldots, S_r\}$ is the set of feasible community structures in a network. Each $f_i \in \{f_1, f_2, \ldots, f_d\}$ is the modularity related to a different network dimension i where d is the number of dimensions in the network.

4 Algorithm Description

In this section we describe the multi-objective algorithm MOGA-MDNet, the genetic representation of community structure, the objective functions selected, and the genetic operators used by the algorithm.

Genetic representation: Our algorithm adopts the locus-based adjacency representation proposed in [15]. According to this representation any individual (chromosome) c, in the population consists of n genes, where n is the number of nodes in the network, and each gene correspond to a node in this network. In individual c each gene g_i can assigned an arbitrary integer value j, where j is in the range of $\{1, 2, \ldots, n\}$. A value j assigned to the gene g_i is interpreted as a link between the nodes i and j which means that, nodes i and j will be in a same community. To figure out community structure, a decoding step is necessary to identify connected components which can be done in linear time [16].

A main advantage of this representation is that there is no need to know in advance the number of communities, as the number of communities denoted by each individual is automatically determined in the decoding process.

Genetic initialization: Our algorithm adopts the initialization process proposed in [17] (safe initialization), which takes in account the effective connections of the nodes in the social network. A random generation of individuals could generate components that are disconnected in the original graph. When generating individual gene g_i could be assigned to value j, but no connection between nodes i and j exists in the original graph, which means that grouping both nodes i and j in the same group is a wrong choice. Using safe initialization such this case is avoided by substituting value j with one of the neighbors of i.

Crossover operator: Uniform crossover is used. Given two parents, a random binary vector is created. Then uniform crossover selects the genes where the vector is a 1 from the first parent, and the genes where the vector is a 0 from the second parent. Then these genes are combined to form the new child.

Mutation: The mutation operator that randomly changes the value of a randomly chosen gene causes a useless exploration of the search space. Therefore, as in the initialization step, we randomly select percent of the genes and for each gene i we randomly change its value to j such that node i and j are neighbors.

Fitness function: In genetic algorithms, the fitness function plays an important role in the evolution process. For the community detection problem, there is a wide variety of methods to measure the quality of discovered community structure [8], each of which could potentially be used as a fitness function. We decided to focus on modularity quality measure as an objective to determine the optimal community structure. Modularity [18] was designed specifically to measure the strength of a community structure for real-world networks. Networks with high modularity have dense connections between the nodes within communities but sparse connections between nodes in different communities. Studies in the literature shown that modularity is effective in various kinds of complex networks [19, 20].

5 Experimental Results

We tested our algorithm on a synthetic data set and real world social networks. For evaluation, we compared the results obtained by our algorithm with the results obtained by the Principal Modularity Maximization (PMM) method proposed in [14]. Normalized mutual information (NMI) [21], which measure the similarity between the discovered community structures and the true ones is adopted to measure the clustering performance in the controlled experiments. For each network we began by applying the single objective genetic algorithm to detect communities in each dimension separately as baselines. We employed standard parameters for the GA: a crossover rate of 0.8, a mutation rate of 0.4, the elite reproduction was 10 % of the population size, the population size was 200, and the number of iterations was 100. The algorithm was implemented in Python environment using DEAP (Distributed Evolutionary Algorithms in Python) [12]. Then in the multi-objective case we used the Nondominated Sorting Genetic Algorithm (NSGA-II) proposed in [22] implemented in the MATLAB Genetic Algorithm and Direct Search Toolbox as the multi-objective GA. We employed standard parameters for the GA; a crossover rate of 0.8, a mutation rate of 0.2, the elite reproduction rate was 10 % of the population size. We also employed a binary tournament selection function. The population size was 200, and the number of generations was 200.

5.1 Synthetic Network Dataset

To conduct some controlled experiments, we used the benchmark proposed in [14]. The network consists of 350 nodes divided into three communities, with each having 50, 100, 200 nodes respectively. There are 4 different types of relationships (dimensions) among these nodes. For each dimension, group members connect with each other following a random generated within-group interaction probability. To discover communities in this network we applied our algorithm ten times. For each run, we investigated the set of Pareto-optimal returned by the algorithm; each solution we calculated the summation of modularity over all dimensions then we selected the community structure with the maximum summation; and compared it with the optimal known structure of the network in terms of the NMI similarity measure. Average NMI over the 10 runs are calculated and reported in Table 1. Table 1 also reports the average performance of PMM method in terms of NMI as declared in [14].

5.2 Real World Social Network Data

We use three real-world data sets to test our algorithm, YouTube network [14], students cooperation network [23], and medical innovation network [24]. Table 2 shows the density of all dimensions of each network.

- **Students Cooperation Social Network**: The students cooperation social network [23] was constructed from the data collected during a "Computer and Network Security" course; a mandatory course at Ben-Gurion University. The students

Table 1 Average NMI results obtained by MOGA-MDNet and PMM algorithms for synthetic network

	Strategy	Average NMI
Single-Dimensional using SGA	D1	0.6587
	D2	0.5176
	D3	0.6544
	D4	0.6617
Multi-Dimensional	PMM	0.935
	MOGA-MDNet	0.946

Table 2 The Density of each dimension for the real world social networks datasets

Network	Dimension	Density
Students Cooperation	Partner	1.41×10^{-2}
	Same computer	1.35×10^{-3}
	Same time	5.69×10^{-3}
Medical Innovation	Advice	1.59×10^{-2}
	Discussion	1.87×10^{-2}
	Friend	1.68×10^{-2}
YouTube	Contact	6.74×10^{-4}
	Co-contact	7.28×10^{-2}
	Co-subscription	4.90×10^{-2}
	Co-subscribed	1.97×10^{-2}
	Favorite	8.91×10^{-2}

cooperation network contained 185 students and 360 links and three different types of connections.

- **Medical Innovation Network**: This data set was prepared by Ron Burt [24]. He dug out the 1966 data collected by Coleman on medical innovation. The medical innovation network contained 246 nodes and three different types of connections.
- **Youtube network**: The Youtube network [14] is constructed from data crawled from the popular video sharing site in December 2008. The crawler collected information about contacts, favorite videos, and subscriptions. In total, it reached 848,003 users, with 15,088 users sharing all of the information types. The network contains five different interaction types.

To discover communities in each network we applied our algorithm 10 times. For each run, the set of Pareto-optimal returned by the algorithm was investigated; and the modularity of each dimension for the best structure was calculated. Average modularity of each dimension for the best structure over the ten runs was calculated and reported. Also the PMM algorithm is applied 10 times to divide the network into different number k of communities, ranging from 2 to 60. For each k average modularity over the 10 runs was calculated, and then the highest modularity value was selected.

Figure 1 summarizes the result of applying Single Objective Genetic algorithm SGA, MOGA-MDNet and PMM algorithms for the Students and Medical Innovation networks. First by analyzing the result of SGA we observed that the result of the optimization process of GA using one dimension detect a modular community structure that successfully maximize the Modularity only on the selected dimension however by calculating the Modularity of the detected community structure on the other dimensions we found that it has a low or a less modular structure in the other dimensions,this is due to that SGA optimizes only using one dimension information i.e. connectivity pattern. Such observation is clearly visible in the result for Student network Fig. 1a.

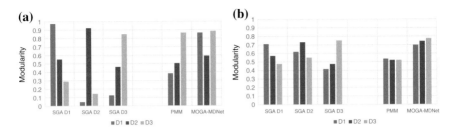

Fig. 1 Average Modularity results obtained by SGA, MOGA-MDNet and PMM for the Students and Medical Innovation networks. **a** Students network. **b** Medical Innovation networks

This problem is overcame by using MOGA in which all dimensions informations are used in the optimization process. As we can observe the MOGA-MDNet is able to detect a community structure that simultaneously maximize Modularity in all network dimensions. Regarding PMM algorithm it was able to find a community structure that maximize Modularity in all dimensions for the Medical network however the result was not high as the result of MOGA-MDNet on the same network. As for the student network, PMM fail to find a community structure that maximize Modularity in all dimensions, as we can observe in Fig. 1a the result of PMM has a high Modularity value in the third dimension and low Modularity in the other dimensions compared to MOGA-MDNet.

A final note regarding the result of MOGA-MDNet, is that the goal of MOGA is to find an optimized structure that maximize the objective function in all dimensions, so when looking at the result of the Student network when applying MOGA-MDNet and SGA on the first dimension D1, we observe that the average Modularity in the first dimension using MOGA-MDNet was 0.699 and it was 0.973 using SGA-D1. Such a drop in the Modularity value in D1 is compensated with an increase in the Modularity value in the other dimensions which is our ultimate goal. Table 3 shows in details the average Modularity values obtained by SGA, MOGA-MDNet and PMM for the Students and Medical Innovation networks.

Regarding Youtube network; by first analyzing the result from SGA using each dimension separately. We observed that some how the network dimension are conflicting each other i.e. when maximizing the modularity on one dimension lead to poor Modularity on the other conflicted dimensions. For example maximizing the Modularity on D1 lead to a low Modularity value in D5 and vice versa also hold true; maximizing the Modularity on D5 lead to a poor Modularity value in D1. Thus making it a challenge for any community detection algorithm to detect a shared community structure that maximize the Modularity in each dimension as possible as it could. Table 4 summarizes the result of SGA, MOGA-MDNet and PMM for the Youtube network. As we can observe MOGA-MDNet able to detect a good community structure that maximize the Modularity in most dimensions.

Table 3 Average Modularity results obtained by SGA, MOGA-MDNet and PMM for the Students and Medical Innovation networks

		D1	D2	D3
Students network	SGA-D1	0.9739	0.551	0.2891
	SGA-D2	0.0466	0.9225	0.144
	SGA-D3	0.1248	0.4612	0.8501
	PMM ($k = 19$)	0.3842	0.5042	0.8657
	MOGA-MDNet	0.8656	0.5933	0.8856
Medical Innovation	SGA-D1	0.712	0.57	0.476
	SGA-D2	0.62	0.732	0.549
	SGA-D3	0.415	0.474	0.751
	PMM ($k = 5$)	0.535	0.52	0.52
	MOGA-MDNet	0.699	0.743	0.776

Table 4 Average Modularity results obtained by SGA, MOGA-MDNet and PMM for Youtube network

	D1	D2	D3	D4	D5
SGA-D1	0.428	0.061	0.071	0.133	0.017
SGA-D2	0.2	0.228	0.052	0.085	0.032
SGA-D3	0.211	0.066	0.193	0.13	0.031
SGA-D4	0.267	0.061	0.088	0.328	0.02
SGA-D5	0.082	0.041	0.046	0.032	0.081
PMM ($k = 10$)	0.47	0.079	0.104	0.153	0.032
PMM ($k = 20$)	0.398	0.055	0.063	0.1	0.025
MOGA-MDNet	0.341	0.108	0.128	0.234	0.028

5.3 Further Analysis

In the previous section our aim was to find the community structure that simultaneously maximize modularity in all network dimensions, however, respect to a certain query, different relations might have different importance. To find a community structure taking into account the importance of a certain relation(s), we first need to identify which relation(s) plays an important role in such structure, then MOGA-MDNet is applied to return the set of Pareto-optimal solution, then this set is investigated and the community structure that best maximize Modularity in the dimension(s) representing this relation(s) is selected.

6 Conclusion and Future Work

This paper presented a multi objective genetic algorithm for detecting communities in multi-dimensional networks. The aim of this method is to find community structure that simultaneously maximizes modularity; as an objective function, in all network dimensions, where the number of objectives needed to be optimized equals the number of dimensions. Experiments in synthetic and real world network showed the ability of this method to correctly detect communities structures that maximize modularity across all network dimensions. The algorithm has the advantage that it does not need any prior knowledge about number of communities; also it guarantees that dimensions with intensive interaction can not affect the structural information in other dimensions. Also using this algorithm we can detect a shared community structure taking into account the importance of a certain relation(s). As the size of real world social networks increases continuously, increasing scalability of our proposed method will be investigated in the future. Also we plan to enhance the capabilities of this algorithm to discover overlapped communities.

References

1. Wasserman, S.: Social Network Analysis: Methods and Applications, vol. 8. Cambridge University Press, Cambridge (1994)
2. Newman, M.E.J., Girvan, M.: Finding and evaluating community structure in networks. Phys. Rev. E **69**(2), 026113 (2004)
3. Blondel, V.D., Guillaume, J.-L., Lambiotte, R., Lefebvre, E.: Fast unfolding of communities in large networks. J. Stat. Mech. Theory Exp. **2008**(10), P10008 (2008)
4. Fortunato, S.: Community detection in graphs. Phys. Rep. **486**(3), 75–174 (2010)
5. Plantié, M., Crampes, M.: Survey on social community detection. In: Social Media Retrieval, pp. 65–85. Springer, Berlin (2013)
6. Duch, J., Arenas, A.: Community detection in complex networks using extremal optimization. Phys. Rev. E **72**(2), 027104 (2005)
7. Liu, J., Liu, T.: Detecting community structure in complex networks using simulated annealing with k-means algorithms. Physica A: Stat. Mech. Appl. **389**(11), 2300–2309 (2010)
8. Hafez, A.I., Al-Shammari, E.T., Hassanien, A.E., Fahmy, A.A.: Genetic algorithms for multi-objective community detection in complex networks. In: Social Networks: A Framework of Computational Intelligence, pp. 145–171. Springer (2014)
9. Berlingerio, M., Coscia, M., Giannotti, F., Monreale, A., Pedreschi, D.: Multidimensional networks: foundations of structural analysis. World Wide Web **16**(5–6), 567–593 (2013)
10. Berlingerio, M., Coscia, M., Giannotti, F., Monreale, A., Pedreschi, D.: Foundations of multidimensional network analysis. In: International Conference on Advances in Social Networks Analysis and Mining (ASONAM), pp. 485–489. IEEE (2011)
11. Berlingerio, M., Coscia, M., Giannotti, F.: Finding and characterizing communities in multidimensional networks. In: International Conference on Advances in Social Networks Analysis and Mining (ASONAM), pp. 490–494. IEEE (2011)
12. Fortin, F.-A., Rainville, D., Gardner, M.-A.G., Parizeau, M., Gagné, C., et al.: Deap: evolutionary algorithms made easy. J. Mach. Learn. Res. **13**(1), 2171–2175 (2012)
13. Cai, D., Shao, Z., He, X., Yan, X., Han, J.: Community mining from multi-relational networks. In: Knowledge Discovery in Databases: PKDD 2005, pp. 445–452. Springer, Heidelberg (2005)

14. Tang, L., Liu, H.: Uncovering cross-dimension group structures in multi-dimensional networks. In: SDM workshop on Analysis of Dynamic Networks (2009)
15. Park, Y., Song, M.S.: A genetic algorithm for clustering problems. In: Proceedings of the Third Annual Conference on Genetic Programming, pp. 568–575 (1998)
16. Cormen, T.H., Leiserson, C.E., Rivest, R.L., Stein, C.: Introduction to Algorithms. Mit Press, Cambridge (2003)
17. Pizzuti, C.: GA-Net: a genetic algorithm for community detection in social networks. In: Parallel Problem Solving from Nature PPSN X. LNCS, pp. 1081–1090. Springer, Heidelberg (2008)
18. Newman, M.E.J.: Modularity and community structure in networks. Proc. Natl. Acad. Sci. **103**(23), 8577–8582 (2006)
19. White, S., Smyth, P.: A spectral clustering approach to finding communities in graph. In: SDM, vol. 5, pp. 76–84. SIAM (2005)
20. Newman, M.E.J.: Finding community structure in networks using the eigenvectors of matrices. Phys. Rev. E **74**(3), 036104 (2006)
21. Danon, L., Diaz-Guilera, A., Duch, J., Arenas, A.: Comparing community structure identification. J. Stat. Mech. Theory Exp. **2005**(09), P09008 (2005)
22. Deb, K., Agrawal, S., Pratap, A., Meyarivan, T.: A fast elitist non-dominated sorting genetic algorithm for multi-objective optimization: Nsga-ii. Lecture Notes in Computer Science, vol. 2000, pp. 849–858 (1917)
23. Fire, M., Katz, G., Elovici, Y., Shapira, B., Rokach, L.: Predicting student exam's scores by analyzing social network data. In: Active Media Technology, pp. 584–595. Springer, Berlin (2012)
24. Samuel, J., Katz, C.E., Menzel, H., et al.: Medical innovation: a diffusion study. Bobbs-Merrill Indianap. (1966)

Creativity in the Era of Social Networking: A Case Study at Tertiary Education in the Greek Context

Evgenia Theodotou and Avraam Papastathopoulos

Abstract This paper investigates the utilization of a social network tool in order to promote creativity in higher education. Buddypress was selected as a social network tool and de Bono's "6 thinking hats" as a creativity strategy. The participants were 17 undergraduate students from a case study in a University in Greece in the field of social sciences. Creativity was defined by Torrance and Ball's [25] factors and the results were analyzed using authentic assessment and a questionnaire. The findings show that the research process was beneficial to students' creativity and that social network tools can be utilized successfully with such a focus. These findings should be treated carefully in terms of generalizing them, as they were derived from a case study. This research could be useful to educators and researchers as a pioneering approach in all levels of education.

Keywords Creativity · Social network tools · 6 thinking hats · Higher education

1 Introduction

Creativity is a very important element of the educational process at all levels. This is even more important in tertiary education where students are trained to develop their professional and academic skills for the rest of their lives.

The Web 2.0 tools offer substantial and remarkable features that can be used beneficially in education and their interface can promote creativity [5, 13]. Social Network Tools (SNT) are some of the most popular Web 2.0 tools that according to

E. Theodotou (✉)
Cass School of Education and Communities, University of East London, London, UK
e-mail: e.theodotou@uel.ac.uk

A. Papastathopoulos
Abu Dhabi University, Abu Dhabi, UAE
e-mail: avraam.p@adu.ac.ae

© Springer International Publishing Switzerland 2016   141
T. Gaber et al. (eds.), *The 1st International Conference on Advanced Intelligent System and Informatics (AISI2015), November 28-30, 2015, Beni Suef, Egypt.*,
Advances in Intelligent Systems and Computing 407, DOI 10.1007/978-3-319-26690-9_13

Zhou et al. [29], provide the opportunity to promote creativity because there is common ground between them. It is worth mentioning that SNT enable users to participate actively and exchange information through a creative collaboration without a specialized knowledge.

So far, researchers have focused on the utilization of SNT in students' cognitive development [16, 18, 22, 26]. There are a lot of arguments that SNT can be used to promote creativity but there is limited, if any, research with such a focus. Therefore there is a need to investigate further this aspect.

This piece of research aims to investigate the utilization of a Social Network Tool to promote creativity in Undergraduate students. Creativity was defined according to the 4 factors that were suggested by Torrance and Ball [25] and these are: Fluency, Flexibility, Elaborate any ideas identified and Originality. The research question and sub-questions are:

1. Can the application of a SNT, which utilizes de Bono's "6 thinking hats", promote creativity?

    (a) What are the effects on students' fluency through this strategy in a SNT?
    (b) What are the effects on students' flexibility through this strategy in a SNT?
    (c) What are the effects on students' elaboration of any ideas identified through this strategy in a SNT?
    (d) What are the effects on students' originality through this strategy in a SNT?

2 Literature Review on Creativity and Social Network Tools

According to Dewey [9, p. 167), students' active involvement is the cornerstone of the learning process especially in order to acquire new knowledge. Web 2.0 tools offer opportunities for active involvement but, by comparing some of them, it becomes obvious that SNT provide more features to users (Table 1).

There are a lot of definitions regarding Social Network Tools [1, 2, 14, 27] but all of them agree that they are tools that provide an electronic and online socialization and interaction among users through the creation of their personal profile. There are a lot of researchers that investigate the effects of SNT on teaching and learning at different levels of education and they all agree on their beneficial outcomes. Bowers-Campbell [1] verify that SNT have positive effects on students' self-efficacy and self-regulation. One year later, Sturgeon and Walker [22] note that SNT make students feel more comfortable as there is an environment of open discussion and some years later, Walker [26] and Reid [19] concluded that SNT contribute positively to students' cognitive development. On the other hand, there are researchers [10, 15] that think differently about the usage of SNT in the educational process as they stressed that SNT might be popular but this does not

Table 1 Comparison of Web 2.0 tools

Web 2.0 Tools			
Features	SNT	Blogs	Wikis
Forum	✓	✓	✓
Chat	✓		
Tagging	✓	✓	✓
Groups	✓		
Friends	✓		
Profile pages	✓	✓	✓
File sharing	✓	✓	✓
Real time activities	✓		
Post/publish	✓	✓	✓
Build virtual communities of practice	✓	✓	

necessarily mean that they have an educational focus. So, for this purpose SNT need to be used with specific educational techniques.

Similar to this discussion, the research concerning creativity in education has been a main topic of discussion in social sciences. The development of creativity is very important to the advancement of science and society [24]. Creativity is the mental process or activities which include the generation of new concepts or theories, or association among them [23]. It is also the ability to generate many unique ideas that solve a problematic situation [17].

There are a lot of creativity strategies that have been suggested with specific steps that can be used at a practical level in order to promote creativity. The "6 Thinking Hats" is a creativity strategy that was put forward by de Bono [6–8] for the development of creative solutions in a problematic situation. Each hat has a different color and represents a different kind of thinking (see Table 2).

This creativity strategy is very popular among researchers at all levels of education. De bono's "6 thinking hats" reinforce active involvement, collaboration and open discussion among the members of the team [11, 12, 18, 20, 21]. It is obvious that this creativity strategy has some common characteristics with SNT, as they both promote open discussion, cooperation and active involvement. Therefore, their combination in a research procedure is an interesting topic to investigate.

Table 2 The 6 thinking hats of de Bono

Hat	Thinking
White hat	Known information, facts, data
Red hat	Emotions, hunches, feelings
Black hat	Difficulties, problems
Yellow hat	Positive, values, benefits
Green hat	Possibilities, ideas, solutions, alternatives
Blue hat	Overview, decision, next steps

3   Research Methodology

This research is a case study in the Greek context which uses a SNT for the development of students' creativity through de Bono's "6 thinking hats". The sample consisted of 17 undergraduate students from a University in Greece in social sciences. A comparison was conducted among different SNT and Buddypress [3] was selected as it was free of charge and had the most features (see Table 3).

It was uploaded to the University's website and access was allowed only to the participants/users. There were 3 types of users with different permissions (Creator, Manager/Teacher, Student) and only the Creator and the Manager/Teacher could invite members to the SNT. Students were divided into 4 groups and they decided the name and the logo of the group. Following this, 4 private groups were formed in the SNT and named after the students' groups. There was, also, one common group "The creativity group" in which the teaching material was. Students were assigned with a real life case scenario in their profession and they had to use the SNT and de Bono's "6 thinking hats" in order to find a solution. Following this they had to submit their proposal as a comic book. Being more specific there were 4 different periods for the students:

1. Induction, groups, presentation of SNT, 6 thinking hats strategy and the case scenario
2. Teacher announces the beginning of each hat, students discuss in the SNT to find a solution according to the thinking of each hat that de Bono suggests (for example in the white hat students were discussing the general information of the allocated problem, while, in the green hat students were suggesting possible solutions to the problem etc.).
3. Students present their ideas in a comic book
4. Assessment and evaluation

Table 3 Comparison of social network tools

Social network tools				
Features	BuddyPress	Ning	Facebook	Diigo
Free	✓		✓	✓
Safety concerns	✓	✓		✓
Server-free		✓	✓	✓
Educational focus	✓	✓		
Users create forum	✓	✓	✓	✓
Forum	✓	✓	✓	✓
Chat	✓	✓	✓	✓
Messages	✓	✓	✓	✓
Group formation	✓	✓	✓	✓
Upload any kind of file	✓	✓		
Select the characteristics	✓			

Table 4 Methodology for the evaluation of creativity

Creativity factors	Scoring criteria	Score awarded
Fluency	The number of different ideas that one can produce	1 point for each idea
Flexibility	The number of categories of ideas that one produces	1 point for each category
Elaboration	The number of categories of ideas that one produces	1 point for each creative elaboration
Originality	The uniqueness of ideas that one produces as compared to the whole sample of ideas suggested	Between 1 and 5 % = 1 Point. If 1 % = 2 points

To measure the outcomes of the procedure we used authentic assessment and a questionnaire. Authentic assessment is the measurement of real abilities, capabilities and knowledge in procedures that can be applied in real situations [28]. The comic book and the discussions in the SNT were evaluated based on the 4 creativity factors using Choon-Keong et al.'s [4] methodology. This methodology takes qualitative data and categorizes it (see Table 4). In addition, at the end of the research procedure, students completed a questionnaire that was based on Ziogkou and Dimitriadis's [30] research in order to evaluate their experience through the SNT.

For the assessment of originality, we need to measure the ideas of each group and divide them with the total number of all ideas. The outcome is multiplied by 100. If the resulting number is less or equal to 1 % then originality is scored with 2 points. If it is greater than 1 % and less than 5 %, originality is scored with 1 point, while if it is greater or equal to 5 % it receives 0 points.

Following this and according to the total score of each team, there is a categorization of the teams in 3 different creativity groups (creative, moderate creative and less creative). These 3 creativity groups are formed by dividing the highest total score by 3.

4 Data Analysis and Discussion

For the authentic assessment, the data from the comic book and the discussion in the SNT were used in order to evaluate creativity. Regarding the data analysis of the discussion of the SNT of the 4 teams, the highest creativity score in the sample of proposed ideas (N = 379) was 646. Regarding the data analysis of the comic book of the 4 teams, the highest creativity score in the sample of proposed ideas (N = 36) was 63. The data from all teams were successfully related to the creativity factors and there were very interesting and pioneering ideas. However, the last two teams had significantly higher scores than the first two in both measurements (see Table 5).

Table 5 Creativity scores

Team	Fluency	Flexibility	Elaboration	Originality	Total score
Creativity scores from the discussion of SNT					
Powerpuff girls	48	24	39	83	194
Sailormoon	61	18	40	104	223
Bad Girls	78	45	155	290	568
Little explorers	192	46	86	322	646
Creativity scores from the comic book					
Powerpuff girls	4	4	1	8	17
Sailormoon	7	7	7	14	35
Bad Girls	11	10	11	22	54
Little explorers	14	12	13	22	61

Table 6 Creativity groups

Creative	Moderate Creative	Less creative
Greativity groups from the discussion in the SNT		
431–646	216-430	0–215
Bad Girls (568) Little explorers (646)	Sailormoon (223)	Powerpuff girls (194)
Greativity groups from the Comic Book		
42–63	21–42	0–21
Bad girls (55) Little explorers (63)	Sailormoon (35)	Powerpuff girls (17)

Based on the total scores of all teams, the 3 different creativity groups were formed. The data analysis from both the discussion in the SNT and the comic book shows that the groups remained in the same creativity group with no significant decrease (see Table 6).

After the authentic assessment, there was one more measurement regarding students' views of the research procedure and the use of the SNT. Data from the questionnaire was categorized in 4 factors regarding the SNT learning, usability, efficiency and utilization. Students had to express their views concerning 5-point Likert scale questions. The data analysis shows that the SNT and the research procedure was well received by students as most of them answered positively (see Fig. 1).

Analyzing the data further, it can be argued that there are some very interesting results which can be verified by the existing literature. To answer the research question and sub-questions, the ideas from all teams covered successfully all creativity factors and there were very good examples that could be implemented in everyday practice in order to meet the upcoming challenges. SNT can offer direct and immediate interaction and this enables communication among the members of the team. This was very important in the successful outcomes of the research procedure and it is also verified by researchers [1, 22] that conclude that SNT make learning more interesting as they offer an open discussion.

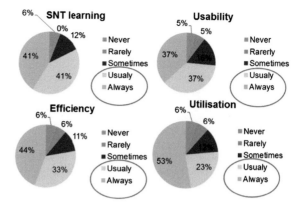

Fig. 1 Data from the questionnaire

The teams displayed the same progress with no significant decrease in the creativity groups in both measurements with authentic assessment. This can be attributed to the fact that the SNT make students to become actively involved as they have attractive features. This is also verified by literature, as Na-songkhla [16] claims that SNT provides the opportunity to present the teaching material in a stimulating way to students. It has to be acknowledged that SNT are widely used in everyday life and thus participants were familiar with the research procedure.

As stated earlier, there are a lot of arguments supporting the idea that SNT can promote creativity but up to now, there is no, if any, research with such a focus. This piece of research shows positive outcomes in the beneficial use of SNT concerning students' creativity. However, the data should be treated carefully as it was from a case study and thus the sample size was limited. This research could be helpful to researchers and to those who seek pioneering approaches to all levels of education.

5 Conclusions

This research examined the use of a SNT to promote creativity in undergraduate students in a case study in a University in Greece. Buddypress was selected as a SNT, in which students utilized de Bono's "6 thinking hats". The data demonstrated the beneficial outcome of the research procedure and that the procedure in general was well received by the students. Findings should be treated carefully as they were derived from a case study and the sample size was small. It would be interesting to further investigate this in a bigger sample and with more Web 2.0 tools. This research could be beneficial to researchers and to those who seek innovative approaches to all levels of education.

References

1. Bowers-Campbell, J.: Cyber "Pokes": motivational antidote for developmental college readers. J. Coll. Read. Learn. **39**(1), 74–87 (2008)
2. Boyd, D.M., Ellison, N.B.: Social network sites: definition, history and scholarship. J. Comput. Med. Commun. **13**(1), 210–230 (2007)
3. BuddyPress.org: http://buddypress.org/
4. Choon-Keong, T., Kean-Wah, L., Baharuddin, A., Jamalludin, H.: Training and measuring creativity using computer-based morphological analysis method. In: Proceedings of the 5th MIT LINC Conference, Boston 23–25 May. USA (2010)
5. Chui, M., Miller, A., Roberts, R.P.: Six ways to make Web 2.0 work. Business Technology. http://www.rachelcolic.com/documents/6%20Ways%20to%20Make%20Web%202.0% 20Work.pdf (2009). Accessed 26 Nov 2011
6. de Bono, E.: Exploring patterns of…serious creativity. J. Qual. Particip. **18**(5), 12–18 (1995)
7. de Bono, E.: Handbook for the Positive Revolution. Penguin Books, England (1992)
8. de Bono, E.: Six Thinking Hats: Revised and Updated. Little, Brown and Company, Boston (1999)
9. Dewey, J.: Democracy and Education: An Introduction to the Philosophy of Education. The Macmillan Company, New York (1916)
10. Greenhow, C.: Tapping the wealth of social networks for professional development. Learn. Learn Technol **36**(8), 10–11 (2009)
11. Kwok, R.C.-W., Cheng, S.H., Ip, H.H.-S., Kong, J.S.-L.: Design of affectively evocative smart ambient media for learning. Comput. Educ. **56**(1), 101–111 (2011)
12. Maroulis, J., Reushle, S.: Bluring of the boundaries: innovative online pedagogical practices in an Australian Faculty of Education. In: 17th Biennial Conference of the Open and Distance Learning Association of Australia. Adelaide, Open Distance Learning Association of Australia (ODLAA) (2005)
13. McLoughlin, C., Lee, M.: Social software and participatory learning: pedagogical choices with technology affordances in the Web 2.0 era. In: Proceedings ascilite Singapore (2007)
14. McVey, M.: To block or not to block? The complicated territory of social networking. Sch. Bus. Aff. **75**(1), 27–28 (2009)
15. Mishra, P., Koehler, M.: Too cool for school? No way! Using the TPCK framework: you can have your hot tools and teach with them, too. Learn. Learn. Technol. **36**(7), 14–18 (2009)
16. Na-songkhla, J.: Flexible learning in a workplace model: blended a motivation to a lifelong learner in a social network environment. In: The Annual Meeting of the Association for the Advancement of Computing in Education (AACE). Melbourne, Australia (2011)
17. Paraskevopoulos, I.: Dimiourgiki skepsi sto sholeio kai stin ikogenia. Athina, Gelebesis (2008)
18. Porcaro, D.: Computer-supported collaborative learning (CSCL) in an Omani undergraduate course: a design-based study. In: The 3rd Annual Forum on e-Learning Excellence: Bringing Global Quality to a Local Context, Dubai, U.A.E. (2010)
19. Reid, J.: "We don't Twitter, we Facebook": an alternative pedagogical space that enables critical practices in relation to writing. Engl. Teach. Pract. Critiq. **10**(1), 58–80 (2011)
20. Ropes, D.C.: Grounding interventions in the service of communities of practice. In: Proceedings of the 5th Conference of the International Human Resource Network. Tilburg, The Netherlands (2007)
21. Schellens, T., Van Keer, H., De Wever, B., Valcke, M.: The effects of two computer-supported collaborative learning (CSCL) scripts on university students' critical thinking. In: The 4th International Conference on Multimedia and Information and Communication Technologies in Education (m-ICTE), Spain (2006)
22. Sturgeon, C.M., Walker, C.: Faculty on Facebook: confirm or deny? In: 14th Annual Instructional Technology Conference. Middle Tennessee State University 29–31 March. Murfreeboro, Tennesse (2009)

23. Sulaiman, F.: The effectiveness of problem-based learning (PBL) online students' creative and critical thinking in physics at tertiary level in Malaysia. PhD thesis. University of Waikato, New Zealand [Online]. http://researchcommons.waikato.ac.nz//handle/10289/4963 (2011). Accessed 25th Oct 2011

24. Theodotou, E.: H dimiourgikotita stin epohi ton neon technologion. Athina, Kritiki (2015). (in Greek)

25. Torrance, E.P., Ball, O.E.: Torrance tests of creative thinking streamlined (revised) manual, figural forms A & B. Scholastic Testing Service, Benseville (1984)

26. Walker, A.: Using social networks and ICTs to enhance literature circles: a practical approach. In: 39th International Association of School Librarianship Conference, Brisbane 27 September–1 October, Australia (2010)

27. Watson, S.W., Smith, Z., Driver, J. Alcohol, sex and illegal activities: An analysis of selected Facebook central photos in fifty states. http://www.eric.ed.gov/PDFS/ED493049.pdf (2006). Accessed 15th Oct 2011

28. Woolfolk, A.: Educational Psychology, 10th edn. Allyn & Bacon, Boston (2007)

29. Zhou, J., Shin, S.J., Brass, D.J., Choi, J., Zhang, Z.X.: Social networks, personal values and creativity: evidence for curvilinear and interaction effects. J. Appl. Psychol. **94**(6), 1544–1552 (2009)

30. Ziogkou, M. Dimitriadis, S.: I chrisi ergalion tipou wiki stin ekpaideusi: Mia meleti periptosis stin tritovathimia ekpaideusi. In: 7o Panellinio Sidedrio me Diethi Simmetohi "I TPE stin ekpaideusi", Tomos II, pp. 321–328. Panepistimio Pelloponisou, Korinthos, 23-26 Septembriou (2010) (in Greek)

OCR System for Poor Quality Images Using Chain-Code Representation

Ali H. Ahmed, Mahmoud Afifi, Mostafa Korashy, Ebram K. William, Mahmoud Abd El-sattar and Zenab Hafez

Abstract The field of Optical Character Recognition (OCR) has gained more attention in the recent years because of its importance and applications. Some examples of OCR are: video indexing, references archiving, car-plate recognition, and data entry. In this work a robust system for OCR is presented. The proposed system recognizes text in poor quality images. Characters are extracted from the given poor quality image to be recognized using chain-code representation. The proposed system uses Google online spelling to suggest replacements for words which are misspelled during the recognition process. For evaluating the proposed system, the born-digital dataset ICDAR is used. The proposed system achieves 74.02 % correctly recognized word rate. The results demonstrate that the proposed system recognizes text in poor quality images efficiently.

1 Introduction

Optical Character Recognition (OCR), is the process of transforming characters in an image (handwritten, printed, or typewritten) text to a machine-encoded format. Two major types of character recognition in computer science are in place: (1) **Optical Character Recognition (OCR)**: techniques based solely on image processing techniques which include extracting features in the image, comparing those features with predefined ones and finally character recognition, and (2) **Intelligent Character Recognition (ICR)**: includes machine learning algorithms within the recognition process.

Although OCR is an old discipline, it still one of the most important and hot topics in image processing. OCR obtains its importance from the wide variety of applications of this field e.g. (1) reading applications for blinds, (2) automatic plate number

A.H. Ahmed · M. Afifi (✉) · M. Korashy · E.K. William · M.A. El-sattar · Z. Hafez
Assiut University, Asyut, Egypt
e-mail: m.afifi@aun.edu.eg

© Springer International Publishing Switzerland 2016
T. Gaber et al. (eds.), *The 1st International Conference on Advanced Intelligent System and Informatics (AISI2015), November 28–30, 2015, Beni Suef, Egypt,*
Advances in Intelligent Systems and Computing 407, DOI 10.1007/978-3-319-26690-9_14

recognition, (3) data entry, (4) archiving of old valuable books and references, (5) playing music notes, (6) CAPTCHA, and (7) office automation.

Many online tools exist to provide a free OCR engines such as: (1) *Free-OCR*[1]: is used to extract text from an uploaded image you to the website using Tesseract engine, and (2) *Newocr*[2]: is also an online tool for optical character recognition in addition to a developer API.

Many researches are debated to OCR, such as [1], in which the authors presented a technique for extracting text captions from news video. In their research the authors addresses two main problems in character recognition from videos namely: low resolution of characters, and high complexity of the backgrounds. The system is suggested to be used in news videos archiving. The captions in the news videos are often be the news itself, names, places or any other relevant information that can be used in indexing and archiving the videos. In [2], the authors presented an OCR platform for embedded and mobile system. The system can be used for developing OCR applications that can be migrated from one operating system to another one without any loose in performance. In [3], the authors distinguished between two types of text overlay text and scene text. Overlay text is the text that is added in the edit (montage) time artificially by the video editor. Scene text is the text captured by the acquisition device at the recording time. Because each type of text has the algorithms and techniques that are used to deal with it, the authors presented a system based on [4] for classifying the text in the video to either overlay or scene text to properly applies the required pre-processing techniques. In [5], the authors explore the feasibility of using general object detection techniques in text recognition. The paper evaluated the performance of two systems. The first one represents the traditional way of text detection and consists of a two ordinary pipeline stages of text detection followed by an OCR engine. The second system based on [6] and used the general purpose object recognition methods. The performance evaluation has revealed that the system based on general computer vision techniques provided a superior performance compared to the traditional one. This result can lead to using the highly optimized and high performance algorithms and techniques in computer vision in the process of text detection. Nguyen et al. [7] presented a method for recognizing text in poor quality images. This method assumes that, the given image consists of two classes of pixels, background and foreground "text" pixels. Working with poor resolution images has been paid attention in characters recognition field. Thus, we present a system for extracting and recognizing text from poor quality images. The rest of the paper is organized as follows: the details of the proposed system are presented in Sect. 2. In Sect. 3, the results of the proposed system are presented. Finally, the paper is concluded in Sect. 4.

---

[1]http://www.free-ocr.com/.

[2]http://www.newocr.com/.

2 The Proposed System

In this work, a system for recognizing text in poor quality images is presented. For a given image, the proposed system extracts the characters, exports each character for performing pre-processing filtering. After that, chain-code is performed for recognizing each segmented character from the given image. Finally, the estimated text is corrected for improving the process of text recognition, see Fig. 1. For extracting characters from a poor quality image, a series of image filtering is performed. This process enhances the given image for exporting only the text from the given image to the next stage, which enhances the extracted character image for removing noise and unwanted pixels. After that, the filtered image is used for recognizing its character. The text consists of the recognized characters using chain-code representation. An auto-correction process is performed for improving the accuracy of recognition process. The rest of this section describes the steps of the proposed system in more details.

2.1 Characters Extraction

For recognizing characters in poor quality images, a set of pre-processing steps is required. Extracting text from poor quality images with the assumption they have bi-modal histogram [7] is not efficient. Where, in some cases the image contains a pattern in the background which contains some colors of the text color, such as images with stochastic texture in the background, see Fig. 2.

Thus, For extracting the text in poor quality images, we perform the following steps. First, bilinear interpolation is applied for the given image to scale it up. Working with grayscale version of the resized image for generating a blurred version of it is performed using a circular averaging filter. The text mask M is generated using the following equation

$$M = G \vee \triangle B, \tag{1}$$

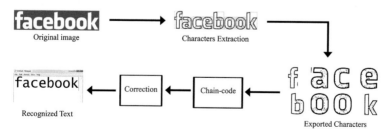

Fig. 1 The proposed system overview

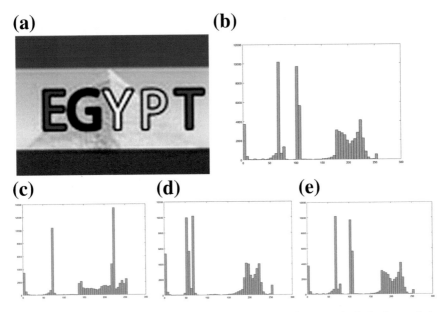

Fig. 2 The histogram of a poor quality image contains stochastic texture in the background. As shown the background contains colors of the text color. **a** The original image. **b** The histogram of the graysclae of the image. **c** The histogram of the red channel. **d** The histogram of the green channel. **e** The histogram of the blue channel

where G is the grayscale version of the given image, and $\triangle B$ is the gradient of the blurred version of G. After applying exclusive disjunction to generate the mask M, we add a border for avoiding problems of characters and borders intersection. The border's color is determined using the following equation

$$B_C = \begin{cases} 1 & \text{if } \sum P_{(x,y)} \in \Omega > \frac{|q|}{2} \\ 0 & otherwise, \end{cases} \qquad (2)$$

where B_C is the binary value of the new border pixels, $P_{(x,y)}$ is the binary value of pixels in the boundary Ω of the mask M, and $|q|$ is the number of pixels in the boundary Ω.

Morphological opening is performed for removing small objects using a circular averaging filter as shown in the following equations

$$M' = \overline{(M \ominus F) \oplus F}, \qquad (3)$$

where F is the circular averaging filter, $\oplus$ and $\ominus$ refer the dilation and erosion, respectively. Labeling connected components is performed for separating each character in a single image that is used by the chain-code representation. For avoiding the

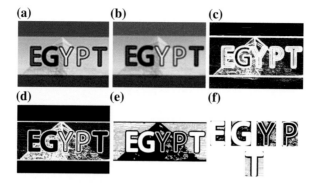

Fig. 3 The internal steps of characters extraction process. **a** The original image. **b** The blurred version of the original image. **c** The gradient of the blurred image. **d** The generated mask M. **e** The final mask M'. **f** After extracting each character in a single image

problem of unwanted small objects that may be considered as a single character, we assume that if the area of the extracted character is smaller than a predefined threshold compared with the area of the image, we ignore it. Figure 3 shows the internal steps of the characters extraction process.

2.2 Characters Recognition

The next step is to erode symbol's borders to extract the contour by subtracting the eroded image from the original image as shown in Fig. 4.

The basic principle of chain-codes is to separately encode each connected component. For each region, a point on the boundary is selected and its coordinates are transmitted. The encoder then moves along the boundary of the region and, at each step, transmits a symbol representing the direction of this movement. This continues until the encoder returns to the starting position, at which point the blob has been completely described, and encoding continues with the next blob in the image. This encoding method is particularly effective for images consisting of a reasonably small number of large connected components. The chain-code vector is obtained

Fig. 4 The morphological erosion and subtraction. **a** The original image. **b** Result of subtraction

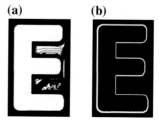

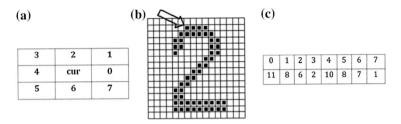

Fig. 5 A chain counted example. **a** Chain code directions. **b** The binary image. **c** The chain vector

from the boundary extracted by subtraction as discussed in the previous step. Chain-code vector is an 8-elements array of integer values, each element stores a counter for the number of connected pixels along the contour in a specified direction. Figure 5 shows the encoding directions for chain-code.

Euclidean distance is used to measure chain codes similarity. In this step the chain-codes for specific domain (numbers, alphabet, or alphanumeric recognition system) is obtained and preprocessed as in the following equation

$$Chain_i = \frac{chain_i}{Max(Chain)}, \tag{4}$$

where *chain* is the symbol chain vector, so that symbol with large space can be recognized and matched with the images in the domain. The next step is to get the Euclidean distance as in Eq. 5 between input character chain vector and every symbol is the domain, the lowest distance means that the input character is similar to the character in the domain.

$$d(p, q) = \sqrt{(q_1 - p_1)^2 + (q_2 - p_2)^2 + \cdots + (q_n - p_n)^2}, \tag{5}$$

where p and q are chain vectors.

2.3 Estimated Text Correction

After the recognition process, the proposed system uses Google's online spelling[3] to suggest replacements for words which are misspelled during the recognition process.

It is highly recommended to use a web of online text in the correction process since using a dictionary would need constantly updates to include new words [8].

---

[3]http://www.google.com.

Fig. 6 Samples of the images that were used for evaluating the proposed OCR

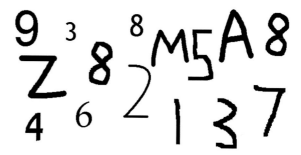

3 Experiment Results

3.1 Datasets

The born-digital dataset "Word Recognition" that was introduced in the first Challenge of ICDAR 2011 Robust Reading Competition [9] was used for evaluating the results of the proposed system. It includes 3583 words in 918 images which were extracted from different types of HTML documents. Another set that consists of 400 images which contains numbers, characters, and symbols was used for testing the character recognition stage using the chain-code representation.

3.2 Results of Chain-Code Representation

In this section, the experimental results of the chain-code in character recognition are presented. The proposed OCR was applied to a dataset consists of 400 images, as described in Sect. 3.1, see Fig. 6. We compared the efficiency of the proposed OCR with other OCR techniques. Some of them are online OCR service, such as onlineocr,[4] free-ocr, and Newocr. In addition; we compared the efficiency of the proposed OCR with embedded OCR in Google Translate App.[5] For each image in the dataset, we applied the presented OCR technique and all of online OCR services which are mentioned above. Figure 7b shows the presented OCR technique after extracting features of the given image. Figure 7c shows the recognized symbol after comparing the extracted features with the features of the dataset. Figure 8b shows the recognized symbol using the presented OCR. Figure 8c shows the recognized symbols using the OCR provided by Google translate. Figure 8d shows the recognized symbol using onlineocr. Figure 8e shows the recognized symbol using

---

[4]http://www.onlineocr.net/.

[5]https://play.google.com/store/apps/details?id=com.google.android.apps.translate.

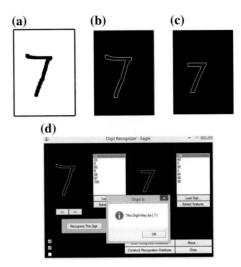

Fig. 7 The results of the presented OCR. **a** Input image. **b** Extracted features of the given image. **c** The recognized symbol. **d** The user interface of the program

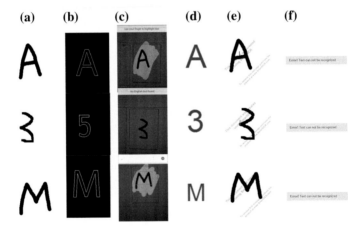

Fig. 8 Comparisons among the performance of OCR techniques and services. **a** Samples of images which were used in the comparison. The recognized symbol is shown using: **b** the proposed OCR, **c** Google translate, **d** onlineocr, **e** free-ocr, and **f** Newocr

free-ocr. Finally, Fig. 8f shows the recognized symbol using Free Online OCR. As shown in Table 1, onlineocr has the highest success recognition rate compared with the others, and Free Online OCR has the lowest recognition rate compared.

Table 1 Comparison of the efficiency of different OCR techniques

OCR provider	Success	Failure
The presented OCR	310	90
OnlineOCR	389	11
Free-OCR	250	150
Free online OCR	190	110
Google translate OCR	280	20

3.3 Results of the Proposed System

The result of the proposed system is considered good with the poor quality images, see Fig. 9. By applying the proposed system on the ICDAR 2011 dataset [9], we get the rate 74.02 % of correctly recognized word. Table 2 shows comparisons among the proposed system and other techniques. Finally our proposed system fails in separating the characters in some cases which the characters is more interrelated together or the background is difficult to be separated as shown in Fig. 10.[6]

Fig. 9 Samples of the results of extracting characters from given images using ICDAR 2011 dataset [9]. **a** Characters extraction without requiring the correction process. **b** Errors in characters extraction process which are handled using the correction process

(a)

(b)

Table 2 Comparisons among the proposed system and other techniques using ICDAR 2011 dataset [9]

Participant	Technique	Correctly Recognized Word Rate (%)
Baseline	Baseline	63.94
Liu et al. [10]	OCR engine TH-OCR 2007	61.98
Su et al. [11]	Proposed_abbyy	72.77
Yang et al. [4]	VOCR	82.03
Kumar et al. [12]	Power-law transformation for enhanced recognition	82.9
Gonzalez et al. [13]	Alvaro Gonzalez	66.88
Nguyen et al. [7]	280513	72.88
Ali H. Ahmed et al.	The proposed system	74.02

---

[6]http://finereader.abbyy.com/.

Fig. 10 An example of
failed cases of extraction and
recognition

4 Conclusions

In this paper, OCR system has been presented. The presented system is based on characters extraction, characters recognition, and text correction to recognize text in poor quality images. The characters extraction preprocessing step is performed to ease the recognition process using chain-code. To improve the misspelled text, we use Google's online suggestions. The performance of the proposed system is evaluated using ICDAR 2011 dataset. The proposed system achieves 74.02 % correctly recognized word rate. The proposed system efficiently recognizes text in poor quality images compared with other techniques.

References

1. Sato, T., Kanade, T., Hughes, E.K., Smith, M.A.: Video ocr for digital news archive. In: IEEE International Workshop on Content-Based Access of Image and Video Database, pp. 52–60. IEEE (1998)
2. Khan, M.N.H., Siddiqui, F., Das, A., et al.: Pin number detection from mobile phone scratch card using OCR on android platform and build an application for balance recharge. Ph.D. thesis, BRAC University (2014)
3. Quehl, B., Yang, H., Sack, H.: Improving text recognition by distinguishing scene and overlay text. In Seventh International Conference on Machine Vision (ICMV 2014), International Society for Optics and Photonics, pp. 944509–944509 (2015)
4. Yang, H., Quehl, B., Sack, H.: A framework for improved video text detection and recognition. Multimed. Tools Appl. **69**(1), 217–245 (2014)
5. Wang, K., Babenko, B., Belongie, S.: End-to-end scene text recognition. In: IEEE International Conference on Computer Vision (ICCV), pp. 1457–1464. IEEE (2011)
6. Wang, K., Belongie, S.: Word Spotting in the Wild. Springer, Berlin (2010)
7. Nguyen, M.H., Kim, S.-H., Lee, G.: Recognizing text in low resolution born-digital images. Ubiquitous Information Technologies and Applications, pp. 85–92. Springer, Berlin (2014)
8. Bassil, Y., Alwani, M.: Ocr post-processing error correction algorithm using google online spelling suggestion (2012). arXiv preprint arXiv:1204.0191
9. Karatzas, D., Mestre, S.R., Mas, J., Nourbakhsh, F., Roy, P.P.: Icdar 2011 robust reading competition-challenge 1: Reading text in born-digital images (web and email). In 2011 International Conference on Document Analysis and Recognition (ICDAR), pages 1485–1490. IEEE (2011)
10. Liu, H., Ding, X.: Handwritten character recognition using gradient feature and quadratic classifier with multiple discrimination schemes. In Eighth International Conference on Document Analysis and Recognition, 2005, pp.19 23. IEEE (2005)

11. Su, B., Lu, S., Phan, T.Q., Tan, C.L.: Character extraction in web image for text recognition. In 2012 21st International Conference on Pattern Recognition (ICPR), pp. 3042–3045. IEEE (2012)
12. Kumar, D., Ramakrishnan, A.G.: Power-law transformation for enhanced recognition of born-digital word images. In 2012 International Conference on Signal Processing and Communications (SPCOM), pp. 1–5. IEEE (2012)
13. Gonzalez, A., Bergasa, L.M.: A text reading algorithm for natural images. Image Vis. Comput. **31**(3), 255–274 (2013)

Human Thermal Face Extraction Based on SuperPixel Technique

Abdelhameed Ibrahim, Tarek Gaber, Takahiko Horiuchi,
Vaclav Snasel and Aboul Ella Hassanien

Abstract Face extraction is considered a very important step in developing a recognition system. It is a challenging task as there are different face expressions, rotations, and artifacts including glasses and hats. In this paper, a face extraction model is proposed for thermal IR human face images based on superpixel technique. Superpixels can improve the computational efficiency of algorithms as it reduces hundreds of thousands of pixels to at most a few thousand superpixels. Superpixels in this paper are formulated using the quick-shift method. The Quick-Shift's superpixels and automatic thresholding using a simple Otsu's thresholding help to produce good results of extracting faces from the thermal images. To evaluate our approach, 18 persons with 22,784 thermal images were used from the Terravic Facial IR Database. The Experimental results showed that the proposed model was robust against image illumination, face rotations, and different artifacts in many cases compared to the most related work.

Keywords Thermal face extraction · Terravic facial IR · Superpixels

A. Ibrahim (✉)
Faculty of Engineering, Mansoura University, Mansoura, Egypt
e-mail: afai79@yahoo.com

T. Gaber
Faculty of Computers and Informatics, Suez Canal University, Ismailia, Egypt

T. Horiuchi
Graduate School of Advanced Integration Science, Chiba University, Chiba, Japan

T. Gaber · V. Snasel
Faculty of Electrical Engineering and Computer Science, IT4Innovations,
VSB-TU of Ostrava, Ostrava, Czech Republic

A.E. Hassanien
Faculty of Computers and Information, Cairo University, Giza, Egypt

A. Ibrahim · T. Gaber · A.E. Hassanien
Scientific Research Group in Egypt, (SRGE), Cairo, Egypt
URL: http://www.egyptscience.net

© Springer International Publishing Switzerland 2016
T. Gaber et al. (eds.), *The 1st International Conference on Advanced Intelligent System and Informatics (AISI2015), November 28–30, 2015, Beni Suef, Egypt,*
Advances in Intelligent Systems and Computing 407, DOI 10.1007/978-3-319-26690-9_15

1 Introduction

There is a great volume of research efforts in the area of the face recognition using the visible spectrum mechanism. However, the visible spectrum-based face recognition suffers from the problem of the variations of light [1]. To address this problem, 3D face recognition [2] or a combination between visible and Infrared (IR) spectrum have been suggested [3]. There is always a need for more robust security systems which does not affect by variations of light. As the IR spectrum is not affected by variations of light, this have increased the rise to develop face recognition system based only on the infrared spectrum.

It is reported that IR spectrum could offer a promising alternative face recognition systems to visible spectrum specifically in case of variations in the face appearance causing by illumination changes [4, 5]. In particular, Jain et al. [6] reported that IR spectrum provides an ability for human identification under different lighting conditions even in the total darkness. Wolff et al. [7] have concluded that IR spectrum is nearly invariant to any change in ambient illumination. Thus, IR-based human recognition systems have the potential to offer simpler and yet robust solutions which achieve a good performance in uncontrolled environment.

Segmentation is an important step in recognition systems. Recognition rates of most recognition approaches can be improved by a good segmentation technique as it enables the utilization of the face shape in the recognition process [8, 9]. There are limited studies about the segmentation approaches for thermal face image. Here, we will give an overview about these studies.

Aglika et al. [10] proposed a segmentation approach using an elliptical mask to be put over the face image to remove the background, align and scale the faces. However, this approach is applicable only for frontal and centered faces. Pavlidis et. al. [11] suggested a face segmentation method based on Bayesian approach. This method is based on both of the models of skin and the background pixel intensities. Thus, clothes pixels was included as skin pixels while ignoring other skin pixels and considering them as a background. In another study, Cho et al. in [12] proposed a segmentation method for the IR face images using contours and morphological operations. The Sobel edge detector was used for the edge detection then the morphological operations was applied to the contour to connect open contours and remove small areas. Recently, Filipe et al. [13] proposed two segmentation methods which make use of the active contour approaches and the statistical modeling of pixel intensities. The two methods are robust against face pose, expression, and rotation. In addition, they addressed the problem of considering the clothes as part of the face, thus enabling the segmentation of the face shape to be used recognition methods.

Superpixels can improve the computational efficiency of algorithms as it reduces hundreds of thousands of pixels to at most a few thousand superpixels. Algorithms for generating superpixels can be categorized as either graph based [14–17] or gradient-ascent based [18–22]. Quick-shift is a common image segmentation method as a gradient-ascent based method [18]. The quick-shift's superpixels are not fixed in size or number and preserve most of the boundaries in the original image. The quick-shift parameters are usually determined by segmenting a few training images. Generating superpixels by quick shift are controlled by three parameters of Ratio, Kernel Size, and Distance.

In this paper, a face extraction model is proposed based on superpixel technique for thermal IR human face images. Superpixels formation using quick-shift helps to get more accurate face extractions. The Quick-Shift parameters' values and automatic thresholding, using a simple Otsu's thresholding, help to produce good results of extracting faces from the thermal images. The Terravic Facial IR Database is used to evaluate our approach. The Experimental results showed that the proposed model was robust against image illumination, face rotations, and different artifacts. Comparing to the most related work, our model was found better in many cases.

2 Theoretical Background

2.1 Quick-Shift Method

The quick-shift method [18] is used to extract superpixels from the thermal face image. The superpixels, in this method, depend on three different parameters of ratio, kernel size, and maximum distance. Determining the quick-shift parameters successfully makes the resulted image more meaningful and easier to be used to extract the thermal face superpixels. In this paper, the three parameters' values are determined by segmenting a few training images by hand until we find a set that shows a good segmentation result for nearly all of the face boundaries and had the largest possible average segment size. In practice, the quick-shift algorithm is not too much sensitive to the choice of parameters, thus a quick tuning by hand is somewhat sufficient for thermal face extraction.

In summary, the superpixels extracted by quick-shift depends on the following parameters:

- Ratio, *Ratio*: It is a tradeoff between spatial and intensity consistency.
- Kernel size, *KernalSize*: It is the parameter that controls the scale at which the density is estimated.
- Max-distance, *MaxDist*: It is the distance between two pixels that the method considers when building the tree.

2.2 Otsu's Thresholding Method

Converting a greyscale image to a binary image is a common task in image processing. Otsu's segmentation method [23] is usually used to automatically perform clustering-based image thresholding [24]. This method converts a grayscale image to binary image. The algorithm assumes that the image contains two classes of pixels (foreground and background pixels). Thresholding tries all possible threshold values to separate the pixels that either fall in foreground or background. The optimum threshold value minimizes the sum of foreground and background spreads.

3 Proposed Thermal Face Extraction Model

A model is proposed to extract human faces based on their thermal images. The model makes use of the Quick-Shift algorithm to produce superpixels and the Otsu's method for automatic thresholding. The proposed model steps are as shown in the Algorithm 1. This model works as follows.

Firstly, a thermal face image I_i is selected, where I_i represents the ith input image of the total number of images N in this group for $i = 1, 2, 3, \ldots, N$. The Terravic Facial IR Database is used for the proposed model. The Quick-Shift method is applied with its initial parameters, ratio, kernel size, and maximum distance, to produce superpixels. The Otsu's thresholding method is then applied to the produced superpixels image. Thus, each superpixels image is converted to a binary image B_i based on the optimum threshold. Finally, the relevant pixel values from the original thermal image are extracted. The Quick-Shift parameters' values with the automatic thresholding one can get the best results of extracting faces from the thermal images.

Algorithm 1 : Thermal Face Extraction

1: Input the thermal face images $\{I\}_{i=1}^{N}$, where N represents the total number of images and I_i represents the ith input image.
2: Determine the superpixels $S_i(I_i)$ using Quick-Shift with initial values for *Ratio*, *KernalSize*, *MaxDist*.
3: Tries all possible threshold values to separate the pixels that either fall in foreground or background for each superpixels image $S_i(I_i)$ using the Otsu's method.
4: Each superpixels' image is converted to a binary image $B_i(S_i)$ based on the optimum threshold.
5: Extract the face pixels from the original thermal image I_i to get $I_i(face)$ based on the binary image $B_i(S_i)$.

---

4 Experimental Results and Discussion

To evaluate our approach, the Terravic facial IR database [25] was used. This database consists of 20 persons each of one of them has a different number of images with various variations (front, left, right; indoor/outdoor; glasses, hat). Its images' format is 8-bit grayscale JPEG with the size of 320×240 pixels. 18 persons with 22,784 thermal images were used from this database in our experiments. Table 1 shows the distribution of the images for each class.

Table 1 Terravic facial IR database images distribution

Face class	Face01	Face02	Face03	Face04	Face07	Face08	Face09	Face10	Face11
Images/class	227	620	592	487	1297	857	1117	283	434
Face class	Face12	Face13	Face14	Face15	Face16	Face17	Face18	Face19	Face20
Images/class	2179	1417	1482	1125	1611	2632	2215	2539	1670

Two main scenarios were designed to evaluate our proposed model:

- The first scenario was to check the accuracy of extracting thermal faces using only Otsu's automatic threshold.
- The second scenario was to evaluate the accuracy of extracting thermal faces using Quick-Shift based automatic threshold.

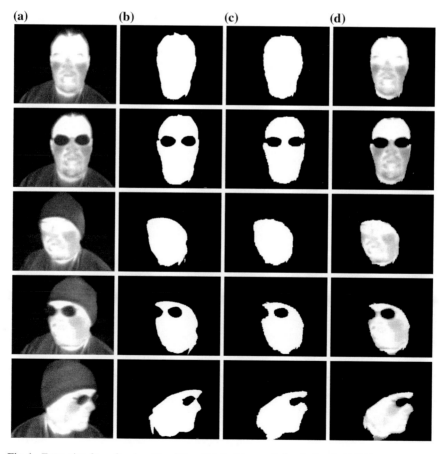

Fig. 1 Extracting faces for class 'face16'; **a** Original image, **b** Otsu's threshold [23], **c** Quick-Shift based threshold, **d** Face Extraction

To show the evaluation of the two scenarios, we use classes (01), (12), and (16) which represents various poses and variations required for face recantation. All of these classes contain different poses (front, left, right). The class of (1) contains indoor images while (12) and (16) includes outdoor images. For the glasses and hat poses, class (1) contains glasses whereas (12) and (16) include glasses and hat.

Figure 1 shows the results for class 'face16' for the two scenarios. This class of the thermal images were captured outdoor from front, left, right direction with glasses, hat, and both. From this figure, its clear that both methods (Otsu's and Quick-Shift) achieve good results for this class because there is clear different in the brightness between the face area and the object of clothes, hate, glass, and other surroundings.

On the other side, Fig. 2 shows the results of the class 'face01' where both methods (i.e. scenarios) are totally different. As the thermal images of class 'face01' were captured indoor with different front, left, right direction with glasses, the Otsu's method did not succeed to extract the face area because of the brightness of the clothes.

For images of class 'face12', which were captured outdoor from front, left, right direction and containing glasses, hat, both the Quick-Shift accomplished some good results and other not good as shown in Fig. 3. The good or the bad the results were noticed that they depend on the face direction.

(a)   **(b)**   **(c)**   **(d)**

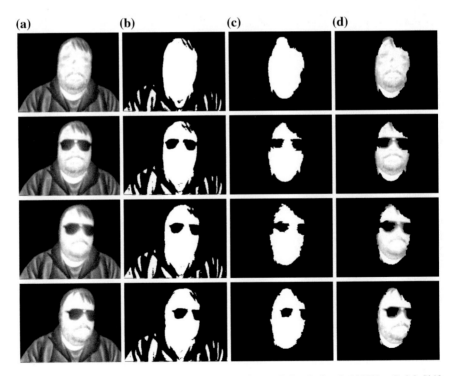

Fig. 2 Extracting faces for class 'face01'; **a** Original image, **b** Otsu's threshold [23], **c** Quick-Shift based threshold, **d** Face Extraction

(a)   (b)   (c)   (d)

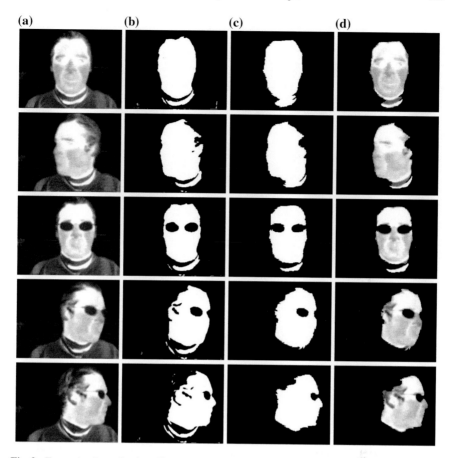

Fig. 3 Extracting faces for class 'face12'; **a** Original image, **b** Otsu's threshold [23], **c** Quick-Shift based threshold, **d** Face extraction

To show the effectiveness of the proposed method, a face segmentation approach based on active counters [26] and suggested by Filipe et al. [13] was implemented and its results were compared with our model's results. This comparison was conducted with the same database (the Terravic facial IR database) and its results are illustrated in Fig. 4. From this figure, it can be noticed that the proposed method was more robust than Filipe's approach. Although both methods could effectively extract human faces when the intensity clothes are high, still our method showed results better than Filipe's one.

From the above results, it can be concluded that the proposed method can successfully extract face from thermal images which were taken indoor/outdoor under various variations, e.g. different directions (left, right, front), glasses, clothes, and hat. On the other side, with high-intensity clothes in the images, the model needs refinement.

(a)   (b)   (c)

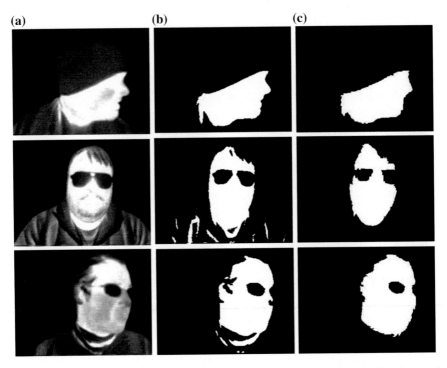

Fig. 4 Comparing the proposed model; **a** Original image, **b** Active Contours [13, 26], **c** Proposed method

5 Conclusion

In this paper, we proposed a face extraction model from IR human face images. This model made use of Otsu and Quick-Shift methods. Based on an extensive experimental results using 18 persons with 22,784 thermal images from the Terravic Facial IR Database, it was concluded that Quick-shift can improve the face extraction results. Our model achieved excellent results for extracting faces from thermal images which were taken under various variations, e.g. different directions, glasses, clothes, and hat. Comparing with the related work, our model's results were better in all different variations. As for the future work, we plan to (a) make a refinement for the case where there is high intensity in clothes in the images and (2) explore the effectiveness of the proposed model for object detection and extracted in different types of thermal databases such as Terravic Motion IR Database, Terravic Weapon IR Database, and Thermal Infrared Video Benchmark for Visual Analysis.

References

1. Ross, A., Nandakumar, K., Jain, A.: Handbook of Multibiometrics (International Series on Biometrics). Springer, New York (2006)
2. Quy, N.H. et al.: 3D human face recognition using sift descriptors of face's feature regions. In: New Trends in Computational Collective Intelligence. Springer International Publishing, pp. 117–126 (2015)
3. Akhloufi, M., Bendada, A., Batsale, J.: State of the art in infrared face recognition. Quant. Infrared Thermogr. J. **5**(1), 3–26 (2008)
4. Ramaiah, N.P., Ijjina, E.P., Mohan, C.K.,: Illumination invariant face recognition using convolutional neural networks. In: IEEE International Conference on Signal Processing, Informatics, Communication and Energy Systems (SPICES), 2015, pp. 1–4 (2015)
5. Chen, C.-L., Jian, B.-L.: Infrared thermal facial image sequence registration analysis and verification. Infrared Phys. Technol. **69**, 1–6 (2015)
6. Jain, A., Bolle, R., Pankanti, S.: Biometrics: Personal Identification in Networked Society. Kluwer Academic Publishers, London (1999)
7. Wolff, L., Socolinsky, D., Eveland, C.: Quantitative measurement of illumination invariance for face recognition using thermal infrared imagery. In: IEEE Workshop on Computer Vision Beyond the Visible Spectrum: Methods and Applications, Hawaii (2001)
8. Pantofaru, C.: Studies in using image segmentation to improve object recognition. Ph.D. thesis, Robotics Institute, Carnegie Mellon University, Pittsburgh (2008)
9. Segundo, M.P., Silva, L., Bellon, O.R.P., Queirolo, C.C.: Automatic face segmentation and facial landmark detection in range images. IEEE Trans. Syst. Man Cybern. Part B Cybern. **40**(5), 1319–1330 (2010)
10. Gyaourova, A., Bebis, G., Pavlidis, I.: Fusion of infrared and visibleimages for face recognition. In: Computer Vision-ECCV 2004, pp. 456–468. Springer, Berlin(2004)
11. Pavlidis, I., Tsiamyrtzis, P., Manohar, C., Buddharaju, P.: Biometrics: face recognition in thermal infrared. In: Biomedical Engineering Handbook, 3rd edn., Chap. 29, pp. 1–15. CRC Press, Boca Raton
12. Cho, S., Wang, L., Ong, W.: Thermal imprint feature analysis for face recognition. IEEE Int. Symp. Ind. Electron, pp. 1875–1880 (2009)
13. Filipe, S., Alexandre, L.A.: Algorithms for invariant long-wave infrared face segmentation: evaluation and comparison. Pattern Anal. Appl. **17**(4), 823–837 (2014)
14. Ren, X., Malik, J.: Learning a classification model for segmentation. IEEE Proc. ICCV **1**, 10–17 (2003)
15. Mori, G., Ren, X., Efros, A., Malik, J.: Recovering human body configurations: combining segmentation and recognition. IEEE Proc. CVPR **2**, 326–333 (2004)
16. Felzenszwalb, P., Huttenlocher, D.: Efficient graph-based image segmentation. Int. J. Comput. Vis. **59**(2), 167–181 (2004)
17. Moore, A., Prince, S., Warrell, J., Mohammed, U., Jones, G.: Superpixel Lattices. IEEE Proc. CVPR, pp. 1–8 (2008)
18. Vedaldi, A., Soatto, S.: Quick shift and kernel methods for mode seeking. In:Proceedings of the European Conference on Computer Vision (ECCV) (2008)
19. Levinshtein, A., Stere, A., Kutulakos, K., Fleet, D., Dickinson, S., Siddiqi, K.: Turbopixels: fast superpixels using geometric flows. IEEE Trans. Pattern Anal. Mach. Intell. **31**(12), pp. 2290–2297 (2009)
20. Arbelaez, P., Maire, M., Fowlkes, C., Malik, J.: From contours to regions: an empirical evaluation. IEEE Proc. CVPR, pp. 2294–2301 (2009)
21. Achanta, R., Shaji, A., Smith, K., Lucchi, A., Fua, P., Ssstrunk, S.: SLIC superpixels compared to state-of-the-art superpixel methods. IEEE Trans. Pattern Anal. Mach. Intell. **34**(11), 2274–2282 (2012)
22. Ren, C.Y., Reid, I.: gSLIC: a real-time implementation of SLIC superpixel segmentation. Department of Engineering Science, University of Oxford (2011)

23. Otsu, N.: A threshold selection method from gray-level histograms. IEEE Trans. Syst. Man Cybern. **9**(1), 62–66 (1979)
24. Sezgin, M., Sankur, B.: Survey over image thresholding techniques and quantitative performance evaluation. J. Electron. Imag. **13**(1), 146–165 (2004)
25. IEEE OTCBVS WS Series Bench; Roland Miezianko, Terravic Research Infrared Database
26. Chan, T.F., Vese, L.A.: Active contours without edges. IEEE Trans. Image Process. **10**(2), 266–277 (2001)

Unsupervised Brain MRI Tumor Segmentation with Deformation-Based Feature

Shichen Zhang, Fei Hu, Shang-Ling Jui, Aboul Ella Hassanien
and Kai Xiao

Abstract Deformation-based features has been proven effective for enhancing brain tumor segmentation accuracy. In our previous work, a component for extracting features based on brain lateral ventricular (LaV) deformation has been proposed. By employing the extracted feature on classifiers of artificial neural networks (ANN) and support vector machines (SVM), we have demonstrated its effect for enhancing brain magnetic resonance (MR) image tumor segmentation accuracy with supervised segmentation methods. In this paper, we propose an unsupervised brain tumor segmentation system with the use of extracted brain LaV deformation feature. By modifying the LaV deformation feature component, deformation-based feature is combined with MR image features as input dataset for the unsupervised fuzzy c-means (FCM) to perform clustering. Experimental results shows the positive effect from the deformation-based feature on FCM-based unsupervised brain tumor segmentation accuracy.

Keywords MRI · Tumor · Feature · Segmentation · Deformation · Unsupervised · Fuzzy C-Means

---

S. Zhang · F. Hu · S.-L. Jui · K. Xiao (✉)
School of Electronic Information and Electrical Engineering,
Shanghai Jiao Tong University, 800 Dongchuan Road, Shanghai, China
e-mail: showkey@sjtu.edu.cn

S. Zhang
e-mail: sjtuzsc@sjtu.edu.cn

F. Hu
e-mail: hu_fei@sjtu.edu.cn

S.-L. Jui
e-mail: shang_ling_jui@hotmail.com

A.E. Hassanien
Scientific Research Group, Faculty of Computers and Information, Cairo University,
Giza, Egypt
e-mail: aboitcairo@gmail.com

© Springer International Publishing Switzerland 2016   173
T. Gaber et al. (eds.), *The 1st International Conference on Advanced Intelligent
System and Informatics (AISI2015), November 28–30, 2015, Beni Suef, Egypt,*
Advances in Intelligent Systems and Computing 407, DOI 10.1007/978-3-319-26690-9_16

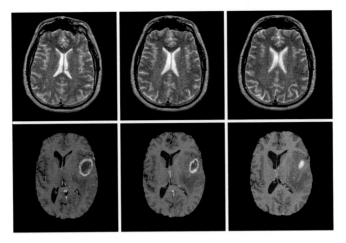

Fig. 1 Consecutive brain MR image T2-weighted scans from a healthy brain (*first row*), and T1-weighted scans from a patient with high-grade glioma (*second row*)

1 Background

Brain tumors normally compress normal intracranial tissues and create deformation on them [1, 2]. In the first row of Fig. 1, the three consecutive scans from healthy brain MR image[1] scans in the axial views show the general symmetry between the left and right hemispheres; while in the second row, where the images are from a patient with high-grade glioma,[2] major intracranial tissues and structures are compressed and deformed towards the opposite side of the tumor growth.

In our previous work [3], following the structure of general medical image segmentation system as shown in Fig. 2, this correlation has been utilized by extracting the LaV deformation feature on supervised support vector machines (SVM) and artificial neural networks (ANN) classifiers. It is proven that the additional deformation-based feature is useful for enhancing brain tumor segmentation accuracy.

---

[1] The healthy brain image data used in this work were obtained from the Whole Brain Atlas (http://www.med.havardedu/aanlib), by K. A. Johnson and J. A. Beker.

[2] The brain tumor image data used in this work were obtained from the MICCAI 2012 Challenge on Multimodal Brain Tumor Segmentation (http://www.imm.dtu.dk/projects/BRATS2012) organized by B. Menze, A. Jakab, S. Bauer, M. Reyes, M. Prastawa, and K. Van Leemput. This database contains fully anonymized images from the following institutions: ETH Zurich, University of Bern, University of Debrecen, and University of Utah. Size of each image in the database is 256×256.

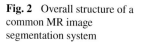 **Fig. 2** Overall structure of a common MR image segmentation system

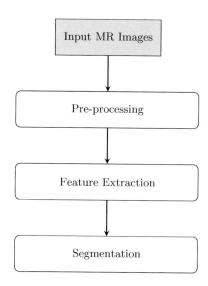

2 Methods

LaV deformation feature extraction component which includes dynamic template image creation and 3-dimensional deformation modeling. Figure 3 illustrates its architecture. In the component, steps of segmentation of lateral ventricles, 3-dimensional alignment and transforming LaV deformation to feature are performed consecutively.

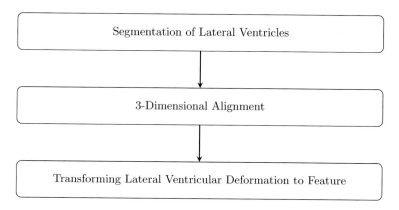

Fig. 3 Structure of lateral ventricular deformation feature extraction component

2.1 Segmentation of Lateral Ventricles

By segmenting LaVs, we can retrieve the shape variation of later ventricles caused by the compression from brain tumor and edema. We use our recently published work of dwFCM algorithm [5] to separate cerebrospinal fluid (CSF) [6] and other tissues. Then based on the knowledge that LaVs are located in the brain center with relatively large volume, the segmentation can be achieved by using a mask in the brain followed by a filter for removing small isolated CSF pixels. Figure 4 illustrates the original MR images and their respective LaV segmentation results.

2.2 3-Dimensional Alignment

3-Dimensional alignment step is further broken down into three processes of creating dynamic template images, obtaining control point pairs and control points registration.

Control points on the skull and LaVs of both template and tumor-affected images are adopted for the deformation modeling. After image binarization, morphological dilation, resizing and convex hull, we can obtain a simulated skull. The midline of the simulated skull can be obtained and treated as the axis which bisects the left and right hemispheres. Figure 5a illustrates this approach. We can then obtain the template image by mirroring the less tumor-affected brain hemisphere.

We then apply a non-linear thin plate splines (TPS) to solve the issue of deviation of central axis of LaVs in tumor-affected MR images. By mirroring this hemisphere

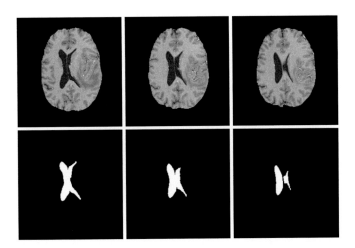

Fig. 4 Original brain MR image T1-weighted contrast-enhanced scans (*first row*), and respective segmentation results of LaVs (*second row*)

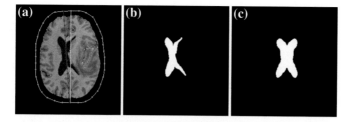

Fig. 5 Visualization of creating a dynamic template image: **a** creating simulated skull and finding its midline; **b** the deformed LaVs; **c** the created template LaVs

Fig. 6 Selected control points on a deformed (*left*) and template LaVs (*right*)

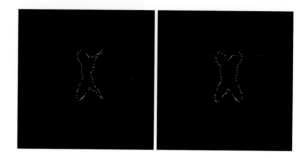

to the other side, we can then create the template image which represents the LaVs in the non-tumor-affected image. Figure 5b, c shows the deformed and created template LaVs images, respectively.

We use Canny edge detector [8] on the adjusted image pairs to get the boundary of LaV and curvature scale space (CSS) corner detect algorithm [9] to locate key control points based on retrieving the maximum curvature. Other control points can be selected on the boundary to provide more coherent alignment. We can finally retrieve a pair of point sets in the 3-dimensional view by combining all the control point sets in each included 2-dimensional image. Figure 6 illustrates the control points on a pair of deformed and template LaVs. Thin plate spline and robust point matching (TPS-RPM) algorithm [10] have been used to automatically perform the control points registration. TPS-RPM algorithm detects the correspondence between template and deformed images, and matches the two point sets according to a non-rigid transformation.

2.3 Transforming Lateral Ventricular Deformation to Feature

After the alignment of LaV control points of template and original images, additional control points on the convex hull to aligned images are included to confine the scope of the deformation. The nonlinear TPS [7] warping function is again employed for

Fig. 7 Tumor-affected brain
MR images (*first row*) and
their respectively extracted
deformation features (*second
row*)

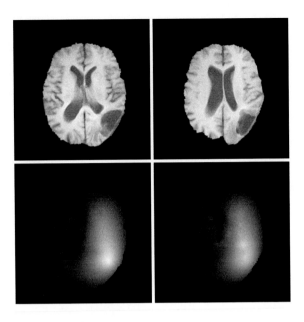

modeling the non-rigid deformation between the pair of point sets. The result of the
modeling can be expressed by the following equation:

$$F \rightarrow F'\qquad(1)$$

where coordinate of each voxel F in the template brain image is displaced to F'.

The LaV deformation feature extraction is achieved by performing a 3-dimen-
sional nonlinear warping followed by calculating and quantifying the displacement
of each voxel generated from the warping process. Figure 7 illustrate the feature
extraction results on two MR image slices from one tumor-affected patient case.

3 Experiments

3.1 Experimental Settings and Evaluation Methods

The common MR image segmentation system [4] shown in Fig. 2 is applied for
tumor segmentation. In order to test the segmentation results from the extracted
deformation-based feature, we have applied our previously published work of modified FCM [11] which automatically adjust feature weights to achieve optimized clustering performance. After clustering, the particular cluster of tumor is selected as the
tumor segmentation result. To evaluate the effect from the extracted feature, we treat
the manual segmentation results from the medical experts as ground truth, and per-

form comparative tumor segmentation experiments on the feature data set, accordingly with or without the extracted LaV deformation feature.

We have employed several statistical methods, i.e., *Accuracy*, *Sensitivity*, *Specificity* and *F-Score* [12, 13], to evaluate the effects from the extracted LaV deformation features. By treating correctly segmented tumor, wrongly segmented tumor, correctly segmented non-tumor and wrongly segmented non-tumor pixel number as *true positive* ($true^+$), *false positive* ($false^+$), *true negative* ($true^-$) and *false negative* ($false^-$), respectively.

3.2 Results and Discussion

We have conducted comparative experiments on 26 MR image cases. Using feature sets with or without this feature, comparative brain tumor segmentation results are illustrated in Fig. 8. It can be observed that, the added LaV deformation feature significantly eliminates the wrongly labeled image pixels and biases the segmentation toward the results obtained by the medical experts.

Experimental results of *Accuracy*, *Sensitivity*, *Specificity* and *F-Score* are tabulated in Table 1. It can be seen that when the feature set is combined with the additional LaV deformation feature, all cases creates the improved or same tumor segmentation results. It should be noted that, the average *Accuracy*, *Sensitivity*, *Specificity* and *F-Score* values increase from 96.8 to 97.8 %, 60.3 to 67.2 %, 99.58

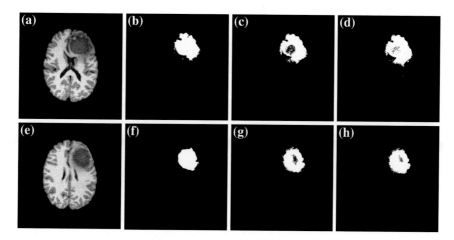

Fig. 8 Two tumor-affected MR image cases for demonstrating the effect from the extracted LaV deformation feature: **a** original MR image case 1; **b** tumor and edema segmentation of (**a**) by medical experts; **c** tumor segmentation of (**a**) without LaV deformation feature; **d** tumor segmentation of (**a**) with LaV deformation feature; **e** original MR image case 2; **f** tumor and edema segmentation of (**e**) by medical experts; **g** tumor segmentation of (**e**) without LaV deformation feature; **h** tumor segmentation of (**e**) with LaV deformation feature

Table 1 *Accuracy, Specificity, Sensitivity* and *F-Score* values of unsupervised brain tumor segmentation with or without deformation-based feature

	Accuracy		Sensitivity		Specificity		F-Score	
	Feature sets							
	Without added feature (%)	With added feature (%)	Without added feature (%)	With added feature (%)	Without added feature (%)	With added feature (%)	Without added feature (%)	With added feature (%)
Case 1	99.0	99.1	92.3	92.5	99.3	99.4	86.7	88.0
Case 2	98.2	98.5	51.6	56.7	99.9	99.9	66.3	70.9
Case 3	91.2	94.8	14.1	22.0	100.0	100.0	24.7	36.0
Case 4	97.4	98.0	80.5	87.9	98.7	98.7	81.4	85.1
Case 5	98.0	98.0	88.4	88.5	98.7	98.7	85.1	85.2
Case 6	98.2	98.2	87.9	88.1	98.9	98.9	86.3	86.5
Case 7	98.1	98.1	83.7	83.8	99.1	99.1	85.1	85.3
Case 8	98.3	98.6	83.3	86.6	99.4	99.4	86.9	88.9
Case 9	98.5	98.5	89.5	89.7	99.1	99.1	88.5	88.6
Case 10	98.3	98.8	84.7	92.2	99.3	99.3	87.6	91.3
Case 11	98.7	98.7	82.6	83.3	99.9	99.9	89.5	89.8
Case 12	98.3	98.9	73.8	84.0	99.5	99.5	79.8	86.1
Case 13	99.1	99.1	88.0	88.0	99.6	99.6	89.1	89.1
Case 14	91.4	94.1	5.2	7.5	100.0	100.0	9.8	13.9
Case 15	91.0	94.3	6.2	10.9	99.9	100.0	11.5	19.6
Case 16	90.4	94.8	4.1	14.8	99.4	100.0	7.5	25.7
Case 17	94.7	95.1	15.1	18.3	99.8	100.0	25.6	31.0
Case 18	94.2	95.5	19.1	24.5	99.9	100.0	31.8	39.4
Case 19	95.3	96.1	26.8	31.1	99.9	100.0	41.8	47.3
Case 20	95.6	96.3	30.0	33.9	99.9	100.0	45.6	50.3
Case 21	96.4	99.9	35.8	96.2	99.9	99.9	52.2	96.5
Case 22	99.8	99.8	96.0	96.5	99.9	99.9	95.4	96.2
Case 23	99.2	99.8	76.9	96.4	99.9	99.9	85.3	96.4
Case 24	99.8	99.8	96.8	96.9	99.9	99.9	95.9	96.6
Case 25	98.6	99.3	70.7	82.3	99.8	99.9	80.1	89.1
Case 26	98.9	99.5	84.3	95.3	99.7	99.7	88.4	94.1

to 99.64 % and 65.7 to 71.8 %, respectively. The enhancement can be attributed to the added relevancy of the LaV deformation feature and the existence of brain tumor and edema that is employed in the brain tumor segmentation.

4 Conclusion

This paper presents the validation of the effect from LaV deformation feature on unsupervised brain tumor segmentation. We apply the newly proposed LaV deformation feature extraction component on clustering method. The positive effect of deformation-based feature is demonstrated by the observation of segmentation results, and then further proven by statistical measures. It is shown that, the deformation-based feature can also be used for unsupervised brain tumor segmentation.

References

1. Deangelis, L.M.: Brain tumors. New Engl. J. Med. **344**, 114–123 (2001)
2. Wen, P.Y., et al.: Updated response assessment criteria for high-grade gliomas: response assessment in neurooncology working group. J. Clin. Oncol. **28**, 1963–1972 (2010)
3. Jui, S.L., Zhang, S., Xiong, W., Yu, F., Fu, M., Wang, D., Hassanien, A.E.: Brain MR image tumor segmentation with 3-Dimensional intracranial structure deformation features. IEEE Intell. Syst. submitted, under review (2015)
4. Clarke, L.P., Velthuizen, R.P., Camacho, M.A., Heine, J.J., Vaidyanathan, M., Hall, L.O., Thatcher, R., Silbiger, M.L.: MRI segmentation: methods and applications. Neuroanatomy **11**(3), 343–368 (1995)
5. Jui, S.L., Lin, C., Guan, H.B., Abraham, A., Hassanien, A.E., Xiao, K.: Fuzzy C-Means with wavelet filtration for mr image segmentation. In: 6th World Congress on Nature and Biologically Inspired Computing, pp. 12–16 (2014)
6. Fix, J.D.: Neuroanatomy, 3rd edn. Lippincott Williams Wilkins, Philadelphia (2002)
7. Bookstein, F.L.: Thin-Plate splines and the decomposition of deformation. IEEE Trans. Pattern Anal. Mach. Intell. **11**, 567–585 (1989)
8. Canny, J.A.: Computational approach to edge detection. IEEE Trans. Pattern Anal. Mach. Intell. **8**(6), 679–698 (1986)
9. Mokhtarian, F., Suomela, R.: Robust image corner detection through curvature scale space. IEEE Trans. Pattern Anal. Mach. Intell. **20**(12), 1376–1381 (1998)
10. Chui, H., Rangarajan, A.: A new point matching algorithm for non-rigid registration. Comput. Vis. Image Underst. **89**(2–3), 114–141 (2003)
11. Xiao, K., Ho, S.H., Hassanien, A.E.: Automatic Unsupervised segmentation methods for mri based on modified fuzzy C-Means. Fundam. Inform. **87**(3–4), 465–481 (2008)
12. Fawcett, T.: An introduction to ROC analysis. Pattern Recognit. Lett. **27**(8), 861–874 (2006)
13. Powers, D.: Evaluation: from precision, recall and F-Measure to ROC, informedness, markedness & correlation. J. Mach. Learn. Technol. **2**(1), 37–63 (2011)

Face Sketch Synthesis and Recognition Based on Linear Regression Transformation and Multi-Classifier Technique

Alaa Tharwat, Hani Mahdi, Adel El Hennawy
and Aboul Ella Hassanien

Abstract Biometric technique becomes essential to identify individuals in different applications. Face sketch is one of the biometric methods, which are used to identify criminals. In this paper, a face sketch synthesis and recognition model is proposed. In this model, the photo images are transformed to pseudo-sketch images using linear regression technique. Moreover, Gabor filters are used to extract the features from three scales of the images. For each scale, a face sketch image is matched with face pseudo-sketches instead of the original photos to identify an unknown individual. Minimum distance classifier is used to match the sketches with pseudo-sketches in each scale. Further, a classification level fusion is used to combine the outputs of the classifiers at three scales namely, decision, rank, and score level fusion. CHUK database images is used in our experiments. The experimental results show that the proposed model is superior to other existed models in terms of identification accuracy. Moreover, matching sketch images with pseudo-sketches achieved accuracy better than matching sketch images with the original photo images. The proposed model achieved a recognition rate ranged from 82.95 to 94.32 % using single scale, while the accuracy increased to 94.32, 96.6 and 97.7 % when the decision, rank, and score level fusion, respectively, are used.

Keywords Face sketch · Classifier fusion · Face sketch synthesis · Gabor filters · Linear Discriminant Analysis (lda) · Biometrics

A. Tharwat (✉)
Faculty of Engineering, Suez Canal University, Ismailia, Egypt
e-mail: engalaatharwat@hotmail.com

A. Tharwat · H. Mahdi · A.E. Hennawy
Faculty of Engineering, Ain Shams University, Cairo, Egypt

A.E. Hassanien
Faculty of Computers & Information, Cairo University, Cairo, Egypt

© Springer International Publishing Switzerland 2016
T. Gaber et al. (eds.), *The 1st International Conference on Advanced Intelligent System and Informatics (AISI2015), November 28–30, 2015, Beni Suef, Egypt,*
Advances in Intelligent Systems and Computing 407, DOI 10.1007/978-3-319-26690-9_17

1 Introduction

Biometrics is the science of measuring and analyzing physiological and behavioral characteristics to identify individuals. The progress in biometric techniques increases the accuracy of security systems. Different biological biometrics have different weak points. For example, fingerprints and iris need to interact cooperatively with the sensors. While face, gait, and ear recognition can be measured from a distance, which is more convenient and allows remote recognition [1].

Face recognition is one of the most common biometrics and has many applications. Face sketch recognition is one of the recent methods of face recognition methods, which is used to identify criminals to assist law enforcement. In the past, face sketch recognition was based on manual measurements. However, the process of automatic face sketch recognition is used to increase the accuracy and decrease the processing time. Automatic face sketch recognition system starts from draw sketches manually by artists and extract robust features from the sketches and match the extracted features with the face images of the criminals in the database to determine the nearest image to the sketch, hence identify the person. Thus, the artists' drawing skills and experience is one of the most important factors to increase the accuracy of the system [2, 3].

Face sketch system has two types, namely, viewed sketches and forensic sketches. Viewed sketches are sketches that are drawn while viewing a photograph of the person or the person himself. On the other hand, forensic sketches are drawn by asking a witness to know a description of the criminal [4]. Due to different modalities, features, and textures between the sketches and faces, many studies proposed to generate a face sketches from the original photos, which is called pseudo-sketches and match the original face sketches with the pseudo-sketches [2].

Zhong et al. synthesized a photo from a sketch by transforming the sketch to a photo, then they used photo-photo recognition to achieve better accuracy. They used 56 color photo-sketch pairs and *Embedded Hidden Markov Model* (EHMM) to map between photos and sketches and used enginspace to identify sketch images. Their proposed model achieved recognition rate 17.6 % when they directly applied photo-sketch recognition while they achieved 88.2 % when they applied photos-to-sketch and sketch-to-photo transformation [5]. Sun et al. used Active Shape Modeling (ASM) to extract the 68 features points to transform the photos to sketches automatically [6]. Yong et al. [7], used hand-drawn face sketch recognition based on *Principal Component Analysis* (PCA). They collected sketches from five different artists and they used 50 subjects each has five images from different five artists. Mahalanobis distance is used as a classifier and score level fusion and achieved 90 % recognition rate. Brendan et al. used *Scale Invariant Feature Transform* (SIFT) and *Multiscale Local Binary Patterns* (MLBP). They achieved accuracy 99.27, 98.6 and 99.47 % using SIFT, MLBP, and SIFT+MLBP, respectively [4].

In this paper, linear regression method is used to transform a photo to a pseudo-sketch. Moreover, Gabor filters are used to extract the features from different scales and then the unknown image is identified using each scale individually and the results of the three scales are combined at the classification level. The rest of this paper is organized as follows, the preliminaries are introduced in Sect. 2. The proposed model is explained in Sect. 3. Experimental results and discussion are introduced in Sect. 4. Finally, concluding remarks and future work are introduced in Sect. 5.

2 Preliminaries

2.1 Gabor Features

Gabor filter method is one of the common methods that are used to extract texture features from grayscale images. Gabor filter is an effective method in texture analysis; hence it is used in many applications such as segmentation, and biometrics [8]. Moreover, Gabor filter method is less sensitive to noise, a small range of translation, rotation, and scaling. A 2D Gabor function $g(x, y)$ is defined as follows [9]:

$$g(x, y) = \frac{1}{2\pi\sigma_x\sigma_y} exp\left[-\frac{1}{2}\left(\frac{x^2}{\sigma_x^2} + \frac{y^2}{\sigma_y^2}\right) + 2\pi jWx\right] \tag{1}$$

where σ_x and σ_y characterize the spatial extent and frequency bandwidth of the Gabor filter, and W represents the frequency of the filter, and $g(x, y)$ is the mother generating function of a Gabor filter family. A set of different Gabor functions, $g_{m,n}(x, y)$, can be generated by rotating and scaling $g(x, y)$ to form non-orthogonal basis set, as follows, $g_{m,n}(x, y) = a^{-2m}g(x', y')$, where $\hat{x} = a^{-m}(x\cos\theta_n + y\sin\theta_n)$, $\hat{y} = a^{-m}(-x\sin\theta_n + y\cos\theta_n)$, $a > 1$, $\theta_n = n\pi/K$, $m = 0, 1, \ldots, S - 1$, and $n = 0, 1, \ldots, K - 1$. The parameter S is the total number of scales, and the parameter K is the total number of orientations. Thus, the total number of generated functions is determined by S and K parameters. Gabor filters features for any image (I) are computed as in Eq. (2).

$$G_{m,n}(x, y) = \sum_{x_1}\sum_{y_1} I(x_1, y_1)g_{m,n}(x - x_1, y - y_1)) \tag{2}$$

2.2 Linear Discriminant Analysis (LDA)

LDA is one of the common feature extraction and dimensionality reduction methods. The main objective of LDA is to find the projection space (i.e. LDA space) that have a good discrimination of the original data by increasing the between-class variance (S_B) and decreasing the within-class variance (S_W). The redundant and unimportant features are neglected when the original data are projected on the LDA space [10].

2.3 Classifier Fusion

Combining independent sources of information may help to determine the most suitable decisions. In machine learning, combining data can be performed in many levels such as sensor, feature, and classification level fusion. In classification level, the performance of the systems may be improved if the classifiers are independent (i.e. diverse classifiers). The outputs of classifiers can be combined in decision, rank, or score level. Fusion in decision level considers the simplest fusion method and easiest one to implement because only decisions are combined in this level. One of the most famous combination methods used in decision level fusion is majority voting (MV) [11]. In rank level fusion, the ranked lists, which represent the output of the classifiers, are combined. Rank level fusion has data more than decision level, thus it may achieve results better than decision level. The score level fusion has information more than the other two levels and achieved good results [12–14].

3 Proposed Model

The proposed model consists of three phases, namely, photo to sketch transformation, training phase, and testing phase. The descriptions of each phase are introduced in the following sections.

3.1 Photo to Sketch Transformation

In this phase, the photo image is transformed into pseudo-sketch using linear regression as shown in Fig. 1. As shown from the figure, the photo and the corresponding sketch image are divided into sub-regions ($n \times n$), where n represents odd number (e.g. 3, 5, 7, ... , etc.). Each pixel in the sketch image represents one label and the corresponding feature vector is extracted from the photos. For example, the first label Y_1 represents the pixel value $S_{2,2}$, while the corresponding feature vector is denoted by X_1, where X_1 represents the first feature vector (i.e. first row) and it is calculated from the square region around the center pixel ($p_{2,2}$) as follows, $X_1 = [p_{1,1}, p_{1,2}, p_{1,3}, p_{2,1}, p_{2,2}, p_{2,3}, p_{3,1}, p_{3,2}, p_{3,3}]$, where $p_{2,2}$ is the value of the pixel in the second row and second column, and X_1 is the first feature vector (i.e. first row). The detailed steps of the proposed transformation model are summarized in Algorithm 1 and Fig. 1.

Algorithm 1 : Transformation algorithm from photo to pseudo-sketch using linear regression technique.

1: Divide Photo image ($P(N \times M)$) and sketch image ($S(N \times M)$) into sub-regions, each region ($n \times n$).

2: **for** (each pixel in sketch image (S) (e.g. first pixel $Y_1 = [S_{2,2}])$) **do**

3: Extract the corresponding pixel from photo image P and its surrounding pixels (e.g. $X_1 = [P_{1,1}, P_{1,2}, \dots, P_{3,3}]$.

4: **end for**

5: $\mathbf{X} \Leftarrow X_1, X_2, \dots, X_N, X(N \times M, n \times n)$.

6: $\mathbf{Y} \Leftarrow Y_1, Y_2, \dots, Y_N, X(N \times M, 1)$.

7: Normalize the feature matrix X.

8: Initialize Learning Rate (α) to zeros.

9: Run Gradient Descent algorithm to compute regression parameters, λ_i.

10: **for** (any new photo image P_{new}) **do**

11: Divide P_{new} into sub-regions ($n \times n$).

12: To estimate the value of pixels in the pseudo-sketch image PS, extract the corresponding pixel in photo image P_{new} and its surrounding pixels.

13: Predict the pixels in pseudo-sketch image (PS) as follows: $PS_{new} = P_{new} \times \lambda_i$.

14: **end for**

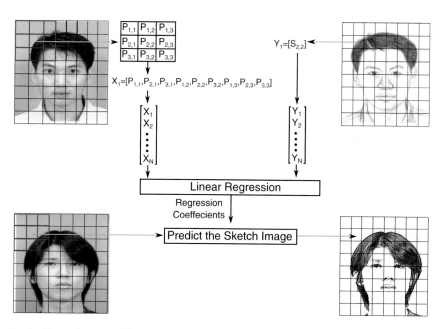

Fig. 1 Block diagram of the proposed transformation method (Photo to sketch transformation using linear regression)

3.2 Training Phase

In this phase, the photo images or pseudo-sketches are collected in different scales
(256×256, 128×128, and 64×64), thus each pseudo-sketch is represented by three
scales ($PS_{256 \times 256}$, $PS_{128 \times 128}$, and $PS_{64 \times 64}$), where $PS_{n \times n}$ is the pseudo-sketch in scale
$n \times n$. For each scale, Gabor features are extracted. LDA dimensionality reduction
method is then used to reduce the dimensions of each scale as shown in Fig. 2.

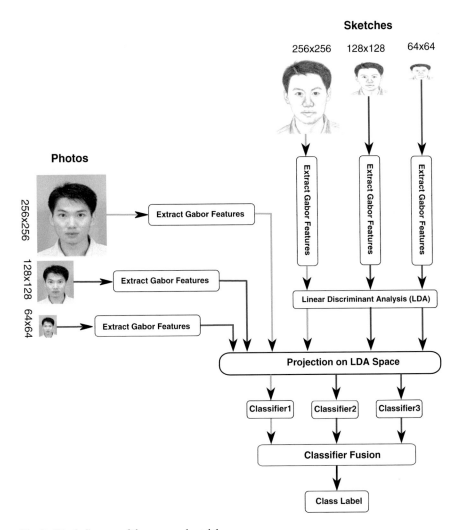

Fig. 2 Block diagram of the proposed model

3.3 Testing Phase

In this phase, an unknown image (i.e. sketch image) is collected and resized to be at three different scales, thus each sketch represented by three images ($S_{256\times256}$, $S_{128\times128}$, and $S_{64\times64}$). Gabor features are then extracted from each scale and project each feature vector on its corresponding LDA space (There are three different LDA spaces for the three scales), which is calculated in the training phase. After that, match the test feature vector of each scale with its corresponding feature matrix to determine the class label of the sketch image at all three different scales as shown in Fig. 2. Finally, combine the outputs of the classifiers at all three fusion levels (i.e. decision, rank, and score) to determine the final class label of the unknown sketch image.

4 Experimental Results and Discussion

In this section, two experiments are performed. The first experiment is performed to transform the photo images into pseudo-sketches. In the second experiment, face sketch recognition is implemented to match an unknown face sketch with the pseudo-sketch images to determine the nearest person.

4.1 Experimental Setup

In this experiment, CHUK dataset is used. CHUK consists of 606 individuals. Each person has a frontal face photo image and a sketch image drawn by an artist. The dataset is divided into two sets, namely, training and testing sets. The training set consists of 306 individuals, while the other images are used as a testing set. Figure 3 shows samples of the used datasets. The training images are used to; (1) build the regression model to transform the photo image to pseudo-sketch image, (2) train or learn the classifiers to identify an unknown sketch image.

4.2 First Experiment (Photo to Pseudo-Sketch Transformation)

In this experiment, a linear regression technique is used to transform the photo image to pseudo-sketch image. In this experiment, some of the photo images are selected to train the regression model to transform the photo images into pseudo-sketches. As shown in Algorithm 1, the photo and sketch images are divided into sub-regions and the size of each region is $n \times n$. Each sub-region is used to predict one pixel in the

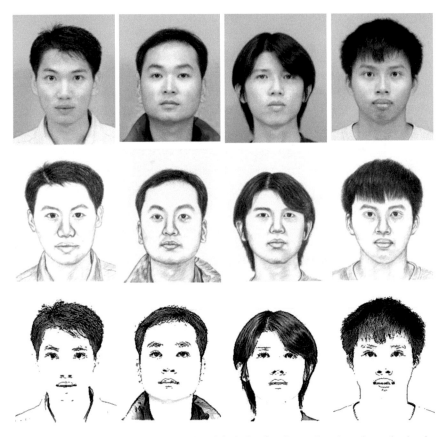

Fig. 3 Samples of face photos (*top row*), original sketches (*second row*), and pseudo-sketches using the proposed transformation model for different four individuals

pseudo-sketch image as shown in Algorithm 1). In this experiment, the size of the sub-regions will be 3×3. Samples of original photos, sketches, and pseudo-sketches are illustrated in Fig. 3.

4.3 Second Experiment (Face Sketch Recognition)

The aim of this experiment is to identify individuals by matching the unknown sketch with the pseudo-sketches. In this experiment, Gabor filters are used to extract the features from the sketches, pseudo-sketches, and photo images. This experiment consists of two sub-experiments. The first sub-experiment aimed to match the sketches with the original photos. While in the second sub-experiment, the sketches are matched with the pseudo-sketches. In this experiment, the images are scaled into three different scales (256×256, 128×128, and 64×64) as mentioned in Sect. 3. Minimum distance classifier is used to match the training (i.e. photos or pseudo-sketches) and testing (i.e. sketches) samples. Moreover, different levels of fusion are used to combine the outputs of the classification of the three scales. In other words, the extracted features from the three different scales are then matched independently (i.e. the photos and sketches of each scale is matched together) and combine the outputs of all classifiers in the three different levels (decision, rank, and score level) as shown in Fig. 2. In decision level, majority voting (MV) is used to combine the final decisions. While Borda count method is used to combine the ranked lists, which result from matching the unknown sketch with the training samples in different scales. Finally, different rules such as sum, min, and max rules are used to combine the scores at score level fusion. Accuracy assessment method is used to measure the performance of this experiment. The accuracy represents the ratio between the correctly classified samples and all samples. A summary of this scenario is shown in Table 1.

4.4 Discussion

Figure 1 shows the pseudo-sketches, which are generated using linear regression technique. The pseudo-sketches are more similar to the face sketches than photos, which may has a great impact on the accuracy of sketch recognition.

From Table 1 many remarks can be seen. First, the accuracy values of the scales are different and the scale of 128×128 achieved the highest accuracy, while the 256×256 achieved the lowest accuracy. Second, combining different scales in classification level (i.e. decision, rank, or score) achieved accuracy better than all single scales. Third, score level fusion achieved the best accuracy, while the decision level achieved the lowest accuracy because the score level has information more than all

Table 1 Accuracy of the proposed model using photo-to-sketch recognition and pseudo-sketch-to-sketch recognition

Single/multi scale	Classifier	Accuracy (%)	
		Using photos	Using pseudo-sketches
64×64	Minimum distance	52	85.23
128×128	Minimum distance	63	94.32
256×256	Minimum distance	63	82.95
Classifier fusion	Abstract- MV	77.45	94.32
	Rank- Borda count	80.25	96.6
	Measurement-sum	82.25	95.5
	Measurement-min	80.5	97.7
	Measurement-max	83.12	93.2

Table 2 Accuracy of some state-of-the-art models

Author	Method	Accuracy (in %)
Himanshu S. Bhatt et al. [15]	EUCLBP+GA	94.12
X. Tang et al. [16]	Eigentransform+PCA	75
X. Tang et al. [16]	Eigentransform+Bayes	81.3
Qingshan Liu et al. [17]	LDA	85
Qingshan Liu et al. [17]	PCA	64.33

other classification levels. Fourth, the accuracy of matching sketches with pseudo-sketches is much higher than matching sketch images with the original photos, which reflects that pseudo-sketch images are more similar to sketches than original photos.

It can also be noticed from Tables 1 and 2 that our proposed model achieved results better than some of the state-of-the-art models, which are listed in Table 2. Finally, we note that the fusion in classification level achieved accuracy better than matching photos with sketches at single scales. Moreover, transforming sketches to pseudo-sketches improves the accuracy of the proposed model.

5 Conclusions and Future Work

In this paper, a face sketch recognition model is proposed. The proposed model is used to transform a face photo image to pseudo-sketch. Two different experiments are performed. The goal of the first experiment is to transform the face photo to pseudo-sketch using linear regression. While in the second experiment, an unknown sketch is identified by matching the unknown sketch with (1) original face photos or (2) pseudo-sketch. In all experiments, Gabor filters method is as a feature extraction method. In the first sub-experiment, the unknown face sketch is matched with the original photos for each scale individually. While in the second sub-experiment, the face sketches are matched with the pseudo-sketch images. In the two sub-experiments, the outputs of the three scales are combined at classification level. The experimental results have shown that the classification level fusion in all levels achieved results better than all other single scales. Moreover, score level fusion achieved the best accuracy while the decision level achieved the worst accuracy. Moreover, matching sketches with pseudo-sketches outperform matching sketches with original photos. Thus, it could be concluded that the proposed model has achieved an excellent accuracy against many state-of-the-art methods. In the future work, we will combine the features of different scales at a feature level. Moreover, more robust feature extraction method is used.

References

1. Tharwat, A., Ibrahim, A.F., Ali, H.A.: Multimodal biometric authentication algorithm using ear and finger knuckle images. In: 2012 Seventh International Conference on Computer Engineering and Systems (ICCES), pp. 176–179 IEEE (2012)
2. Tang, X., Wang, X.: Face sketch recognition. IEEE Trans. Circ. Syst. Video Technol. **14**(1), 50–57 (2004)
3. Wang, X., Tang, X.: Face photo-sketch synthesis and recognition. IEEE Trans. Pattern Anal. Mach. Intell. **31**(11), 1955–1967 (2009)
4. Klare, B.F., Li, Z., Jain, A.K.: Matching forensic sketches to mug shot photos. IEEE Trans. Pattern Anal. Mach. Intell. **33**(3), 639–646 (2011)
5. Zhong, J., Gao, X., Tian, C.: Face sketch synthesis using e-hmm and selective ensemble. In: IEEE International Conference on Acoustics, Speech and Signal Processing, 2007. ICASSP 2007. vol. 1, pp. I–485. IEEE (2007)
6. Sun, Y., Miao, Z., Wang, Y., Wang, Z.: Automatic personalized facial sketch based on asm. In: International Conference on Image Analysis and Signal Processing (IASP), pp. 123–126. IEEE (2010)
7. Zhang, Y., McCullough, C., Sullins, J.R., Ross, C.R.: Hand-drawn face sketch recognition by humans and a pca-based algorithm for forensic applications. IEEE Trans. Syst. Man Cybern. Part A: Syst. Hum. **40**(3), 475–485 (2010)
8. Tharwat, Alaa, Gaber, Tarek, Hassanien, Aboul Ella: Cattle identification based on muzzle images using gabor features and SVM classifier. In: Hassanien, Aboul Ella, Tolba, Mohamed F., Taher Azar, Ahmad (eds.) AMLTA 2014. CCIS, vol. 488, pp. 236–247. Springer, Heidelberg (2014)
9. Han, J., Ma, K.K.: Rotation-invariant and scale-invariant gabor features for texture image retrieval. Image Vision Comput. **25**(9), 1474–1481 (2007)
10. Tharwat, A., Gaber, T., Hassanien, A.E., Hassanien, H.A., Tolba, M.F.: Cattle identification using muzzle print images based on texture features approach. In: Proceedings of the 5th International Conference on Innovations in Bio-Inspired Computing and Applications IBICA 2014, pp. 217–227. Springer (2014)
11. Kuncheva, L.I.: Combining pattern classifiers: methods and algorithms, 2nd edn. Wiley, Hoboken (2014)
12. Tharwat, A., Ibrahim, A., Ali, H.A.: Personal identification using ear images based on fast and accurate principal component analysis. In: 8th International Conference on Informatics and Systems (INFOS), pp. 56–59 (2012)
13. Tharwat, Alaa, Ibrahim, Abdelhameed, Hassanien, Aboul Ella, Schaefer, Gerald: Ear recognition using block-based principal component analysis and decision fusion. In: Kryszkiewicz, Marzena, Bandyopadhyay, Sanghamitra, Rybinski, Henryk, Pal, Sankar K. (eds.) PReMI 2015. LNCS, vol. 9124, pp. 246–254. Springer, Heidelberg (2015)
14. Semary, N.A., Tharwat, A., Elhariri, E., Hassanien, A.E.: Fruit-based tomato grading system using features fusion and support vector machine. In: Intelligent Systems' 2014, pp. 401–410. Springer (2015)
15. Bhatt, H.S., Bharadwaj, S., Singh, R., Vatsa, M.: On matching sketches with digital face images. In: 4th IEEE International Conference on Biometrics: Theory Applications and Systems (BTAS), pp. 1–7. IEEE (2010)
16. Tang, X., Wang, X.: Face sketch synthesis and recognition. In: Proceedings of 9th IEEE International Conference on Computer Vision, pp. 687–694. IEEE (2003)
17. Liu, Q., Tang, X., Jin, H., Lu, H., Ma, S.: A nonlinear approach for face sketch synthesis and recognition. In: IEEE Computer Society Conference on Computer Vision and Pattern Recognition, 2005. CVPR 2005. vol. 1, pp. 1005–1010. IEEE (2005)

Part III
Swarms Optimization and Applications

A New Learning Strategy for Complex-Valued Neural Networks Using Particle Swarm Optimization

Mohammed E. El-Telbany, Samah Refat and Engy I. Nasr

Abstract QSAR (Quantitative Structure-Activity Relationship) modelling is one of the well developed areas in drug development through computational chemistry. This kind of relationship between molecular structure and change in biological activity is center of focus for QSAR modelling. Machine learning algorithms are important tools for QSAR analysis, as a result, they are integrated into the drug production process. In this paper we will try to go through the problem of learning the *Complex-Valued Neural Networks*(CVNNs) using *Particle Swarm Optimization*(PSO); which is one of the open topics in the machine learning society. We presents CVNN model for real-valued regression problems. We tested such trained CVNN on two drug sets as a real world benchmark problem. The results show that the prediction and generalization abilities of CVNNs is superior in comparison to the conventional real-valued neural networks (RVNNs). Moreover, convergence of CVNNs is much faster than that of RVNNs in most of the cases.

Keywords Particle swarm optimization · Complex-valued neural networks · Drug design

1 Introduction

The problem of drug design is to find drug candidates from a large collection of compounds that will bind to a target molecule of interest. The development of a new drug is still a challenging, time-consuming and cost-intensive process and due to the enormous expense of failures of candidate drugs late in their development. Designing '*drug-like*' molecules using computational methods can be used to assist and speed

M.E. El-Telbany (✉)
Computers and Systems Department, Electronics Research Institute, Cairo, Egypt
e-mail: telbany@eri.sci.eg

S. Refat · E.I. Nasr
Faculty of Women for Arts, Science and Education, Ain Shams University,
Cairo, Egypt

© Springer International Publishing Switzerland 2016
T. Gaber et al. (eds.), *The 1st International Conference on Advanced Intelligent System and Informatics (AISI2015), November 28–30, 2015, Beni Suef, Egypt,*
Advances in Intelligent Systems and Computing 407, DOI 10.1007/978-3-319-26690-9_18

up the drug design process [1–3]. The major bottlenecks in drug discovery ware addressed with computer-assisted methods, such as QSAR models [4], where the molecular activities are critical for drug design. The QSAR models used to predict the drug activity within a large number of chemical compounds using their descriptors that are often generated with high-noise in high-dimensional space. Nowadays, *machine learning* algorithms have been used in the modelling of QSAR problems [5–7]. They extract information from experimental data by computational and statistical methods and generate a set of rules, functions or procedures that allow them to predict the properties of novel objects that are not included in the learning set. Formally, a learning algorithm is tasked with selecting a hypothesis that best supports the data. Considering the hypothesis to be a function f mapping from the data space X to the response space Y; i.e., $f : X \rightarrow Y$. The learner selects the best hypothesis f^* from a space of all possible hypotheses $\mathcal{F}$ by minimize errors when predicting value for new data, or if our model includes a cost function over errors, to minimize the total cost of errors. As shown in Fig. 1, the QSAR modelling is heavily dependent on the selection of molecular descriptors; if the association of the descriptors selected to biological property is strong the QSAR model can identify valid relations between molecular features and biological property/activity. Thus, uninformative or redundant molecular descriptors should be removed using some feature selection methods during (*filters*) or before (*wrappers*) the learning process. Subsequently, for tuning and validation of the predictively of learned QSAR model, one of the validation strategy can be applied likes cross-validation, leave-one-out or the full data set is divided into a training set and a testing set prior to learning (See [8] for a survey).

Actually, the machine learning field [9–11] have versatile methods or algorithms such as decision trees (DT), lazy learning, k-nearest neighbours, Bayesian methods, Gaussian processes, artificial neural networks (ANN), support vector machines (SVM), and kernel algorithms for a variety of tasks in drug design. These methods are alternatives to obtain satisfying models by training on a data set. However, the prediction from most regression models - be it multiple regression, ANN, SVM, DT, etc. is a point estimate of the conditional mean of a response (i.e., quantity being predicted), given a set of predictors. Currently, the complex numbers are very actively

Fig. 1 General steps of developing QSAR models

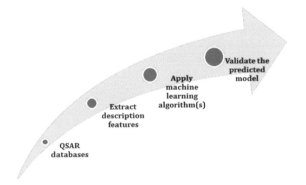

used in modern engineering and in modern physics, to describe real-life phenomena which allow *feasible predictions* and *efficient control strategies*. Operations such as integration and optimization are only *feasible* when the corresponding functions can be extended to smooth functions in a complex domain and consequently feasible algorithms become possible. Moreover, sometimes, even in the situations when a real-valued feasible algorithm is possible, the use of complex numbers can speed up computations. A typical example of such a situation is the use of *Fast Fourier Transform*(FFT). In recent years, complex-valued neural networks have widened the scope of application in telecommunications, imaging, remote sensing, time-series, spatio-temporal analysis of physiological neural systems, and artificial neural information processing. Also, multi-valued neural network is a special type of CVNN, its has a threshold function of multi-valued logic and complex-valued weights is considered [18]. The CVNN which has complex number processing structure that made the network have stronger learning ability, better generalization ability [12], superior reducing power [13], faster convergence [14], lower computational complexity and less data is needed for network training. There are many previously successful attempts to implement the PSO in generating QSAR models. One of the attempts is the using of Binary PSO for feature selection followed by a neural network which is trained using back propagation (BP) for the construction of the QSAR models [15, 16]. The major disadvantage of BPSO-BP is the difficulty in choosing parameters for the *back-propagation* that can ensure efficient network training. Recently, [17] uses the PSO for training the RVNN to overcome such drawbacks. The main objective of this work is to exploit the complex-valued characteristic by apply the CVNN model that trained by PSO for real world forecasting problem (e.g. drug design). The key contributions of this paper are concluded as follows:

- We propose a new strategy for training the CVNN using PSO in QSAR modelling for drug design.
- We evaluate the proposed strategy by determine its prediction accuracy and convergence rate for real data through experimental results.

The rest of the paper is organized as follows. In Sects. 2 and 3 an introduction to the CVNN and PSO algorithm are presented. Section 4 briefly introduces the problem formulation. Section 5 describes the data set, their descriptions and processing step. Section 6 describes an evaluation of QSAR modelling and prediction results using the proposed strategy. Finally Sect. 7 presents the findings and conclusions.

2 Complex-Valued Neural Networks

The CVNNs are simply the generalization of the RVNN in the complex valued domain as shown in Fig. 2, where all the parameters including weights, biases, inputs, outputs could be complex variables, and the activation function and its

Fig. 2 The CVNN model

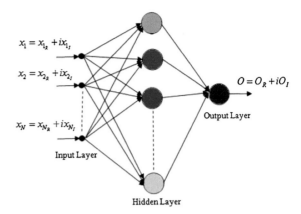

derivatives have to be well behaved everywhere in the complex plane. The complex back-propagation (BP) algorithm, which is the complex-valued version of the real valued back propagation algorithm, is widely used to train the CVNNs [20, 21]. Let us take the CVNN which has n inputs, m neurons in the hidden layer and k outputs. The o network's output could be calculated as follows:

$$O_k = f(B_k + \sum_{i=1}^{m} W_{km} \times H_m)$$ (1)

and the output of each hidden neuron m is given like:

$$H_m = f(B_m + \sum_{i=1}^{n} W_{mn} \times X_n)$$ (2)

Where X_n complex valued input and B_k, W_{km}, B_m, W_{mn} are biases and the weights from the input to hidden and from the hidden to output layers, respectively. $f(.)$ is a complex valued activation function. According to the *Liouville's theorem*, in which the analytic and bounded functions on entire complex plane are constant, the $f(.)$ takes a great attention and several complex activation functions proposed in the literature [21]. In this paper the split sigmoid function is taken and it is given as follows:

$$f = \frac{1}{1 + e^{-Re(z)}} + j\frac{1}{1 + e^{-Im(z)}}$$ (3)

where $z = x + jy$. It should be noted that the use of the split sigmoid function rather than the non-split function could avoid the problem of function's singularity [21].

3 Particle Swarm Optimization

The particle swarm optimization algorithm [22] is based on two socio-metric principles. Particles fly through the solution space and are influenced by both the best particle in the particle population and the best solution that a current particle has discovered so far. The best particle in the population is typically denoted by *gobal best*, while the best position that has been visited by the current particle is donated by *local best*. The *global best* individual conceptually connects all members of the population to one another. That is, each particle is influenced by the very best performance of any member in the entire population. The *local best* individual is conceptually seen as the ability for particles to remember past personal success. The PSO makes use of a velocity vector to update the current position of each particle in the swarm. The position of each particle is updated based on the social behaviour that a population of individuals adapts to its environment by returning to promising regions that were previously discovered. The PSO is one of the methods that can be used to update the weights of a neural network during training. Weight adjustment between processing units in PSO is carried out according to the difference between the target value and the output value of the neural model. At the end of each iteration, the smallest fitness value is remembered by PSO, and the corresponding particle is retained as global best.

4 Problem Formulation

QSAR models are in essence a mathematical function that relates features and descriptors generated from small molecule structures to some experimental determined activity or property [15]. The structure-activity study can indicate which features of a given molecule correlate with its activity, thus making it possible to synthesize new and more potent compounds with enhanced biological activities. QSAR analysis is based on the assumption that the behaviour of compounds is correlated to the characteristics of their structure. In general, a QSAR model is represented as follows:

$$activity = \beta_0 + \sum_{i=1}^{n} \beta_i x_i \qquad (4)$$

where the parameters x_i are a set of measured (or computed) properties of the compounds and β_0 through β_i are the calculated coefficients of the QSAR model. Using both RVNN and CVNN for QSAR modelling where a network of nodes and connecting weights is used to represent the interaction between input and output parameters in a prediction model. The primary function of a *QSAR neural network* model is to assign appropriate weights to the input nodes of a network so that a weighted function of the input nodes predicts the outputs. By formulating the QSAR neural network design problem as an optimization and search problem. This objective function $Q(S)$ needs to be optimized. Where $Q(S)$ represents the quality measurement for a solution S_i given $\forall S_i Q(S_i) \geqslant 0$. The problem is to find the *best solution* (i.e., QSAR model) $\hat{S}$ such that:

$$Q(\hat{S}) = \max_{S} Q(S) \tag{5}$$

The validation of a QSAR relationship is probably the most important step of all. The validation estimates the reliability and accuracy of predictions before the model is put into practice. Poor predictions misguide the direction of drug development and turn downstream efforts meaningless. To verify QSAR model quality in regression tasks, we employ the commonly used mean squared error (MSE) given by,

$$\mathbf{MSE} = \frac{1}{n} \sum_{i=1}^{n} (y_i - \hat{y}_i)^2 \tag{6}$$

where, $\hat{y}_i$, values of the predicted values, and y_i, values of the actual values. However, it is necessary to get a large number of testing compounds in order to draw statistically convincing conclusion. The main steps of the QSAR optimization using PSO algorithm listed in Algorithm 1, where there are three main steps (1) Initialize a swarm of particles (position and velocities); (2) Updating velocities; (3) Updating positions.

Input: Particles, a set of QSAR models
          : Fitness-Fn, a function that calculate the QSAR model error function
Output: Predictive QSAR model
1. Initialize swarm $P(t)$, each position $X_i(t)$ of each particle $P_i \in P(t)$ (random)
2. Evaluate performance $F(X_i(t))$ of each particle (using current position $X_i(t)$)
3. Compare performance of each individual to its best performance
4. Compare performance of each particle to the global best
while *(error $\geq \epsilon$)* **do**
    for *(i = 1 to Size(Particles))* **do**
        Change velocity vector of each *particle*

$$v_i(t+1) = w v_i(t) + \eta_1 \cdot rand() \cdot (p_i(t) - x_i(t))$$
$$+ \eta_2 \cdot rand() \cdot (p_g(t) - x_i(t))$$

        Move each *particle*

$$x_i(t+1) = x_i(t) + v_i(t)$$

    end
end
return *Best QSAR*
Algorithm 1: QSAR optimization using PSO algorithm

5 Data Description and Processing

The chemical structure is susceptible of many numerical representations, commonly known as molecular descriptors. These molecular descriptors map the structure of the molecules into a set of numerical or binary values that characterize specific molecular properties which can explain an activity. The datasets used in this study are obtained from the UCI Data Repository [23]. For this study, two data sets including 74 instances of Pyrimidines with 28 features and 186 instances of Triazines with 61 features were collected. Features were arranged according to the positions of possible substitutions and contained molecular descriptors like polarity, size, flexibility, hydrogen-bond donor, hydrogenbond acceptor, π donor, π acceptor, polarizability, s effect, branching and biological activity. In the first dataset the aim of a learning algorithm is to predict the inhibition of dihydrofolate reductase by pyrimidines with low probability of error. In the second dataset the aims is to predict the inhibition of dihydrofolate reductase by triazines. In order to reproduce cancer cells triazines inhibit dihydrofolate enzymes. To make data useful to the CVNNs, it should be transformed into the complex valued domain. In this paper, each real-valued input is encoded into a phase between 0 and π of a complex number of unity magnitude [24]. One such mapping for each element of the vector x can be done by the following transformation:

$$Let \quad x \in [a, b] \quad where \quad a, b \in \mathfrak{R} \quad then \quad \theta = \frac{\pi(x - a)}{b - a} \qquad (7)$$

and

$$z = \cos \theta + i \sin \theta \qquad (8)$$

This transformation can be regarded as a preprocessing step. The preprocessing is commonly used even in RVNNs in order to map input values into a specified range, such as Min-Max normalization.

6 Experimental Results

In order to evaluate the fittest neural network model, the training algorithms are conducted through several pre-experiments to determine the parameters setting per algorithm that yields the best performance with respect to all data sets. For PSO, all swarm particles start at a random position (i.e., weights). The velocity of each particle is randomized to a small value to provide initial random impetus to the swarm. The swarm size was limited to 60 particles. The most important factor is maximum velocity parameter, which affects the convergence speed of the algorithm, is set to 0.1. For the BP the learning rate is 0.1 and activation function is sigmoid. The two

Table 1 Ten fold cross
validation MSE on the
Pyrimidine and Triazine
datasets

	RVNN	CVNN	BP
Triazines	0.016	0.015	0.04
Pyrimidines	0.0033	0.002	0.012

algorithms are runs of 500 objective function evaluations. Both of the CVNN and RVNN models consists of three-layer neural networks with one input layer, one hidden layer containing 3 processing nodes, and one output layer. The must important factor is maximum velocity parameter w which affect the convergence speed of the algorithm is set to 0.2. The η_1 and η_2 are 2.0 and 2.0 respectively. Using 10-fold cross-validation accuracy to measure prediction quality for each of the QSAR models in the two datasets in order to establish their true learning and generalization capabilities. The results of runs on the two drug data set summarized in Table 1 in terms of the *mean square error*(MSE).

As shown in Figs. 3, 4 and 5, the actual activity of Pyrimidines data set and predicted activity using both CVNN and RVNN as QSAR models and their learning rate, that CVNN is superior for a complex non-linear prediction in comparison with RVNN. Moreover, CVNN has a faster learning rate. From these results it is concluded that the neural network model that trained by PSO is superior for a complex non-linear prediction in comparison with BP which trapped in local minima. This due to the PSO training algorithm escape from local minimal, explore interesting areas of the search space in parallel and maintain multiple solutions during the search.

Also, the CVNN is get better results than RVNN and its learning rate is faster than the RVNN as shown in Fig. 5.

Fig. 3 The predicted
activity for Pyrimidines data
set using QSAR models for 3
hidden nodes

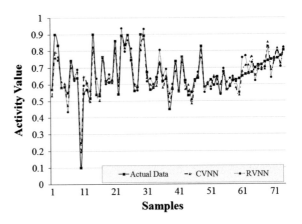

Fig. 4 The effect of number of particles one the predicted error with QSAR model has 3 hidden nodes

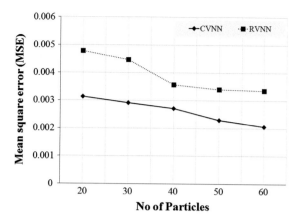

Fig. 5 The learning curve of both PSO for both CVNN and RVNN

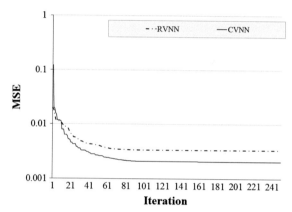

7 Conclusions

One important problem in modern drug design is to predict the activity of a compound of the drug to a binding target using its descriptors, which can be accomplished using machine learning approaches. Computationally, using efficient algorithms in the implementation, since drug datasets are noisy and has high dimensional space. In this work, a new strategy for training the CVNN using PSO as an efficient learning algorithm for real-valued regression problems is implemented. We tested such trained CVNN on two drug datasets as a real world benchmark problems and compared with BP algorithm and RVNN trained by PSO. It was found that the BP has a poor performance in a comparison with CVNN and RVNN. This because the BP uses a predefined way of updating the weights and usually fall in local minima, while the weights in PSO is evolved and influenced by random factors make it escape from local minimal. Moreover, the PSO explores a larger candidate solution space than BP. The RVNN is suffer from this slow rate of learning and gave worst results than CVNN due fact that the complex-valued representation can make the network more flexible in the mapping process from input to output. This idea inspired in kernel-based learning algorithms that use non-linear mappings from input spaces to high dimensional feature spaces should be mentioned.

References

1. Lipinski, C.: Lead- and drug-like compounds: the rule-of-five revolution. Drug Discov. Today: Technol. **1**(4), 337–341 (2004)
2. Leeson, P., Davis, A., Steele, J.: Drug-like properties: guiding principles for design or chemical prejudice? Drug Discov. Today: Technol. **1**(3), 189–195 (2004)
3. Li, A.P.: Preclinical in vitro screening assays for drug-like properties. Drug Discov. Today: Technol. **2**(2), 179–185 (2005)
4. Hansch, C.: A quantitative approach to biochemical structure-activity relationships. Acc. Chem. Res. **2**, 232–239 (1969)
5. Duch, W., Swaminathan, K., Meller J.: Artificial Intelligence Approaches for Rational Drug Design and Discovery Current Pharmaceutical Design, 13, (2007)
6. Chin, L., Chun, Y.: In: Dehmer, M., Varmuza, K., Bonchev, D. (eds.) Current Modelling Methods Used in QSAR/QSPR, in Statistical Modelling of Molecular Descriptors in QSAR/QSPR. Wiley-VCH Verlag GmbH & Co, Weinheim (2012)
7. Gertrudesa, J., Maltarollob, V., Silvaa, R., Oliveiraa, P., Honrioa, K., da Silva, A.: Machine learning techniques and drug design. Curr. Med. Chem. **19**, 4289–4297 (2012)
8. El-Telbany, M.: The predictive learning role in drug design. J. Emerg. Trends Comput. Inf. Sci. JETCIS **5**(3) (2014)
9. Bishop, C.: Pattern Recognition and Machine Learning, 2nd edn. Springer, Berlin (2006)
10. Marsland, S.: Machine Learning—An Algorithmic Perspective. Chapman and Hall, CRC, Boca Raton (2009)
11. Mohri, M., Rostamizadeh, A., Talwalkar, A.: Foundations of Machine Learning. MIT Press (2012)
12. Hirose, A.: Nature of complex number and complex-valued neural networks. Front. Electr. Electron. Eng. China **6**, 171–180 (2011)
13. Kobayashi, M.: Exceptional reducibility of complex-valued neural networks. IEEE Trans. Neural Netw. **21**, 1060–1072 (2010)
14. Hirose, A., Yoshida, S.: Generalization characteristics of complex-valued feedforward neural networks in relation to signal coherence. IEEE Trans. Neural Netw. Learn. Syst. **23**, 541–551 (2012)
15. Agrafiotis, D., Cedeno, W.: Feature selection for structure-activity correlation using binary paricle swarms. J. Med. Chem. **45**, 1098 (2002)
16. Cedeno, W., Agrafiotis, D.: A comparison of particle swarms techniques for the development of quantitative structure-activity relationship models for drug design. In: Proceedings of the IEEE Computational Systems Bioinformatics (2005)
17. El-Telbany, M., Refat, S., Nasr, E.: Particle swarm optimization for drug design. In: Proceedings of the 1st International Workshop on Mechatronics Education, Taif, Saudi Arabia, pp. 117–122 (2015)
18. Aizenberg, I., Moraga, C.: Multilayer feedforward neural network based on multi-valued neurons (MLMVN) and a backpropagation learning algorithm. Soft Comput. **11**(2), 169–183 (2007)
19. Kitajima, T., Yasuno, T.: Output Prediction of Wind Power Generation System Using Complex-valued Neural Network Proceedings SICE Annual Conference, pp. 3610–3613 (2010)
20. Nitta, T.: An extension of the back-propagation algorithm to complex numbers. Neural Netw. **10**(8), 1391–1415 (1997)

21. Ozdemir, N., Iskender, B., Ozgur, N.: Complex valued neural network with Möbius activation function. Commun. Nonlinear Sci. Numer. Simul. **16**(12), 4698–4703 (2011)
22. Kennedy, J., Eberhart, C.: Particle swarm optimization. In: Proceedings of the IEEE International Conference on Neural Networks, pp. 1942–1948 (1995)
23. Newman, D., Hettich, S., Blake, C., Merz, C.: UCI Repository of Machine Learning Databases. Department of Information and Computer Science, University California, Irvine (1998)
24. Faijul, A., Murase, K.: Single-layerd complex-valued neural network for real-valued classification problems. Neurocomputing **72**, 945–955 (2009)

A XOR-Based ABC Algorithm
for Solving Set Covering Problems

Ricardo Soto, Broderick Crawford, Sebastián Lizama,
Franklin Johnson and Fernando Paredes

Abstract The set covering problem is a classical problem in the subject of combinatorial optimization that consists in finding a set of solutions that cover a range of needs at the lowest possible cost. The literature reports various techniques to solve this problem, ranging from exact algorithms to approximate methods. In this paper, we present a new XOR-based artificial bee colony algorithm for solving set covering problems. We integrate a XOR operator to binarize the solution construction in order to cope with the binary nature of set covering problems. We also incorporate preprocessing phases and dynamic ABC parameters so as to improve solving time. We report interesting and competitive experimental results on a set of 65 benchmarks from the Beasley's OR-Library.

R. Soto (✉) · B. Crawford · S. Lizama · F. Johnson
Pontificia Universidad Católica de Valparaíso, Valparaíso, Chile
e-mail: ricardo.soto@ucv.cl

B. Crawford
e-mail: broderick.crawford@ucv.cl

S. Lizama
e-mail: sebastian.lizama.a@mail.pucv.cl

F. Johnson
e-mail: franklin.johnson@upla.cl

R. Soto
Universidad Autónoma de Chile, Temuco, Chile

R. Soto
Universidad Científica Del Sur, Lima, Peru

B. Crawford
Universidad Central de Chile, Santiago, Chile

B. Crawford
Universidad San Sebastián, Region Metropolitana, Chile

F. Johnson
Universidad de Playa Ancha, Valparaíso, Chile

F. Paredes
Escuela de Ingeniería Industrial, Universidad Diego Portales, Santiago, Chile
e-mail: fernando.paredes@udp.cl

© Springer International Publishing Switzerland 2016
T. Gaber et al. (eds.), *The 1st International Conference on Advanced Intelligent System and Informatics (AISI2015), November 28–30, 2015, Beni Suef, Egypt,*
Advances in Intelligent Systems and Computing 407, DOI 10.1007/978-3-319-26690-9_19

Keywords Bio-inspired systems · Artificial bee colony algorithm · Metaheuristics ·
Set covering problem

1 Introduction

The Set Covering Problem (SCP) is a classical problem in the subject of combinatorial optimization that consists in finding a set of solutions that cover a range
of needs at the lowest possible cost. This problem can be applied in various real
scenarios such as in line balancing, location or installation of services, simplification of Boolean expressions, and crew scheduling in airlines, among others [17, 18].
During the last twenty years, different techniques has been proposed to efficiently
solve this problem. For instance, metaheuristics or approximate algorithms consist
on iterative procedures that guide a subordinate heuristic, combining different concepts for exploring and properly exploiting the search space in an intelligent way.
Even though this has some advantages over exact algorithms, it must be emphasized
that do not necessarily deliver optimal solutions. Some examples solving SCPs in
this context include genetic algorithms [4], simulated annealing [5], tabu search [6],
and so on [8]. These algorithms are used when there is no exact method of resolution, or when a lot of calculation time or a large amount of computational resources
is required.

On the other hand, exact algorithms are generally based on branch-and-bound
and branch-and-cut algorithms [2]. The problem is that these algorithms are capable
only to solve instances with limited size. For instance, classical greedy algorithms
can rapidly be implemented, but due to their deterministic nature they cannot always
find high quality solutions [10]. Compared with classical greedy algorithms [15],
Lagrangian relaxation-based heuristics are much more effective such as the ones
proposed in [7, 9].

In this paper, we present a new Artificial Bee Colony (ABC) algorithm for solving
SCPs, we employ pre-processing phases as well as dynamic ABC parameters that
vary depending on the instance size in order to improve solving time. We additionally
introduce a XOR operator to binarize the solution construction in order to adapt them
to the binary nature of SCPs. This allows one to avoid the use of transfer functions
facilitating implementation, while keeping performance. We report interesting and
competitive experimental results on a set of 65 benchmarks from the OR-Library.

The paper is organized as follows: In Sect. 2, the mathematical model of the SCP
is presented. Next section provides an overview of ABC. Then, the pre-processing
phases, the introduction of the xor operator, and the dynamic parameters are
explained. Finally, the experimental results are given followed by conclusions.

2 Problem Description

The Set-Covering Problem aims to find a set of solutions to cover a set of needs at the lowest possible cost. Formally, we define the problem as follows: Let $A = (a_{ij})$ be a $m - rows \times n - columns$ binary matrix $(0,1)$ and let $C = (c_j)$ be a vector representing the cost of each column j, where each element of C is a real non-negative value. Considering $M = \{1, \ldots, m\}$ and $N = \{1, \ldots, n\}$, as the sets of rows and columns respectively, it is possible to say that a column j covers a row i contained in M if $a_{ij} = 1$. The SCP aims at finding a subset of S columns of minimum cost, such that each row i existing in M, is covered by at least a j column existing in S. The corresponding mathematical model is as follows:

$$min\,(z) = \sum_{j=1}^{n} c_j X_j$$

Subject to:

$$\sum_{j=1}^{n} a_{ij} x_j \geq 1 \qquad \forall i \in M$$

$$x_j = \begin{cases} 1 \; j \in S \\ 0 \; if \; not \end{cases} \forall j \in N$$

3 Artificial Bee Colony Algorithm

The Artificial Bee Colony Algorithm (ABC) is one of the most recent algorithms in the domain of collective intelligence. It was developed by Karaboga [11, 13] for resolving numerical problems from the optimization domain, inspired by the dance and feeding process behavior of real honey bee colonies. For the sake of clarity about bees theoretical background, we introduce the following concepts:

Employed bees: Those that are associated in particular to one food source. The function of these bees is to stay in one food source, storing and sharing the information about their neighborhood.

Unemployed bees: These agents play the role of exploration in the algorithm, and split into two sub-types. The *scout bees*, which have the function of seeking new food sources around the hive, and the *onlooker* or *curious bees*, which have the function to remain in the hive, waiting for the information provided by the employed bees, with the objective of going out to search for a food source with potential, based on the knowledge gained through the collected information.

Colony: The colony is equal to the total number of agents that participate in the searching process, i.e. *employed bees* + *onlooker bees*. The scout bees are not counted, since they correspond to the state of the other bees. This will be explained in the next paragraphs.

Food sources: The food sources correspond to the possible solutions of the problem to be solved, i.e. the initial population that is randomly generated. Strictly in ABC, the initial population corresponds to the half of the colony. This is because the relation between food sources and employed bees is always 1:1, therefore the half of the colony would be employed bees and remaining ones will be unemployed bees.

The search process associated to ABC can be described generally in 4 phases [12]:

Algorithm 1 Main steps of ABC

1: *cycles* = 1
2: Initialize food sources x_i, $i = 1 \ldots SN$
3: Evaluate the amount of nectar of food sources
4: **repeat**
5:     Employed bee phase
6:     Onlooker bee phase
7:     Scout bee phase
8:     Memorize the best solution
9: **until** *cycle* = *Max number of cycles*

Initialization phase: For every employed bee, a feasible, random and binary solution is produced, with a drop counter that is initialized for each food source. This counter increases if the produced solution is worse or equal to the stored one. To evaluate each solution, the fitness value is calculated using Eq. 1.

$$fit_i = \begin{cases} \frac{1}{1+f_i} & if \ (f_i \geq 0) \\ 1 + abs \ (f_i) & if \ (f_i < 0) \end{cases} \tag{1}$$

Employed bee phase: At this stage, the employed bee attempts to improve its solution, using the equation of motion. Because of the need to work with binary solutions, we replace the classic ABC motion Eq. (2) by Eq. 3 where the xor ($\oplus$) operator has been introduced as proposed in [14]. In both Eqs. (2) and (3) N is the number of bees used and D the dimensionality of the problem.

$$X_i^j = X_i^j + \varphi \times \left(X_i^j - X_k^j \right) \ i,k \in \{1, \ldots, N\}, j \in \{1, \ldots, D\} \ and \ i \neq k \tag{2}$$

$$V_i^j = X_i^j \oplus \left[\varphi \left(X_i^j \oplus X_k^j \right) \right] i,k \in \{1, \ldots, N\}, j \in \{1, \ldots, D\} \ and \ i \neq k \tag{3}$$

Onlooker bee phase: The onlooker bee waits for the information to be shared from the other food sources. Each food source has a probability of being selected, depending on the potential for improvement. The equation used is not strict, as long as the use of priorities is respected at the time of selection of the food sources. After selecting a food source, the solution is updated with Eq. 3. If the new solution is better, it is replaced and the drop counter is reset, otherwise the counter is incremented by one.

Scout bee phase: This phase begins when the drop counter of a bee exceeds the set limit. In this case, the employed bee becomes a scout bee and generates a new random solution. Once the new solution is generated and if it is feasible, the corresponding drop counter is started and the scout bee becomes an employed bee again [13].

4 Improving Solution Quality and Solving Speed

In addition to the adaptation used to cope with binary solutions [14], there are certain elements that have been added to the classic ABC to obtain better quality solutions and/or reduce solving time.

Parameter Configuration: One of the fundamentals for improving solution quality in metaheuristics is the fine configuration of their parameters. In this work, we have observed that making dynamic some parameters (*limit* (maximum tries to improve a food source) and *maximum cycles*(stop criteria)) in function of the problem size we improve convergence and as a consequence solving time. After carrying out convergence tests (see a summary in Table 1 and detail of instances in Table 2), it was

Table 1 Convergence tests after pre-process

Instance	Best convergence	Worst convergence	$(f * 10) + 2000$
4.1	1650 cycles	2850 cycles	4000 cycles
5.1	1800 cycles	2950 cycles	4000 cycles
6.1	550 cycles	800 cycles	4000 cycles
A.1	2500 cycles	3900 cycles	5000 cycles
B.1	3100 cycles	4750 cycles	5000 cycles
C.1	3300 cycles	4750 cycles	6000 cycles

Table 2 Details of tested instances

Instance set	No. of instances	n	m	Cost range	Density (%)	Optimal solution
4	10	200	1000	[1,100]	2	Known
5	10	200	2000	[1,100]	2	Known
6	5	200	1000	[1,100]	5	Known
A	5	300	3000	[1,100]	2	Known
B	5	300	3000	[1,100]	5	Known
C	5	400	4000	[1,100]	2	Known
D	5	400	4000	[1,100]	5	Known
NRE	5	500	5000	[1,100]	10	Unknown
NRF	5	500	5000	[1,100]	20	Unknown
NRG	5	1000	10000	[1,100]	2	Unknown
NRH	5	1000	10000	[1,100]	5	Unknown

determined that $f * 10$ (f being the number of rows of the SCP) plus a margin error equivalent to 2000 cycles is a good value for the maximum of the cycles, since after that, the optimum reached does not improve. Moreover, a suitable value for the limit is $SN * D$, where SN is the number of food sources used in the implementation [1].

Pre-process: One of the most used methods to reduce execution times is the pre-process. There are a large number of methods proposed, of which two have been used. The first of them is *Column Domination*, which states that is viable to eliminate any column j whose rows can be covered by other columns at lower cost that c_j, where c_j corresponds to the cost vector at the position j. In addition the method *Column inclusion* was used, which indicates that if a row is covered only by a column after using the previous method, this column should be included in the optimal solution [16].

Repairing: One way to treat the generation of infeasible solutions is by repairing them. In current research two methods of repair have been used. The first corresponds to the basic repair of SCP, in which the elements of a solution are revised, searching for the first element that allows this solution to be feasible, forcing compliance with the restriction by inserting a 1. On the other hand, we have a more elaborated fix, proposed by Beasley [4], which attempts to make the genetic algorithm more effective and proposes a step of additional local optimization to the requirement to maintain the viability of the generated solutions.

Basically the idea is to remove redundant columns in the problem, these columns being ones that we can remove while keeping the solution feasible. The selection of the columns is carried out based on an element called *ratio*, which allows sorting the columns based on the result obtained from Eq. 4.

$$Cost\ of\ a\ column\ j/number\ of\ covered\ rows\ by\ column\ j \qquad (4)$$

5 Experiments and Results

The algorithm was implemented in Java and launched on a 2.3 GHz Intel Core i5 with 8GB RAM DDR3 1333 MHz PC running OS X 10.9.2. In order to evaluate the performance and precision of the proposed algorithm, we have employed 65 non-unicost instances from the Beasleys OR-Library [3], divided into 11 groups of 5 or 10 instances each. Table 2 shows in detail the characteristics of each group of test instances before pre-processing.

The algorithm was executed 30 times per instance, with number of food sources = 150, maximum cycles and limit variables according to parameter configuration described in Sect. 4.

Tables 3, 4 and 5 show the results obtained for all instances of SCP, presented from left to right: the instance number, the optimal or best known value for each instance, the best value encountered from the 30 runs of each instance, the average

Table 3 Computational results for groups 4, 5 and 6

Instance	Optimum	Best value found	Average	RPD (%)
4.1	429	430	431	0.23
4.2	512	513	517	0.20
4.3	516	519	524	0.58
4.4	494	495	499	0.20
4.5	512	514	517	0.39
4.6	560	561	564	0.18
4.7	430	431	434	0.23
4.8	492	493	496	0.20
4.9	641	649	656	1.25
4.10	514	517	528	0.58
5.1	253	254	257	0.40
5.2	302	309	312	2.32
5.3	226	229	229	1.33
5.4	242	242	242	0.00
5.5	211	211	211	0.00
5.6	213	214	221	0.47
5.7	293	298	301	1.71
5.8	288	289	292	0.35
5.9	279	280	281	0.36
5.10	265	267	270	0.75
6.1	138	142	145	2.90
6.2	146	147	149	0.68
6.3	145	148	149	2.07
6.4	131	131	132	0.00
6.5	161	165	168	2.48

of 30 runs, and finally the *RPD* depicts the relative percentage difference between the best known optimum value and the best solution found in all executions of the instance, calculated as follow:

$$RDP = \frac{(Z - Z_{opt})}{Z_{opt}} \times 100$$

where Z_{opt} is the best known optimum value and Z is the best optimum value reached by CS.

Figure 1 shows the evolution of the instances and their level of convergence. Instances are presented preceded by three letters that indicate the name of the problem, in this case SCP. In SCPA.2, convergence is achieved exactly at iteration 3357.

Table 4 Computational results for groups A, B, C and D

Instance	Optimum	Best value found	Average	RPD (%)
A.1	253	254	255	0.40
A.2	252	257	259	1.98
A.3	232	235	238	1.29
A.4	234	236	238	0.85
A.5	236	236	237	0.00
B.1	69	70	73	1.45
B.2	76	78	81	2.63
B.3	80	80	81	0.00
B.4	79	80	81	1.27
B.5	72	72	72	0.00
C.1	227	231	231	1.76
C.2	219	222	224	1.37
C.3	243	254	259	4.53
C.4	219	231	233	5.48
C.5	215	216	217	0.47
D.1	60	60	60	0.00
D.2	66	68	69	3.03
D.3	72	76	77	5.56
D.4	62	63	64	1.61
D.5	61	63	63	3.28

Fig. 1 Evolution of mean best values for SCPA.2, SCPB.2 and SCPC.2

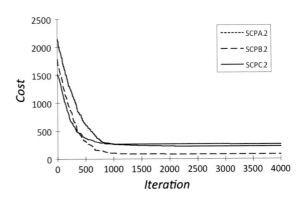

In the cases of instances SCPB.2 and SCPC.2, the convergence is achieved in iterations 3751 and 3395, respectively. In the three cases, the convergence is quite fast providing reasonable good results. We have also observed that pre-processing allowed us to accelerate convergence, due to improvement of initial solutions.

Table 5 Computational results for groups NRE, NRF, NRG and NRH

Instance	Optimum	Best value found	Average	RPD (%)
NRE.1	29	29	29	0.00
NRE.2	30	32	32	6.67
NRE.3	27	29	29	7.41
NRE.4	28	29	31	3.57
NRE.5	28	29	29	3.57
NRF.1	14	14	15	0.00
NRF.2	15	16	16	6.67
NRF.3	14	16	16	14.29
NRF.4	14	15	15	7.14
NRF.5	13	15	15	15.38
NRG.1	176	183	186	3.98
NRG.2	154	162	163	5.19
NRG.3	166	174	176	4.82
NRG.4	168	175	178	4.17
NRG.5	168	179	180	6.55
NRH.1	63	70	72	11.11
NRH.2	63	69	71	9.52
NRH.3	59	66	68	11.86
NRH.4	58	64	65	10.34
NRH.5	55	60	60	9.09

6 Conclusions

In this paper, we have presented a XOR-based ABC algorithm for solving SCPs. We have introduced a XOR operator to the classic ABC approach in order to binarize the solution construction. This allows one to avoid the use of transfer functions facilitating implementation, while keeping performance. We have also incorporated pre-processing phases as well as parameters that vary depending on the instance size in order to improve solving time. We have reported interesting and competitive experimental results where the RPD is equal (or very near) to 0 for various instances, demonstrating that the presented approach is a good candidate for solving this kind of problems. It is also worth mentioning that the modifications made to the basic algorithm can be improved in future research, considering that although each modification to ABC is considered a theoretical base of previous research, the decisions have been made based mainly on trial-and-error processes, since there are no general rules to follow for achieving the expected results in approximate methods domain.

Acknowledgments Ricardo Soto is supported by Grant CONICYT/FONDECYT/INICIACION/ 11130459, Broderick Crawford is supported by Grant CONICYT/FONDECYT/ 1140897, and Fernando Paredes is supported by Grant CONICYT/FONDECYT/ 1130455.

References

1. Akay, B., Karaboga, D.: Parameter tuning for the artificial bee colony algorithm. In: Proceedings of the 1st Conference on Computational Collective Intelligence, volume 5796 of LNCS, pp. 608–619. Springer (2009)
2. Balas, E., Carrera, M.C.: A dynamic subgradient-based branch-and-bound procedure for set covering. Oper. Res. **44**(6), 875–890 (1996)
3. Beasley, J.E.: A lagrangian heuristic for set-covering problems. Naval Res. Logist. **37**(1), 151–164 (1990)
4. Beasley, J.E., Chu, P.C.: A genetic algorithm for the set covering problem. Eur. J. Oper. Res. **94**(2), 392–404 (1996)
5. Brusco, M.J., Jacobs, L.W., Thompson, G.M.: A morphing procedure to supplement a simulated annealing heuristic for cost- and coverage-correlated set-covering problems. Ann. Oper. Res. **86**, 611–627 (1999)
6. Caserta, M.: Tabu search-based metaheuristic algorithm for large-scale set covering
7. Caprara, A., Fischetti, M., Toth, P.: A heuristic method for the set covering problem. Oper. Res. **47**(5), 730–743 (1999)
8. Caprara, A., Toth, P., Fischetti, M.: Algorithms for the set covering problem. Ann. Oper. Res. **98**, 353–371 (2000)
9. Ceria, S., Nobili, P., Sassano, A.: A lagrangian-based heuristic for large-scale set covering problems. Math. Program. **81**(2), 215–228 (1998)
10. Chvatal, V.: A greedy heuristic for the set-covering problem. Math. Oper. Res. **4**(3), 233–235 (1979)
11. Karaboga, D., Basturk, B.: A powerful and efficient algorithm for numerical function optimization: artificial bee colony (ABC) algorithm. J. Glob. Optim. **39**(3), 459–471 (2007)
12. Karaboga, D., Akay, B.: Artificial, bee colony (ABC), harmony search and bees algorithms on numerical optimization. IPROMS: Innovative Production Machines and Systems Virtual Conference. Cardiff, UK (2009)
13. Karaboga, D., Akay, B.: A survey: algorithms simulating bee swarm intelligence. Artif. Intell. Rev. **31**(1–4), 61–85 (2009)
14. Kiran, M.S., Gunduz, M.: XOR-based artificial bee colony algorithm for binary optimization. Turkish J. Electr. Eng. Comput. Sci. **21**(2), 2307–2328 (2013)
15. Lan, G., DePuy, G.W.: On the effectiveness of incorporating randomness and memory into a multi-start metaheuristic with application to the set covering problem. Comput. Ind. Eng. **51**(3), 362–374 (2006)
16. Lan, G., Depuy, G.W., Whitehouse, G.E.: An effective and simple heuristic for the set covering problem. Eur. J. Oper. Res. **176**(3), 1387–1403 (2007)
17. Vasko, F.J., Wilson, G.R.: Using a facility location algorithm to solve large set covering problems. Oper. Res. Lett. **3**(2), 85–90 (1984)
18. Vasko, F.J., Wolf, F.E., Stott, K.L.: Optimal selection of ingot sizes via set covering. Oper. Res. **35**(3), 346–353 (1987)

Abdominal CT Liver Parenchyma Segmentation Based on Particle Swarm Optimization

Gehad Ismail Sayed and Aboul Ella Hassanien

Abstract Image segmentation is an important task in the image processing field. Efficient segmentation of images considered important for further object recognition and classification. This paper presents a novel segmentation approach based on Particle Swarm Optimization (PSO) and an adaptive Watershed algorithm. An application of liver CT imaging has been chosen and PSO approach has been applied to segment abdominal CT images. The experimental results show the efficiency of the proposed approach and it obtains overall accuracy 94 % of good liver extraction.

Keywords Watershed · Preprocessing · Segmentation · Particle swarm optimization · Post-processing · Morphological operators · CT · Parenchyma

1 Introduction

Image segmentation is the process of partitioning the image into meaningful objects that have same visual characteristic. It's considered an important basic task in analyzing and understanding of the image [1]. There are many image segmentation techniques. Clustering is one of the most important techniques used for image segmentation. In image segmentation, Clustering is the process of identifying natural groupings based on some similarity measure like Euclidean distance [2]. Clustering algorithms are used in many applications, such as data mining, image

G.I. Sayed (✉) · A.E. Hassanien
Faculty of Computers and Information, Cairo University, Giza, Egypt
e-mail: darkspot_1993@yahoo.com
URL: http://www.egyptscience.net

A.E. Hassanien
Faculty of Computers and Information, Benisuef University, Beni Suef, Egypt

G.I. Sayed · A.E. Hassanien
Scientific Research Group in Egypt (SRGE), Cairo, Egypt

© Springer International Publishing Switzerland 2016
T. Gaber et al. (eds.), *The 1st International Conference on Advanced Intelligent System and Informatics (AISI2015), November 28–30, 2015, Beni Suef, Egypt,*
Advances in Intelligent Systems and Computing 407, DOI 10.1007/978-3-319-26690-9_20

segmentation, machine learning, etc. [3]. Particle Swarm Optimization (PSO) is a population based stochastic optimization technique proposed by Dr. Eberhart and Dr. Kennedy in 1995. The idea of the algorithm inspired by social behavior of bird flocking or fish schooling [4]. It consists of a number of particles (solutions) that move in the search space (pixels in the image) in order to search for global optima (maximizing the distribution of intensity levels in the image [5]. It has been used to optimize parameters for medical images as in [18]. Many PSO versions have recently proposed like in [19]. In this paper, we will focus on PSO that will used to segment Liver for CT image. A fully automatic liver segmentation approach from abdominal CT images based on PSO and an adaptive watershed is presented in this paper. The remainder of this paper is ordered as follows. Section 2 explains the basic concepts of PSO. Section 3 shows the proposed PSO and watershed approach. In Sect. 4 shows the experimental results and analysis with details of the datasets. Finally, conclusion and future work are discussed in Sect. 5.

2 Preliminaries

PSO algorithm originally developed by Eberhart and Kennedy in 1995 [4], which is taking advantage of the swarm intelligence concept, for example bird flocks and fish schools. The usual aim of the particle swarm optimization (PSO) algorithm is to solve continuous and discrete optimization problems. It is a population-based method, that is iteratively changed until a termination criterion is satisfied. The population of feasible solutions $P = P_1, P_2 \ldots, p_n, p_2$ pain in PSO is often called a swarm. And each P is called particle. These particles travel through the search space to find the optimal solution. The PSO segmentation based algorithm has been one of the most recently used. It has been compared with GA-based algorithms [8]. The results show that it gives better results in less time also it needs only few parameters to adjust. As PSO has no parameter of "mutation", "recombination", and no notion of the "survival of the fittest". At the beginning of the PSO algorithm, particle position velocities are set to zero and their positions are randomly set within the boundaries of the search space. Global, local and neighborhood are initialized with small values. Population size and stopping criteria are very important parameters that need to optimize in order to get an overall good solution with acceptable time limit. In each step of PSO algorithm fitness function is evaluated and each particle's position and velocity are updated according to Eqs. (1) and (2). The fitness function is used to indicate how close a particle to the optimal solution.

$$v_{i+1}^n = wv_i^n + \phi_1 r_1 (\overline{g}_i^n - x_i^n) + \phi_2 r_2 (\overline{p}_i^n - x_i^n) + \phi_3 r_3 (\overline{n}_i^n - x_i^n) \qquad (1)$$

$$x_{i+1}^n = x_i^n + v_{i+1}^n \qquad (2)$$

where w, ϕ_1, ϕ_2 and ϕ_3 coefficients are assigned weights. w Is inertial weight which represents the memory of previous velocity and ϕ_1, ϕ_2 and ϕ_3 are acceleration coefficients which represent cognitive (personal), social (neighborhood) component usually set between 0 and 4. The symbols r_1, r_2 and r_3 represent random variables with uniform distribution between 0 and 1. $\overline{g}_i^n$ is global best information while $\overline{n}_i^n$ is neghborhood best and $\overline{p}_i^n$ is local best.

3   The Proposed Segmentation Approach

The proposed abdominal CT Liver parenchyma segmentation approach is comprised of five fundamental building phases. These five phases are described in detail in the following section along with the steps involved and the characteristics feature for each phase. The proposed algorithm can be summarized as in Algorithm (1). And the overall architecture of the introduced approach is described in Fig. 1.

Algorithm 1 Proposed Approach

1: Read CT image $I(i, j)$ and resize it 256*256

2: Apply median filter with 3*3 window size to $I(i, j)$

3: Apply PSO clutser-based algorithm with number of level = 3

4: **for** each clutser image of $C_k(i, j)$ do

5:     calculate mean

6: **end for**

7: Select best $C(i, j)$ with maximum mean value

8: Apply morphology operators like open and close to enhance $C(i, j)$

9: Apply adaptive watershed on enhanced $C(i, j)$ for final final liver region extraction

---

3.1   Preprocessing Phase

When image transforms from one form to another like scanning some degradation may occur. Image noise is a kind of these degradations. It is the random variation of brightness or color information in images [9]. This noise is always undesirable. So removing noise with preserving edges of the image plays a vital role in image processing. Median filter is one of the simplest and most popular approaches for removing noise like Salt and pepper [10]. It is calculated by sorting all surrounding neighborhood pixel values in numerical order and then replacing the specified pixel being with the middle pixel value. In this paper, CT image will be resized to 256 * 256 in order to reduce computation time. Then, Median filter with window size 3 * 3 is applied to enhance the CT image.

Fig. 1 The proposed automatic CT liver segmentation architecture

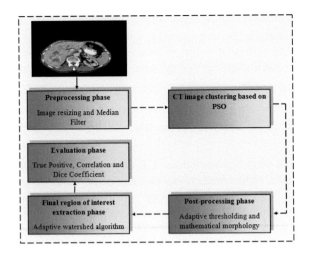

3.2 CT Image Clustering Based on PSO Phase

In this phase the modified image from previous phase will be clustered by using PSO. PSO is used in order to search for the n − 1 optimal n level thresholds that maximize the objective function (fitness function). The initial parameters of PSO are shown in Table 1 which showed the best results obtained.

3.3 Post-Processing Phase: Morphological Operators

In this phase the best cluster image produced from PSO will be selected. Selection criteria depend on the maximum mean value obtained from each cluster. Then the image will be converted to binary with thresholding equal to 0.4 (found to be optimal thresholding value), then open morphological operation used to enhance the clustered image and to focus on liver parenchyma, then the maximum region

Table 1 Initial PSO Parameters

Population size	150
Number of iteration	10
ϕ_1	0.6
ϕ_2	0.6
ϕ_3	0
x_{max}	255
x_{min}	0
v_{max}	2
v_{min}	−2
w	0.4
Number of levels	3

area of the image extracted, then close morphological operation used to fill in holes and small gaps of the image.

3.4 Region of Interest Extraction Phase

Watershed is one of image segmentation techniques. It used to segment the image into several homogeneous regions with similar gray levels [11]. The input image treated as a topographic map. Three types of points are used to express a topographic interpretation points that belong to regional minimum "minimum", points at which a drop of water will fall to a single minimum "catchment basin", and lines that separate catchment basin "watershed line" [12]. The aim of watershed is to search for regions of high intensity gradients (watersheds) which divide neighbored local minima (basins) [11]. In this paper, watershed is used to final segment liver parenchyma from abdominal CT image. Also watershed has a great advantage that it is very fast as No seed is needed, it has some drawbacks. Some of these are very sensitive to noise and over segmentation problem. So in order to overcome these problems. An adaptive approach of watershed used.

3.5 Evaluation Phase

Three measurements are used to evaluate the performance of the presented approach. These measurements are Dice Coefficient, Correlation and True Positive. The true positive ratio measure calculated by dividing the number of true positive (which that is mean pixels that actually belong to liver region) by the total number of liver region pixels. Dice coefficient is statistical validation metric to evaluate spatial overlap accuracy between two binary images. It is commonly used in reporting the performance of segmentation results. Its values range between 0 and 1, where 0 indicate no overlap and 1 perfect agreement [17]. It is calculated using Eq. (3). Correlation is another measurement to indicate how strong relationship between two binary images. 1 indicates a strong positive relationship, -1 indicates a strong negative relationship and 0 indicates no relationship at all. It is calculated using Eq. (4).

$$Dice(x, y) = \frac{2 * |x \cap y|}{|x + y|} \tag{3}$$

$$Corr(x, y) = \frac{\sum_m \sum_n (x(m, n) - \bar{x})(y(m, n) - \bar{y})}{\sqrt{(\sum_m \sum_n (x(m, n) - \bar{x})^2)(\sum_m \sum_n (y(m, n) - \bar{y})^2)}} \tag{4}$$

where $\bar{y}$ is mean of y and $\bar{x}$ is mean of x.

4 Experimental Results and Discussion

CT scanning is a diagnostic imaging procedure that uses X-rays in order to present cross-sectional images (slices) of the body. The proposed approach will be applied to a complex dataset. The dataset is divided into seven categories, depending on the tumor type: Benign (Cyst (CY), Hemangioma (HG), Hepatic Adenoma (HA), and Focal Nodular Hyperplasia (FNH)); or Malignant (hepatocellular carcinoma (HCC), Cholangiocarcinoma (CC), and Metastases (MS)). Each of these categories has more than 15-patients, each patient has more than one hundred slices, and more than one phase of CT scans (arterial, delayed, portal venous, non-contrast). The dataset includes a diagnosis report for each patient. All images are in JPEG format, selected from a DICOM file, and have dimensions of 630630, with horizontal and vertical resolution of 72 DPI, and bit depth of 24 bits [13]. The proposed approach was tested on 43 abdominal CT images from different patients. The accuracy of the proposed approach is measured using Correlation, Dice Coefficient and True Positive Ratio. The proposed CT image segmentation approach was programmed in MATLABR 2007 on a computer having Intel Core I3 and 2 GB of memory.

The following figures will demonstrate the result of each step in the proposed approach. Figure 2 shows the results obtained from preprocessing phase and PSO. Fig. 2a shows the original CT image, Fig. 2b shows CT image after applying median filter in order to remove noise, Fig. 2a shows the first cluster image obtained from PSO, where Fig. 2b shows the second cluster image and Fig. 2c shows the third cluster image. Figure 3 shows post-processing phase results, Fig. 3a shows image after converted to binary image using ostu' thresholding method, Fig. 3b shows the results obtained from applying opening morphology, Fig. 3c shows image after selecting the largest area in the image, Fig. 3d shows image after applying close morphology and Fig. 3e shows image after filling holes. Figure 4

(a)   **(b)**   **(c)**   **(d)**   **(e)**

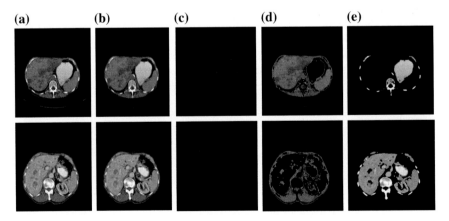

Fig. 2 Preprocessing and PSO results **a** Original Image, **b** Image after applying median filter, **c** Cluster-1, **d** Cluster-2 and **e** Cluster-3

(a)　　　　(b)　　　　(c)　　　　(d)　　　　(e)

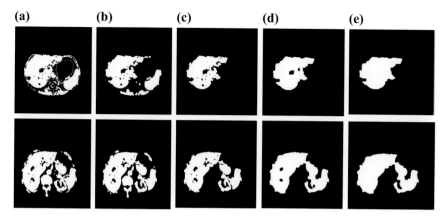

Fig. 3 Post-processing phase results. **a** Binarized clustered image, **b** Image after applying Open Morphology, **c** Image after selecting largest region, **d** Image after applying Close Morphology, **e** Image after fillling holes

(a)　　　　(b)　　　　(c)　　　　(d)　　　　(e)

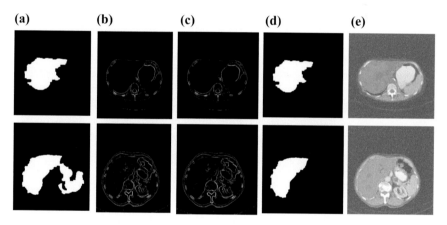

Fig. 4 Results of using modifed watershed. **a** CT image after enhancement, **b** Gradient Image, **c** Gradient image after normalization, **d** Image after remove local maxima regions, **e** Image after apply watershed and take maximum region

shows results of using modified watershed. Fig. 4a the modified watershed use CT image after enhancement in post-processing phase as input image, Fig. 4b shows gradient image calculated from the original image, Fig. 4c shows gradient image after normalization, Fig. 4d shows image after remove local maxima regions and Fig. 4e shows the final segmented liver image after applying watershed and take maximum region.

Table 2 compares the results of the proposed approach with other previous works. As it can be seen, the proposed approach gives better results and it can extract liver in less than 4 s. Table 3 shows the importance of using watershed in the proposed approach in terms of Dice Coefficient, and Correlation for the used

Table 2 Comparison with existing work on liver segmentation

Authors	Year	Accuracy (%)
Jeongjin et al. [16]	2007	70
Ruchaneewan et al. [15]	2007	86
M. Abdallal. [14]	2012	84
Z. Abdallal. [6]	2012	92
M. Anter [13]	2013	93
Nidaa Aldeek [7]	2014	87
Proposed approach	2014	94

Table 3 Comparison between using watershed in proposed approach and without using it in terms of dice coefficient, correlation and true positive

	Dice (%)	Correlation (%)	True positive (%)
Using watershed	91.89	90.62	94.62
Without using watershed	89.12	87.94	90.23

dataset. As we can see, the accuracy increase to almost 3 %. Figure 5 shows the best similarity indices for the used dataset of the proposed approach in terms of Dice Coefficient, and Correlation.

Table 4 compares the results obtained from different levels in terms of Dice Coefficient, True Positive and Correlation. From this table we can prove that the optimal level number is equal to three. Table 5 shows the total CPU process time of the whole data set for different levels. From this table, we can see as the number of levels increasing the CPU process time is increasing too. Table 6 compares the

Fig. 5 The best similarity indices obtained for the used dataset of the proposed approach in terms of dice coefficient, and correlation

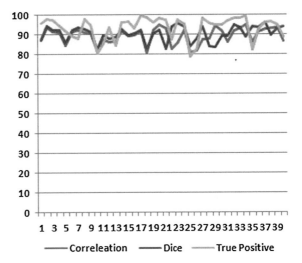

Table 4 Comparison between the results obtained from different levels in terms of dice coefficient, true positive and correlation

No. of levels	Dice (%)	Correlation (%)	True positive (%)
2	78.80	76.51	71.71
3	91.89	90.62	94.62
4	87.98	86.83	93.36
5	80.48	80.62	88.44

Table 5 CPU process time for different levels

No. of levels	Total CPU process time in seconds
2	50.62
3	56.66
4	62.39
5	75.69

Table 6 Comparison between the proposed approach and other methods

	Dice (%)	Correlation (%)	True positive (%)
Active contour	71.87	69.22	72.43
Global threshold	81.34	79.41	81.19
Proposed approach	91.89	90.62	94.62

results from the proposed approach and other methods in terms of Dice Coefficient, Correlation and True Positive. From these results, the proposed approach gives better results than other methods.

5 Conclusion and Future Works

We proposed an integrated approach based on using PSO and watershed algorithm for automatic extract liver from abdominal CT images. The experimental results show that the proposed approach gives better result and obtained over all accuracy about 94 % of good liver extraction. This results from proposed approach can help for further diagnosis and treatment planning. In the future work, we plan to increase the number of CT images to evaluate the performance of the proposed PSO and watershed algorithm. And test new versions of PSO.

References

1. Aly, A., Deris, S., Zaki, N.: Research review for digital image segmentation techniques. Int. J. Comput. Sci. Inf. Technol. **3**(5), 99–106 (2011)
2. Islam, S., Ahmed, M.: Implementation of image segmentation for natural images using clustering methods. Int. J. Emerg. Technol. Adv. Eng. **3**(3), 175–180 (2013)
3. Jain, A., Duin, R., Mao, J.: Statistical pattern recognition: a review. IEEE Trans. Pattern Anal. Mach. Intell. **22**(1), 4–37 (2000)
4. Kennedy, J., Eberhart, R.: Particle swarm optimization. Proc. IEEE Int. Conf. Neural Netw. **4**, 1942–1948 (1995)
5. Ghamisi, P., Couceiro, M., Benediktsson, J., Ferreira, N.: An efficient method for segmentation of images based on fractional calculus and natural selection. Expert Syst. Appl. Elsevier **39**(16), 12407–12417 (2012)
6. Abdalla, Z., Neveen, I.G., Aboul Ella, H., Hesham, A.H.: Level set-based CT liver image segmentation with watershed and artificial neural networks, IEEE, pp 96–102 (2012)
7. Aldeek, N., Alomari, R., Al-Zoubi, M., Hiary, H.: Liver segmentation from abdomen ct images with bayesian model. J. Theor. Appl. Inform. Technol. **60**(3), 483–490 (2014)
8. Hammouche, K., Diaf, M., Siarry, P.: A comparative study of various metaheuristic techniques applied to the multilevel thresholding problem. Eng. Appl. Artif. Intell. **23**, 676–688 (2010)
9. Sakthivel, N., Prabhu2, L.: Mean median filtering for impulsive noise removal. Int. J. Basics Appl. Sci. **2**(4), 47–57 (2014)
10. Leavline, EJ., Singh, DAAG.: Salt and pepper noise detection and removal in gray scale images: an experimental analysis. Int. J. Signal Process. Image Process. Pattern Recognit. 6(5), 343–352 (2013)
11. Salman, N.: Image segmentation based on watershed and edge detection techniques. Int. Arab J. Inf. Tech. **3**(4), 104–110 (2006)
12. Kaur, A., Aayushi.: Image segmentation using watershed transform. Int. J. Soft Comput. Eng. (IJSCE) **4**(1), 5–8 (2014)
13. Anter, A., Azar, A., Hassanien, A., El-Bendary, N., El-Soud, M.: Automatic computer aided segmentation for liver and hepatic lesions using hybrid segmentations techniques. In: IEEE Proceedings of Federated Conference on Computer Science and Information Systems, pp.193–198 (2013)
14. Abdalla, M., Hesham, H., Neven, I.G., Aboul Ella, H., Gerald, S.: Evaluating the effects of image filters in CT liver CAD system. In: proceeding of IEEE-EMBS International Conference on Biomedical and Health Informatics, The Chinese University of Hong Kong, Hong Kong (2012)
15. Ruchaneewan, S., Daniela, S., Jacob, F.: A hybrid approach for liver segmentation, intelligent multimedia processing laboratory, 3d segmentation in the clinic: a grand challenge, pp 151–160 (2007)
16. Jeongjin, L., Namkug, K., Ho, L., Joon, B., Hyung, J., Yong, M., Yeong, S., Soo Hong, K.: Efficient liver segmentation using a level-set method with optimal detection of the initial liver boundary from level-set speed images, Elsevier, computer 1 methods and programs in biomedicine, vol. 88, pp. 26–38 (2007)
17. Dice, L.R.: Measures of the amount of ecologic association between species. Ecology **26**, 297–302 (1945)
18. Chakraborty, S., Samanta, S., Biswas, D., Dey, N., Chaudhuri, S.S.: Particle swarm optimization based parameter optimization technique in medical information hiding. In: IEEE International Conference on Computational Intelligence and Computing Research (ICCIC) pp. 1–6 (2013)
19. Hu, W., Yen, G.G.: Adaptive multiobjective particle swarm optimization based on parallel cell coordinate system, IEEE Trans. Evol. Comput. **19**(1), 1–18 (2013)

Grey Wolf Optimizer and Case-Based Reasoning Model for Water Quality Assessment

Asmaa Hashem Sweidan, Nashwa El-Bendary, Aboul Ella Hassanien,
Osman Mohammed Hegazy and A.E.-K. Mohamed

Abstract This paper presents a bio-inspired optimized classification model for assessing water quality. As fish gills histopathology is a good biomarker for indicating water pollution, the proposed classification model uses fish gills microscopic images in order to asses water pollution and determine water quality. The proposed model comprises five phases; namely, case representation for defining case attributes via pre-processing and feature extraction steps, retrieve, reuse/adapt, revise, and retain phases. Wavelet transform and edge detection algorithms have been utilized for feature extraction stage. Case-based reasoning (CBR) has been employed, along with the bio-inspired Gray Wolf Optimization (GWO) algorithm, for optimizing feature selection and the k case retrieval parameters in order to asses water pollution. The datasets used for conducted experiments in this research contain real sample microscopic images for fish gills exposed to copper and water pH in different histopathological stages. Experimental results showed that the average accuracy achieved by the proposed GWO-CBR classification model exceeded 97.2 % considering variety of water pollutants.

A.H. Sweidan
Faculty of Computers and Information, Fayoum University, Fayoum, Egypt

N. El-Bendary (✉)
Arab Academy for Science, Technology, and Maritime Transport, Cairo, Egypt
e-mail: nashwa.elbendary@ieee.org

A.E. Hassanien · O.M. Hegazy
Faculty of Computers and Information, Cairo University, Cairo, Egypt

A.E. Hassanien
Faculty of Computers and Information, Beni-Suef University, Beni-suef, Egypt

A.H. Sweidan · N. El-Bendary · A.E. Hassanien
Scientific Research Group in Egypt (SRGE), Cairo, Egypt
URL: http://www.egyptscience.net

A.E.-K. Mohamed
Department of Zoology, Faculty of Science, Fayoum University, Fayoum, Egypt

© Springer International Publishing Switzerland 2016
T. Gaber et al. (eds.), *The 1st International Conference on Advanced Intelligent System and Informatics (AISI2015), November 28–30, 2015, Beni Suef, Egypt,*
Advances in Intelligent Systems and Computing 407, DOI 10.1007/978-3-319-26690-9_21

Keywords Classification · Features extraction · Color histogram · Fish gills · Histopathology · Water pollution · Case-based reasoning · Gray Wolf Optimization (GWO) · Microscopic image

1 Introduction

Water pollution is a challenging issue for water quality management as it endangers development, living conditions of humans, and sustainable development of the economy. Also, assessment of water quality is one of the most important ways to manage and monitor the quality of water resources. There are two main methods for assessing and detecting water quality: (1) physical and chemical analysis, and (2) biological monitoring methods. Continuous chemical analysis is a complex and expensive process, and also it provides limited data about the chemical compounds [1, 2].

On the other hand, biological methods, which are valuable for detecting pollution, consider exposing changes in water quality. The most familiar criteria used to assess pollution of water is related to the health of ecosystems. So, fish has been used as biomarker for water quality in order to present kind of detection of biological changes due to exposure to chemical pollutants [3, 4].

Tilapia fish is a major protein source in many of the developing countries and it is the most edible fish in Egypt. It can persist in a highly polluted habitat and has the potential for the development as a biological monitor of environmental water pollution. Moreover, histopathological assessment is a sensitive biomonitoring tool to indicate the effect of toxicants on fish health in polluted aquatic ecosystems. It also allows early warning signs of disease and detection of long-term injury in cells, tissues, or organs. Therefore, several studies used fish gills as a tool to assess the presence of contaminants, in natural aquatic systems or in laboratory experiments, to determine the toxicity of heavy metals. In research, a number of approaches tackling the problem of biomarkers based water quality classification have been recently proposed [5–8]. These proposed researches utilized optimization algorithms combined with the implemented classifiers for enhancing the achieved classification accuracy and accordingly enhancing the overall performance.

The aim of this paper is to use fish gills as biomarker to detect and classify water pollution.Thus, it presents bio-inspired GWO-CBR optimized parameters classification system for identifying the water quality and accordingly the degree of water pollution; in Sharkia Governorate-Egypt, based on microscopic images of fish gills. The selection of CBR suitable matching and similarity measures as well as feature subsets and their weights has been achieved via employing the GWO algorithm [9–11]. The goal of this optimization is to find the best areas of space complex search through the interaction between individuals in the population. Furthermore, the gray wolf optimization proves too much durability against configuration compared to Practical Swarm and Genetic Algorithm optimizers.

The rest of this paper is organized as follows. Section 2 describes the different phases of the proposed classification system. Section 3 introduces the tested fish gills microscopic images datasets and discusses the obtained experimental results. Finally, Sect. 4 presents conclusions and discusses future work.

2 The Proposed GWO-CBR Hybrid Classification Model

This paper proposes an approach for classifying water pollution based on fish gills microscopic image using cased-based reasoning (CBR) [12–15] combined with the bio-inspired grey wolf optimizer (GWO) [9–11] for parameter optimization. Tilapia fish can persist in a highly polluted habitat and has the potential for the development as a biological monitor of environmental water pollution. So, the datasets used for experiments were constructed based on real sample images for Tilapia fish gills, in different histopathlogical stages, exposed to copper and water pH. The datasets contain colored JPEG images as 155 images and 55 images were used as training and datasets, respectively. Training dataset is divided into 4 classes representing the different histopathlogical changes and water quality degrees. Digital image processing techniques used as case representation for proposed system. The proposed CBR classification system utilizes GWO to optimize the feature selection and the number of neighbors that combine in the CBR system. The proposed system comprises five main phases; namely, case representation or defining case attributes (pre-processing, feature extraction), retrieve, reuse/adapt, revise, and retain phases, as depicted in Fig. 1.

2.1 Case Representation Phase

During this phase, case representation is being used to find an appropriate structure for defining case attributes and describing case contents. So, when the user enters image (new case), the system extracts features to define case attributes. This phase consist of two steps, which are pre-processing and feature extraction.

Pre-processing During this step, the proposed approach prepares images for the features extraction. So, input microscopic images are resized, image background are removed to get region of interest (RoI), and also images are converted from RGB to gray scale level. Also, contrast enhancement is applied, so that the contrast of the microscopic images in a given gray level represents descriptors models for the spatial relationship of a pixel and its neighbors.

Feature Extraction In this step, wavelets transform [16, 17] and edge detection algorithm [18, 19] have been utilized. So, after applying pre-processing, the resulted gray scale image is decomposed using wavelets into fourth components as approximation, horizontal, vertical component and diagonal component. Also, edge infor-

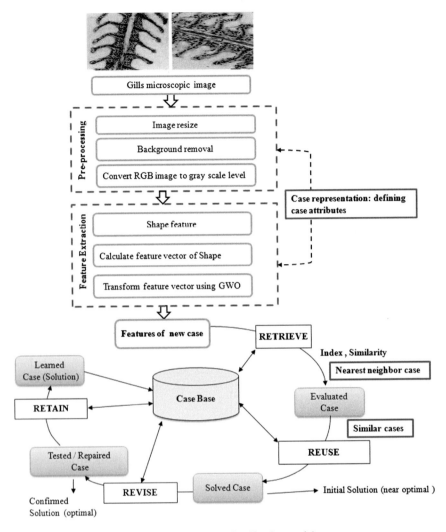

Fig. 1 Architecture of the proposed GWO-CBR classification model

mation has been used for all the four components. The proposed system defines four maps edge by multiplying four masks with the approximation components though implementing the following steps: (1) First and second masks are obtained by placing two different thresholds on the horizontal, vertical, and diagonal components. Two different thresholds are used in this approach to get more edge information, (2) Third mask is obtained by looking at the maximum pixel value of the horizontal, vertical, and diagonal components, and (3) Forth mask is obtained by finding the max intensity pixels among h, v and d components and by multiplying with approximation component. Algorithm 1 shows the details of the feature extraction process.

1: After pre-processing, decompose the image using wavelets.
2: Get approximation a, horizontal y, vertical v and diagonal d components.
3: Filter out the strong edges of horizontal, vertical and diagonal components by using T1γ and T2γ where T is threshold and γ is the standard deviation of respected image.
4: Obtain first edge map by applying T1γ on h, v and d components and combining them and then multiplying the resulting mask with the approximation component.
5: Repeat step5 on T2γ
6: Repeat step5 on T1γ and T2γ
7: Repeat step5 on finding the max intensity pixels among h, v and d components and by multiplying with approximation component

Algorithm 1: Wavelets based feature extraction for case representation phase

T1, it is set to zero (a non edge). If the magnitude is above the high threshold, it is made an edge. And if the magnitude is between the 2 thresholds, then it is set to zero unless there is a path from this pixel to a pixel with a gradient above T2. In this paper, moment invariants are used to represent shape. Feature vectors are calculated for input image and image dataset. Shape feature vector are given by Eqs. (1)–(7).

$$\phi_1 = \eta_{20} + \eta_{02} \tag{1}$$

$$\phi_2 = (\eta_{20} - eta_{02})^2 + 4\eta_{11}^2 \tag{2}$$

$$\phi_3 = (\eta_{30} - 3\eta_{12})^2 + (3\eta_{21} - \eta_{03})^2 \tag{3}$$

$$\phi_4 = (\eta_{30} + \eta_{12})^2 + (\eta_{21} + \eta_{03})^2 \tag{4}$$

$$\phi_5 = (\eta_{30} - 3\eta_{12})(\eta_{30} + \eta_{12})$$
$$[(\eta_{30} + \eta_{12})^2 - 3(\eta_{21} + \eta_{03})^2] +$$
$$(3\eta_{21} - \eta_{03}(\eta_{21} + \eta_{03})$$
$$[3(\eta_{30} + \eta_{12})^2 - (\eta_{21} + \eta_{03}^2] \tag{5}$$

$$\phi_6 = (\eta_{20} - \eta_{02})[(\eta_{30} + \eta_{12}^2 - (\eta_{21} + \eta_{03})^2]$$
$$+ 4\eta_{11}(\eta_{30} + \eta_{12})(\eta_2 1 + \eta_{03}) \tag{6}$$

$$\phi_7 = (3\eta_{21} - \eta_{03})(\eta_{30} + \eta_{12})[(\eta_{30} + \eta_{12})^2$$
$$- 3(\eta_{21} + \eta_{03})^2] + (e\eta_{12} = \eta_{30})(\eta_{21} + \eta_{03})$$
$$[3(\eta_{30} + \eta_{12})^2 - (\eta_{21} + \eta_{03})^2] \tag{7}$$

where $\phi(1)$ is analogous to the moment of inertia around the image's centroid, considering that the pixels' intensities are analogous to physical density, and $\eta_{i,j}$ stands for moments. Then, a feature vector will be formed as a shape feature for gills image (case attributes). Gray wolf optimizer (GWO) [9–11] has been used to choose the optimal subset features. In the end, good sites should be converged, and probably the best ones. GWO makes repetition of exploring new areas in the solution space and exploits the advantage until a near optimal solution has been found. The solution space here represents all the possible selection of features, functions, and thus gray positions represents selection of feature sets. Each feature represents an individual dimension ranging from -2 to 2, in order to decide whether to choose the feature or not.

2.2 Retrieve Phase

The main steps in the CBR process are retrieving similar cases in the case-base, then measuring case similarity to match the best case. The similarity distance is the most visible measure of similarities between two cases. After case attributes are known from the input, an initial inspect is done on the case-base for similarity matching. When establishing the case-base, firstly, the character attributes of fish gills are summarized and extracted. When adding a new case to base, it must extract features in order to be placed at the case-base. When extracting the features of a new case, they are compared to the cases already recorded in the case-based and the best match is chosen. In this paper, distance measures; namely (*Euclidean distance, City Block Distance, Canberra distance, Squared Chord, Squared Chi-square*) are being used. This method determines the case that is most similar to the input case then retrieve the k most similar cases. Moreover, in this paper the k similar cases have been optimized by GWO to enhance the performance of the CBR system. The dataset has been divided into three subsets, as follows:

1. Training subset: used to search for the optimal or near-optimal k parameter and used as a case-base for retrieval.
2. Testing subset: used to measure the success of a candidate for k improves the accuracy of the proposed system.
3. Cross-validation cases: used to verify the classification accuracy.

The GWO algorithm makes iterations of investigation of new regions and exploiting solutions until reaching near-optimal solution. Algorithm 2 shows the details of GWO based retrieve phase.

1: Initialize population.
2: Evaluate the fitness functions

    2.1 Extract attributes of the input image, system searches cases and retrieves case similar to the input case.
    2.2 Use the weighted average of distance functions for each feature as a similarity measure.
    2.3 Put the best accuracy in C population as "*accuracy*".
    2.4 If accuracy satisfies the termination condition, go to step3, otherwise update population and go to step 2.1.

3: Apply the selected kth optimal number of cases.

Algorithm 2: GWO based CBR retrieve phase

2.3 Reuse/Adapt Phase

For the reuse/adapt phase, the proposed system selects the K most similar case from past cases in the case-base that has been obtained through the retrieval phase. Then, it checks the solution of the retrieved similar case and reuses the solution to inform the input case. After being shown the solutions, the system checks the case against the cases in the database to see if a similar case is available there or it should save the new case.

2.4 Revise Phase

For the revise phase, if the case that have been created in the previous reuse/adapt phase contains solution, the system checks if it's an acceptable solution in this first generation. Then, it will be informed and the following retain phase will keep the starting phase of the work. Also, the retain is only used if the case reuse does not occur. In that case, the system had to adapt to a given situation to the list. This adjustment is used to solve a new problem.

2.5 Retrain Phase

For this phase, after the solution is confirmed. It is then saved to the case-base, which will be updated with the new case when new problem solving takes place.

3 Experimental Analysis and Discussion

Nile Tilapia "Oreochromis niloticus" is superlative to be used as bio-indicator for water pollution. The most notable changes of the gills that could be associated with water pollution were restricted to the epithelium of the gill filaments and the secondary lamellae. The datasets used for experiments were constructed based on real sample microscopic images for fish gills in different histopathlogical change stages exposed to copper and water pH. Fish images were collected from Abbassa farm, Abo-Hammad, Sharkia Governorate, Egypt. Some samples of both training and testing datasets are shown in Fig. 2. Training dataset is divided into 4 classes representing the different water quality degrees; namely *excellent quality*, *good quality*, *moderate quality*, and *bad quality* [20], as shown in Fig. 3. (1) **Excellent water quality**: showing primary filament (F) and secondary lamellae (L) arising from these, parallel to each other and perpendicular to the filament axis, as shown in Fig. 3a, (2) **Good water quality**: where fish exposed to copper at $pH = 9$ were more or less

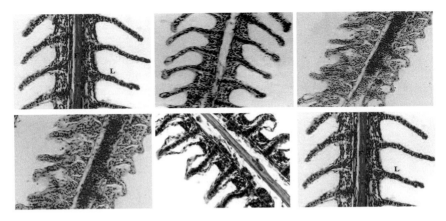

Fig. 2 Examples of training and testing fish gills microscopic images

(a)   **(b)**   **(c)**   **(d)**

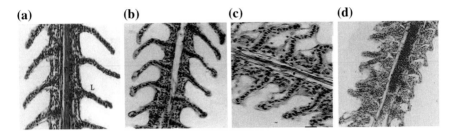

Fig. 3 Examples of water quality degrees

similar to those of control group, as shown in Fig. 3b, (3) **Moderate water quality**: where fish exposed to copper at $pH = 7$ showing hyperplasia of the primary lamellae with complete fusion of the secondary lamellae and shortened, as shown in Fig. 3c, and (4) **Bad water quality**: where fish exposed to copper at $pH = 5$, drooping and shortening of some of the secondary lamellae, as shown in Fig. 3d.

The proposed system was tested using different initialization numbers. As previously stated, the GWO has been employed to optimize the feature selection and the number of neighbors that combine in the CBR algorithm. Also, features have been extracted using shape feature extraction based on (edge detection and wavelet transform). Moreover, CBR was employed with different similarity functions to asses water quality degree. Three different scenarios have been investigated, along with the 10-fold-cross-validation, for the proposed system. These scenarios were applied via retrieving the five, ten, and fifteen most similar cases for a given case, respectively. Figure 4 depicts the obtained experimental results that show accuracy achieved, considering the 3 tested scenarios, via applying each similarity function with optimizing the feature selection for different number of initialization training images per class. The results of the proposed GWO-CBR classification system were evaluated against human expert assessment for measuring obtained accuracy. As shown in Fig. 4,

Fig. 4 Accuracy of
GWO-CBR system using
distance measure functions

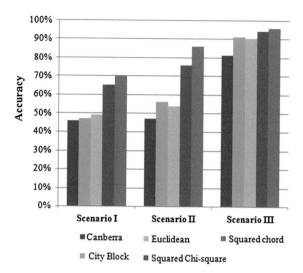

the best accuracy achieved was via applying the proposed GWO-CBR system with
the Squared chord function. Also, as shown in Fig. 4, the performance of the CBR
algorithm increased according to the increase in the number of retrieved cases.

Moreover, Fig. 5 shows comparative analysis for classification accuracy achieved
by CBR against GWO-CBR. The depicted results showed that the proposed GWO-
CBR system outperformed the typical CBR algorithm, via achieving a maximum

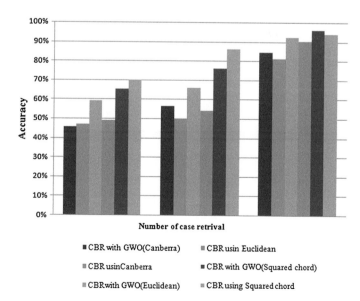

Fig. 5 Comparative analysis for classification accuracy achieved by CBR against GWO-CBR

accuracy of 97.2 %, considering all tested functions, via optimizing the feature selection and the number of neighbors cases.

4 Conclusions and Future Work

This paper proposes an system for classifying water pollution degree based on fish gills microscopic images as biomarker. The proposed optimized case-based reasoning (CBR) system used the bio-inspired grey wolf optimizer (GWO) for feature selection and the number of neighbors that combine in the CBR. Based on the obtained results, the proposed GWO-CBR water quality classification system has achieved an accuracy of 97.2 % that outperforms the accuracy achieved by the CBR without optimization. Also, it has been observed that the classification accuracy increased for all kernel functions as the number of initialization training images per class increases. For future work, it is planned to enhance the approach proposed in this paper via considering other bio-inspiring optimization algorithms and different fitness functions to assess water quality.

References

1. Reddy, P., Rawat, S.S.: Assessment of aquatic pollution using histopathology in fish as a protocol. Int. Res. J. Environ. Sci. **2**(8), 79–82 (2013)
2. Hauser-Davis, R.A., de Campos, R.C., Ziolli, R.L.: Fish Metalloproteins as Biomarkers of Environmental Contamination, vol. 218, pp. 101–123. Springer, US (2012)
3. Omar, W.A., Zaghloul, K.H., Abdel-Khalek, A.A., Abo-Hegab, S.: Risk assessment and toxic effects of metal pollution in two cultured and wild fish species from highly degraded aquatic habitats. Arch. Environ. Contam. Toxicol. **65**(4), 753–764 (2013)
4. Jahanbakhshi, A., Hedayati, A.: Gill histopathological changes in great sturgeon after exposure to crude and water soluble fraction of diesel oil. Comp. Clin. Pathol. **22**(6), 1083–1086 (2013)
5. Zheng, H., Liu, R., Zhang, R., Hu, Y.: A method for real-time measurement of respiratory rhythms in medaka (Oryzias latipes) using computer vision for water quality monitoring. J. Ecotoxicol. Environ. Saf. **100**(4), 76–86 (2014)
6. Liao, Y., Jian-yu, X., Wang, Z.: Application of biomonitoring and support vector machine in water quality assessment. J. Zhejiang Univ. **13**(4), 327–334 (2012)
7. Su, J., Wang, X., Zhao, S., Chen, B., Li, C., Yang, Z.: A structurally simplified hybrid model of genetic algorithm and support vector machine for prediction of chlorophyll a in reservoirs. J. Water **7**(4), 1610–1627 (2015). Multidisciplinary Digital Publishing Institute
8. Sweidan, A.H., El-Bendary, N., Hassanien, A.E., Hegazy, O.M., Mohamed, A.B.D.E.-K.: Machine learning based approach for water pollution detection via fish liver microscopic images analysis. In: The 9th IEEE International Conference on Computer Engineering and Systems (ICCES 2014), Cairo, Egypt, pp. 253–258, 21–23 Dec 2014
9. Mirjalili, S., Mirjalili, S.M., Lewis, A.: Grey wolf optimizer. J. Adv. Eng. Softw. **69**, 46–61 (2014). Elsevier
10. El-Gaafary, A.A.M., Mohamed, Y.S., Hemeida, A.M., Mohamed, A.A.A.: Grey wolf optimization for multi input multi output system. Univers. J. Commun. Netw. **1**(1), 1–6 (2015)

11. Emary, E., Zawbaa, H.M., Grosan, C., Hassenian, A.E.: Feature subset selection approach by gray-wolf optimization. In: Abraham, A., Krömer, P., Snasel, V. (eds.) Afro-European Conf. for Ind. Advancement. AISC, vol. 334, pp. 1–14. Springer, Heidelberg (2015)
12. Vanitha, L., Venmathi, A.R.: Classification of medical images using support vector machine. In: International Conference on Information and Network Technology, IACSIT, vol. 4 (2011)
13. Ashok Kumar, K., Bhaskar Reddy, Y.V.: Content based image retrieval using SVM algorithm. Int. J. Electr. Electron. Eng. (IJEEE) 1(3), 2231–5284 (2012)
14. Kumar, E., Khan, Z., Jain, A.: A review of content based image classification using machine learning approach. Int. J. Adv. Comput. Res. 2(5), 55–60 (2012)
15. Suralkar, S.R., Karode, A.H., Pawade, P.W.: Texture image classification using support vector machine. Int. J. Comput. Appl. Technol. 3(1), 71–75 (2012)
16. Adnan, A., Gul, S., Ali, M., Dar, A.H.: Content based image retrieval using geometrical-shape of objects in image. Int. Conf. Emerg. Technol. ICET 8(9), 222–225 (2007)
17. Akansu, A.N., Serdijn, W.A., Selesnick, I.W.: Wavelet transforms in signal processing: a review of emerging applications. Phys. Commun. 3(1), 1–18 (2010)
18. Suganthy, M., Ramamoorthy, P.: Principal component analysis based feature extraction, morphological edge detection and localization for fast iris recognition. J. Comput. Sci. 8(9), 1428–1433 (2012)
19. Xue, L.-Y., Pan, J.-J.: Edge detection combining wavelet transform and canny operator based on fusion rules. In: International Conference Wavelet Analysis and Pattern Recognition, pp. 324–228 (2009)
20. Mohamed, A.B.D.E.-K., El-sayed, N., El-Shershaby, A.-F., Zaghloul, K.H.H.: Taxicological and histopathological studies on Nile Tilapia Oreochromis niloticus exposed to copper individually and in mixture with zinc at different pH value, Ph.D. thesis, Cairo univeristy (2006)

New Rough Set Attribute Reduction Algorithm Based on Grey Wolf Optimization

Waleed Yamany, E. Emary and Aboul Ella Hassanien

Abstract In this paper, we propose a new attribute reduction strategy based on rough sets and grey wolf optimization (GWO). Rough sets have been used as an attribute reduction technique with much success, but current hill-climbing rough set approaches to attribute reduction are inconvenient at finding optimal reductions as no perfect heuristic can guarantee optimality. Otherwise, complete searches are not feasible for even medium sized datasets. So, stochastic approaches provide a promising attribute reduction technique. Like Genetic Algorithms, GWO is a new evolutionary computation technique, mimics the leadership hierarchy and hunting mechanism of grey wolves in nature. The grey wolf optimization find optimal regions of the complex search space through the interaction of individuals in the population. Compared with GAs, GWO does not need complex operators such as crossover and mutation, it requires only primitive and easy mathematical operators, and is computationally inexpensive in terms of both memory and runtime. Experimentation is carried out, using UCI data, which compares the proposed algorithm with a GA-based approach and other deterministic rough set reduction algorithms. The results show that GWO is efficient for rough set-based attribute reduction.

W. Yamany (✉)
Faculty of Computers and Information, Fayoum University, Fayoum, Egypt
e-mail: wsy00@fayoum.edu.eg
URL: http://www.egyptscience.net

E. Emary · A.E. Hassanien
Faculty of Computers and Information, Cairo University, Cairo, Egypt

A.E. Hassanien
Faculty of Computers and Information, Beni-Suef University, Beni-Suef, Egypt

W. Yamany · A.E. Hassanien
Scientic Research Group in Egypt (SRGE), Cairo, Egypt

© Springer International Publishing Switzerland 2016
T. Gaber et al. (eds.), *The 1st International Conference on Advanced Intelligent System and Informatics (AISI2015), November 28–30, 2015, Beni Suef, Egypt,*
Advances in Intelligent Systems and Computing 407, DOI 10.1007/978-3-319-26690-9_22

1 Introduction

Attribute reduction is one of the essential problems in the fields of data mining, machine learning, and pattern recognition [1]. Attribute reduction is mainly concerned with selecting the smallest subset of attributes for a given problem while preserving a suitably high accuracy in representing the original attributes [2]. In real world problems, attribute reduction is often a necessity due to the abundance of noisy, misleading or irrelevant attributes [3], while attribute reduction enables data processing techniques such as machine learning algorithms to yield good performance [2].

One of the well-known approaches for attribute reduction is based on rough set theory [4] which has the inherent ability to deal with vagueness and uncertainty in data analysis. Roughs set have been extensively used in data mining [5], machine learning [6] and other fields. Here, knowledge is considered as a kind of discriminability, which can also be employed as a scientific instrument to reduce attribute dimensionality and establish data dependencies. While various rough set-based algorithms for attribute reduction have been proposed, the main idea of these methods is to detect minimal reducts by creating every conceivable reduct and subsequently selecting the one with the smallest length. This can be performed by developing a kind of detectability capacity from a given dataset and then streamlining it [6]. On the other hand, the number of possible subsets is typically very large and considering all attributes subsets for choosing the optimal one is considered an NP-hard problem. To address this, rough sets can be integrated with optimisation approaches such as genetic algorithms [7], ant colony optimisation [8], or particle swarm optimisation [9].

Wroblewski [10] integrated a genetic algorithm (GA) with a greedy algorithm to produce small reducts however, could not demonstrate that the produced subset is a reduct. ElAlami [11] also made use of GAs to locate ideal relevant attributes. Zhai [7] proposed an incorporated attribute extraction approach that uses both rough sets and GAs. Jensen and Shen [8] applied the ant colony optimisation to find rough set reducts, while Wang [9] developed a method for attribute reduction using particle swarm optimisation hybridised with rough sets. Although, stochastic techniques can yield strong solutions for global optimisation, this is accomplished at the expense of computational cost [2].

In this paper, we propose a new attribute reduction technique. In our approach, we make use of the grey wolf optimizer (GWO) [12] to discover ideal attribute subsets (reducts). GWO is a recently evolutionary computation algorithm [12] and is based on the imitation of the social leadership and hunting behaviour of grey wolves in nature. GWO can adaptively search the attribute space for optimal attribute combinations that maximise a given fitness function with the fitness function used in our work being rough set-based classification. Experimental results on various benchmark datasets from the UCI repository confirms our technique to perform well in comparison to competing methods.

The organization of this paper as follows: Sect. 2 presents the background of the rough set and gray wolf optimizer. The proposed system section describes in Sect. 3. Experimental results with discussions presents in Sect. 4. Conclusions provide in Sect. 5.

2 Preliminaries

2.1 Rough Set Theory

In this section, we present some of the necessary fundamentals for rough set (RS) theory and RS-based feature selection. For more exhaustive descriptions of the theory, we refer to [4] and other publications.

Let $I = (O, S, B, f)$ be an information system, such that O is a finite non-empty set of instances, S is a finite non-empty set of attributes and B is the set union of attribute scopes such that $B = \bigcup_{s \in S} B_s$ for B_s indicates the value scope of attribute $s. f : O \times S \longrightarrow B$ is an information function that associates a unique magnitude of each attribute with every instance in O such that $f(x, s) \in B_s$ for any $s \in S$ and $x \in O$.

For any $P \subseteq S$ there exists an associated indiscernibility relationship $IND(Z)$

$$IND(Z) = \{(x, y) \in O \times O | \forall s \in p, f(x, s) = f(y, s)\}. \tag{1}$$

The partition of O, deduced by $IND(Z)$ is indicated by $O/INP(Z)$ and can be computed as

$$O/INP(Z) = \otimes \{s \in Z : O/INP(\{s\})\}, \tag{2}$$

where

$$R \otimes K = \{X \bigcap Y : \forall R \in X, \forall K \in Y, X \bigcap Y \neq \emptyset\}. \tag{3}$$

For subset $X \subseteq O$ and equivalence relationship $IND(Z)$, the $Z - lower$ and $Z - upper$ approximations of X are determined as

$$Z_*(X) = \{x \in O : [x]_Z \subseteq X\} \tag{4}$$

and

$$Z^*(X) = \{x \in O : [x]_Z \bigcap X \neq \emptyset\}, \tag{5}$$

respectively

For two subsets $E, W \subset S$ which give raise to two equivalence relationship $IND(E)$ and $IND(W)$, the $E - positive$ and $E - negative$ region of W can be determined as

$$POS_E(W) = \bigcup_{X \in O/IND(W)} E_*(X) \tag{6}$$

and

$$NEG_E(W) = O - \bigcup_{X \in O/IND(W)} E^*(X), \tag{7}$$

respectively, while the $E - boundary$ region of W is defined as

$$BND_E W = \bigcup_{X \in O/IND(W)} E^*(X) - \bigcup_{X \in O/IND(W)} E_*(X). \tag{8}$$

A significant problem in data analysis is finding dependencies among attributes. Dependencies can be determined in the following manner. For $E, W \subset S$, E fully depends on W, if and only if $IND(E) \subset IND(W)$. We say that W depends on E to a degree $0 \leq \mu_E(W) \leq 1$, if

$$\mu_E(W) = \frac{|POS_E(W)|}{|O|}. \tag{9}$$

If $\mu_E(W) = 1$, then W fully depends on P; if $0 \leq \mu_E(W) \leq 1$, then W depends partially on E; and if $\mu_E(W) = 0$ then W does not depend on E. The dependence degree $\mu_E(W)$ can be used as a heuristic in greedy algorithms for calculating attribute selection.

$I = (O, C \cup D, B, f)$ is called a decision table if $S = C \cup D$ and $C \cap D = 0$ in an information table, where C is the set of condition attributes, and D is the decision attributes set. The dependency degree among condition and decision attributes, $\mu_C(D)$, is named the classification accuracy, induced by the decision attributes set.

The goal of attribute reduction is to discard redundant attributes so that the reduced set has the same classification accuracy as the original set. A reduct is resolved as a subset RED of the condition attribute such that

$$\mu_C(D) = \mu_{RED}(D) \text{ and } \forall P \subset RED, \mu_p(D) \neq \mu_C(D). \tag{10}$$

2.2 Grey Wolf Optimization

Generally, grey wolves want to live in a pack. The gathering size is 5 to 12 by and large. They have extremely strict standards in the social predominant progressive system. As per [12] grey wolf pack comprises of the accompanying:

- The alphas are leading the pack, the alpha wolves are in charge of deciding. The alphas choices are directed to the pack
- The betas are subordinate wolves that help the alpha in choice making or different exercises. The beta can be either female or male, and she/he is most likely the best possibility to be the alpha

- The omega plays the role of scapegoat. Omega wolves dependably need to submit to the various overwhelming wolves. They are the last wolves that are permitted to eat.
- The deltas need to submit alphas and betas, yet they rule the omega. Scouts, sentinels, elders, hunters, and caretakers have a place with this class. Scouts are in charge of viewing the limits of the domain and cautioning the pack if there should arise an occurrence of any risk. Sentinels protect and ensure the security of the pack. Elders are the accomplished wolves who utilized to be alpha or beta. Hunters help the alphas and betas when chasing prey and giving sustenance to the pack. Finally, the caretakers are in charge of administering to the feeble, sick, and injured wolves.

In the mathematical model for the GWO, the fittest solution is known as the alpha (α). The second and third best solutions are called beta (β) and delta (δ) respectively. Whatever remains of the hopeful solutions are supposed to be omega (ω). The hunting is guided by α, β, and δ and the ω follow these three candidates.

All together for the pack to chase a prey they first circling it. Keeping in mind the end goal to mathematically model circling conduct the accompanying mathematical Eqs. 11–14 are utilized

$$\vec{S}(t+1) = \vec{S}_p(t) + \vec{A}.\vec{D} \tag{11}$$

where $\vec{D}$ is characterised in Eq. 2 and t is the number of iteration, $\vec{A}, \vec{C}$ are coefficient vectors, $\vec{S}_p$ is the position of prey and $\vec{S}$ is the position of the grey wolf.

$$\vec{D} = |\vec{C}.\vec{S}_p(t) - \vec{S}(t)| \tag{12}$$

The $\vec{A}, \vec{C}$ vectors are computed in Eqs. 3 and

$$\vec{A} = 2\vec{a}.\vec{r_1} - \vec{a} \tag{13}$$

$$\vec{C} = 2\vec{r_2} \tag{14}$$

where a component of $\vec{a}$ is decreasing linearly from 2 to 0 over the course of iterations, and r_1, r_2 are vectors with random values in [0,1]. The chase is generally guided by the alpha. The beta and delta may additionally take part in chasing once in a while. So as to mathematically recreate the chasing conduct of grey wolves, the best candidate solution (alpha), second best candidate solution (beta), and third best candidate solution (delta) are expected to have better learning about the potential area of prey. The initial three best solutions acquired so far and oblige the other search agents (counting the omegas) to overhaul their positions as per the position of the best search agents.

So the updating for the positions of wolves is as in Eqs. 15–17.

$$\vec{D_\alpha} = |\vec{C_1}.\vec{S_\alpha} - \vec{X}|, \vec{D_\beta} = |\vec{C_2}.\vec{S_\beta} - \vec{X}|, \vec{D_\delta} = |\vec{C_3}.\vec{S_\delta} - \vec{X}| \tag{15}$$

$$\vec{S_1} = |\vec{S_\alpha} - \vec{A_1}.\vec{D_\alpha}|, \vec{S_2} = |\vec{S_\beta} - \vec{A_2}.\vec{D_\beta}|, \vec{S_3} = |\vec{S_\delta} - \vec{A_3}.\vec{D_\delta}| \qquad (16)$$

$$\vec{S}(t+1) = \frac{\vec{S_1} + \vec{S_2} + \vec{S_3}}{3} \qquad (17)$$

A final remark about the GWO is the updating of the parameter $\vec{a}$ that controls the tradeoff between exploitation and exploration. The parameter $\vec{a}$ is linearly updated in each iteration to range from 2 to 0 according to the Eq. 18.

$$\vec{a} = 2 - t.\frac{2}{Max_iter} \qquad (18)$$

where t is the number of iteration and Max_iter is the number of total iteration allowed for the optimization.

3 Rough Set Based on Gray Wolf for Attribute Reduction (GWORSAR)

In this paper, we present the proposed GWO optimizer based on the rough set for attribute reduction; see Fig. 1. We used the principles of gray wolf optimization for the optimal attribute reduction problem. Each attribute subset can be seen as a position in such a space. For N attributes, then there will be 2^N different attribute subset.

Fig. 1 The overall Attribute reduction algorithm

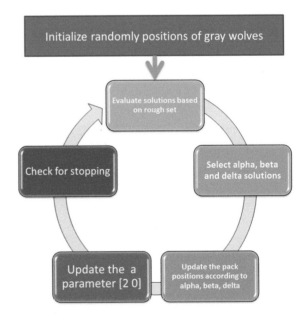

The optimal position is the subset That yields *highest classification accuracy* and *smallest length*. We utilize the exploration and optimisation characteristics of grey wolf optimization for selecting the optimal attribute set. Eventually, they should converge on good, possibly optimal, positions. The GWO makes iterations of exploration of new regions in the attribute space and exploiting solution until reaching the satisfactory solution.

For our GWO, the solution space represents all possible selections of attributes and hence Wolves positions represent the selection of attribute sets. To decide if an attribute subset will be selected or not, we will utilize the used fitness function as in Eq. 19:

$$Fitness = \alpha^* \gamma_R(D) + \beta^* \frac{|C - R|}{|C|} \qquad (19)$$

where $\gamma_R(D)$ is the classification quality of condition attribute set R relative to decision D, $|R|$ is the length of chosen attribute subset. $|C|$ is the total number of attributes. α and β are two corresponding parameters for the importance of classification quality and attribute subset length, $\alpha \in 0, 1$ and $\beta = 1 - \alpha$.

This formula means that the quality of classification and attribute subset length have different significance for attribute reduction task. In our experiment, we suppose that quality classification is more important than the length of attribute subset and set $\alpha = 0.9$, $\beta = 0.1$. The high assures that the best position is at least a real rough set reduct. The goodness of each position is evaluated by this fitness function. The criteria are to maximize fitness values.

4 Experimental Results and Discussions

We performed extensive classification experiments on 11 datasets from the UCI Machine Learning Repository [13] listed in Table 1. We use the LEM2 technique [14] to obtain rules from the data and the global strength [9, 15] for rule parley in classification. For evaluation, we perform standard ten-fold cross validation. For comparison, we also obtain results for conditional entropy-based attribute reduction (CEAR) [16], discernibility matrix-based attribute reduction (DISMAR) [17] and GA-based attribute reduction (GAAR) [15].

The parameters used by each meta-heuristic optimisation algorithms are showed in Table 2. It is significant to shed light over such values have been empirically set. Notice for all optimization algorithms we used 25 agents and 100 iterations.

In Table 1, we list reduct information for the different algorithms. For the evaluated datasets, some may have more than one ideal reduct, while some have only one (exclusive) best reduct. From our results, we can observe that in some cases hill-climbing methods can locate the optimal solution. For example, DISMAR identifies the optimal solution for the Soybean-small and Vote datasets. However, for the other datasets, only sub-optimal solutions are found, leading to redundant attributes. For CEAR, we can see that it often includes even more redundant attributes than

Table 1 Reduct sizes found by different attribute reduction methods

Dataset	Attributes	Instances	CEAR	DISMAR	GAAR	GWORSAR
Breastcancer	9	699	4	5	4	2
M-of-N	13	1000	7	6	6	6
Exactly	13	1000	8	6	6	6
Mushroom	22	8124	5	6	5	3
Vote	16	300	11	8[a]	9	5[a]
Zoo	16	101	10	5	6	5
Lymphography	18	148	8	7	8	5[a]
Led	24	2000	12	18	8	5[a]
Soybean-small	35	47	2	2	6	3
Lung	56	32	5	4	6	4
DNA	57	318	6	6	7	5

[a]indicate optimal solution

Table 2 The Parameters Used by each Optimization algorithm

Algorithm	Parameters	Value
GWO	$\vec{a}$	Linearly decreased from 2 to 0
GA	Crossover probability	0.6
	Mutation probability	0.4

DISMAR. As the experimental results confirm, GWORSAR performs better than GAAR. As we can see, GWORSAR effectively discovers the best reducts on almost all the datasets compared to the competing methods. GWORSAR finds the optimal reduct for the Mushroom data dataset, and is the only method to identify the ideal reduct for the led, Lymphography and Vote datasets.

GWO has strong exploration capabilities. For some of the datasets with more properties, for instance Lung and DNA, the GA-based approach, being influenced by the number of attributes, after having found a sub-optimal solution could not successfully identify a superior one. On the other hand, our GWO-based algorithm is shown to effectively search the attribute domain and to acquire the optimal solution.

Table 3 gives the resulting numbers of decision rules number and classification accuracies based on various reducts. As we can see from there, our approach not only successfully yields compact reducts, the resulting classification accuracy is also superior compared to the other methods. Except for three datasets (Breastcancer and Mushroom), our GWO-based approach gives the best classification performance.

The evaluation criteria or fitness function is a critical issue in the implementation of stochastic algorithms. In the fitness function we use, because the classification

Table 3 Obtained classification accuracies for different reducts

Dataset	CEAR		DISMAR		GAAR		GWORSAR	
	NR	CA	NR	CA	NR	CA	NR	CA
Breastcancer	75	94.20	67	95.94	64	95.65	43	94.64
M-of-N	35	100	35	100	35	100	35	100
Exactly	50	100	50	100	50	100	50	100
Mushroom	61	90.83	19	100	19	100	17	99.8
Vote	25	92.33	28	93.67	25	94.0	15	95.67
Zoo	13	94.0	13	94	13	92.0	10	97.0
Lymphography	42	72.14	40	74.29	38	70.0	47	78.57
Led	228	83.10	257	78.85	10	100	10	100
Soybean-small	4	100	4	100	4	97.50	5	100
Lung	13	73.33	14	73.3	12	70.0	9	93.3
DNA	192	26.45	191	36.45	191	33.87	191	36.45

NR = number of rules
CA = classification accuracy

accuracy parameter outclasses that of subset length (a= 0.9, ? = 0.1), the optimal solution certainly to be a real rough set reduct or super reduct. So, the fitness function in GWO is efficient.

GWO comprises a very simple concept, and the ideas can be implemented in a few lines of computer code. It requires only primitive mathematical operators, and is computationally inexpensive in terms of memory requirements. This optimization technique does not suffer, however, from some of the difficulties of GAs; interaction in the group enhances rather than detracts from progress toward the solution. Further, a grey wolf optimization algorithm system has memory, which the genetic algorithm does not have Changes in genetic populations result in the destruction of preceding knowledge of the problem. In grey wolf optimisation algorithm, individuals who prey past optima are tugged to return towards them; knowledge of good solutions is retained by all wolves [12]. The comparison of the classification quality and the number of decision rules with different reducts are shown in Table 3.

We also conducted a further comparison with the rough set framework RSES [18] and show the results in Table 4. We use the exhaustive calculation to acquire reducts in RSES. RSES could not discover all reducts in all datasets (Soybean-small, Lung and DNA) by exhaustive search due memory constraints. Consequently, we utilise a quick genetic algorithm to obtain 10 reducts. We use the decision rules classifier, LEM2 algorithm for global rules generation and ten-fold cross-validation evaluation technique as in our other experiments. As we can see from Table 3, for almost all cases, reducts obtained by GWORSAR are smaller and exhibit higher classification accuracy.

Table 4 Experimental results by RSES, where TR is the total reducts, MS is the minimal size of reducts, NR is the number of rules and CA is the classification accuracy

Dataset	TR	MS	NR	CA
Breastcancer	20	4	455	94.4
M-of-N	1	6	19	92.9
Exactly	1	6	41	85.90
Exactly2	1	10	242	76.4
Vote	2	8	67	92.93
Zoo	33	5	109	96.8
Lymphography	424	6	353	85.9
Led	140	5	6636	100
Soybean-small	–	2	31	92.5
Lung	–	4	100	75
DNA	–	5	2592	74.3

5 Conclusion

In this paper, we have proposed an algorithm based on rough sets and grey wolf optimisation for attribute reduction. GWO has robust search capabilities and can effectively find smallest attribute reducts based on a suitable definition of a fitness function that combines both classification accuracy and attribute set size. Experimental results prove competitive performance for our GWO-based approach showing that GWO combined with rough set is a useful technique for the attribute reduction problem.

References

1. Laura, M.C.: A framework for feature selection in high-dimensional domains. Ph.D. Thesis, University of Cagliari (2012)
2. Chen, Y., Miao, D., Wang, R., Wu, K.: A rough set approach to feature selection based on power set tree. Knowl.-Based Syst. **24**, 275–281 (2011)
3. Jensen, R.: Combining rough and fuzzy sets for feature selection. Ph.D. Thesis, University of Edinburgh (2005)
4. Pawlak, Z.: Rough sets. Int. J. Comput. Inf. Sci. **11**(5), 341–356 (1982)
5. Guo, Q.L., Zhang, M.: Implement web learning environment based on data mining. Knowl.-Based Syst. **22**(6), 439–442 (2009)
6. Swiniarski, R.W., Skowron, A.: Rough set methods in feature selection and recognition. Pattern Recogn. Lett. **24**, 833–849 (2003)
7. Zhai, L.Y., Khoo, L.P., Fok, S.C.: Feature extraction using rough set theory and genetic algorithms: an application for the simplification of product quality evaluation. Comput. Ind. Eng. **43**(4), 661–676 (2002)
8. Jensen, R., Shen, Q.: Semantics-preserve dimensionality reduction: rough and fuzzy-rough-based approaches. IEEE Trans. Knowl. Data Eng. **16**, 1457–1471 (2004)

9. Wang, X., Yang, J., Teng, X., Xia, W., Jensen, R.: Feature selection based on rough sets and particle swarm optimization. Pattern Recogn. Lett. **28**, 459–471 (2007)
10. Wroblewski, J.: Finding minimal reducts using genetic algorithms. In: Proccedings of the Second Annual Join Conference on Infromation Science, pp. 186–189 (1995)
11. ElAlami, M.E.: A filter model for feature subset selection based on genetic algorithm. Knowl.-Based Syst. **22**(5), 356–362 (2009)
12. Mirjalili, S., Mirjalili, S.M., Lewis, A.: Grey wolf optimizer. Adv. Eng. Softw. **69**, 4661 (2014)
13. Bache, K., Lichman, M.: UCI Machine Learning Repository. Irvine, CA: University of California, School of Information and Computer Science. http://archive.ics.uci.edu/ml (2013)
14. Stefanowski, J.: On rough set based approaches to induction of decision rules. In: Skowron, A., Polkowski, L. (eds.) Rough Sets in Knowledge Discovery, vol. 1, pp. 500–529. Physica Verlag, Heidelberg (1998)
15. Bazan, J., Nguyen, H.S., Nguyen, S.H., Synak, P., Wroblewski, J.: Rough set algorithms in classification problem. In: Polkowski, L., Tsumoto, S., Lin, T.Y. (eds.) Rough Set Methods and Applications, pp. 49–88. Physica-Verlag, Heidelberg, New York (2000)
16. Wang, G.Y., Yu, H., Yang, D.C.: Decision table reduction based on conditional information entropy. Chin. J. Comput. **25**(7), 759–766 (2002)
17. Hu, K., Lu, Y.C., Shi, C.Y.: Feature ranking in rough sets. AI Commun. **16**(1), 41–50 (2003)
18. Skowron, A., Bazan, J., Son, N.H., Wroblewski, J.: RSES, 2.2 User's Guide. Institute of Mathematics, vol. 19. Warsaw University, Warsaw (2005)

Solving Manufacturing Cell Design Problems Using a Shuffled Frog Leaping Algorithm

Ricardo Soto, Broderick Crawford, Emanuel Vega, Franklin Johnson and Fernando Paredes

Abstract The manufacturing Cell Design Problem (MCDP) is a well-known problem for lines of manufacture where the main goal is to minimize the inter-cell moves. To solve the MCDP we employ the Shuffled Frog Leaping Algorithm (SFLA), which is a metaheuristic inspired on the natural memetic features of frogs. The frog tries to leap all over the search space for a better result until the stopping criteria is met. The obtained results are compared with previous approaches of the algorithm to test the real efficiency of our proposed SFLA. The results show that the proposed algorithm produces optimal solutions for all the 50 studied instances.

Keywords Bio-inspired systems · Metaheuristics · Optimization · Manufacturing.

R. Soto (✉) · B. Crawford · E. Vega · F. Johnson
Pontificia Universidad Católica de Valparaíso, Santiago, Chile
e-mail: ricardo.soto@ucv.cl

B. Crawford
e-mail: broderick.crawford@ucv.cl

E. Vega
e-mail: emanuel.vega@usm.cl

F. Johnson
e-mail: franklin.johnson@upla.cl

R. Soto
Universidad Autónoma de Chile, Temuco, Chile

R. Soto
Universidad Científica del Sur, Lima, Peru

B. Crawford
Universidad Central de Chile, Santiago, Chile

B. Crawford
Universidad San Sebastián, Santiago, Chile

F. Johnson
Universidad de Playa Ancha, Valparaiso, Chile

F. Paredes
Escuela de Ingeniería Industrial, Universidad Diego Portales, Santiago, Chile
e-mail: fernando.paredes@udp.cl

© Springer International Publishing Switzerland 2016
T. Gaber et al. (eds.), *The 1st International Conference on Advanced Intelligent
System and Informatics (AISI2015), November 28–30, 2015, Beni Suef, Egypt,*
Advances in Intelligent Systems and Computing 407, DOI 10.1007/978-3-319-26690-9_23

253

1 Introduction

The design of manufacturing cells has emerged in the last two decades as innovation for manufacturing strategy, this strategy involves the creation of an optimal design of production plans. The manufacturing cell design problem (MCDP) considers grouping similar parts into part-families. Ideally, each of these families is processed by a dedicated cluster of manufacturing facilities called manufacturing cell, where the main goal is to minimize movement and exchange of material between cells. In this context, the cell formation problem has been matter of considerable research, where Burbidge with his production flow analysis in 1963, becomes one of the first to propose a process to solve this concern [2].

In this paper, we propose a discrete Shuffled Frog Leaping Algorithm (SFLA) to solve the MCDP. The SFLA is a modern metaheuristic, which has been introduced in 2003 by Eusuff and Lansey [1]. The SFLA is based on the natural behavior of frogs, where they tend to group themselves in what are called memeplexes. Those memeplexes are considered as different cultures of frogs, each memeplex is formed in a way that it can influence another frog of the same group or later in the shuffle process to other frogs from different memeplexes. The standard SFLA works on continuous search spaces but we adapt this approach in order to handle the binary domains of the MCDP by using different transfer functions and discretization methods. We illustrate promising results where the proposed approach noticeable competes with previous reported techniques for solving manufacturing cell design problems.

The rest of this paper is organized as follows. In Sect. 2, we present the related work. Section 3 describes the mathematical model of the MCDP. The SFLA and the corresponding transfer functions and discretization methods are explained in Sect. 4. Finally, Sect. 5 illustrates the experimental results, followed by conclusions and future work.

2 Related Work

The problem of formulating cells has been the subject of considerable research, where Burbidge, with his production flow analysis in 1963, becomes one of the first to solve this problem [2]. Other approaches tried to solve the MCDP by attempting to determine the part families, finding only partial solutions [3, 4]. Most of these methods are based on the incidence machine-part matrix, and can be divided into hierarchical and non-hierarchical clustering. For instance Shargal et al. [5] presented search algorithms and clustering efficiency measures for machine-part matrix, Seifoddini and Hsu [6] worked with clustering algorithms in cellular manufacturing, and Srinivasan [7] used clustering algorithm for machine cell formation in group technology using minimum spanning tree. Also, graph theoretical mathematical programming methods have been employed, for instance Deutsch et al. [8] used an improved p-median model for cell formation, Atmani et al. [9] presented a mathematical programming approach to joint cell formation and operation allocation in cellular manufacturing. Adil et al. [10] proposed a mathematical model for cell formation considering investment and operational costs. Purcheck [11] reported a linear-programming method for the combinatorial grouping of incomplete sets. Olivia Lopez and Purcheck [12] worked on load balancing for group technology planning and control. Finally, Boctor [4, 13, 14] focused mostly on cell formation. The implementation of approximate methods, such as metaheuristics, has also been material of work for researchers devoted to these kind of problems. For instance Durán et al. [15] combined particle swarm optimization and a discrete position update scheme technique for manufacturing cell design. Wu et al. [16] introduced a simulated annealing (SA) approach, and Venugopal and Narendran [17] proposed the use of genetic algorithms (GA), which will be later used by Gupta et al. [18] for multi-objective optimization. In this paper, we employ the modern SFLA metaheuristic for solving the MCDP, which is able to reach encouraging results compared to approximate methods previously reported in the literature.

3 Manufacturing Cell Design Problem

The manufacturing cell desing strategy consists in creating an optimal design of production plants, which are composed of manufacturing cells and machines that process subsets of parts forming families, determined according to the similarity of them. The objective is to minimize movement and exchange of material between cells, in order to reduce production costs and increase productivity. We represent the processing requirements of machine parts through an incidence matrix called machine-part. This matrix contains a binary domains and is denoted as A, where $a_{ij} = 1$ means that machine i is necessary to process part j and $a_{ij} = 0$ otherwise. A rigorous mathematical formulation of machine-part grouping problem with these objectives is given by Boctor [13] and its as follows, let:

- M, the number of machines,
- P, the number of parts,
- C, the number of cells,
- i, the index of machines (i = 1, ... , M),
- j, the index of parts (j = 1, ... , P),
- k, the index of cells (k = 1, ... , C),
- $A = [a_{ij}]$ the binary machine-part incidence matrix M × P,
- M_{max}, the maximum number of machines per cell. We selected as the objective function to be minimized the number of times that a given part must be processed by a machine that does not belong to the cell that the part has been assigned to. Let:

$$y_{ik} = \begin{cases} 1 \text{ if } \texttt{machine } i \in \texttt{cell } k; \\ 0 \qquad\qquad \texttt{otherwise}; \end{cases}$$

$$z_{jk} = \begin{cases} 1 \text{ if } \text{part } j \in \text{family } k; \\ 0 \qquad\qquad \text{otherwise;} \end{cases}$$

The problem is represented by the following mathematical model:

$$Minimize \sum_{k=1}^{C} \sum_{i=1}^{M} \sum_{j=1}^{P} a_{i_j} z_{j_k} (1 - y_{i_k})$$

Subject to

$$\sum_{k=1}^{C} y_{ik} = 1 \qquad \forall i$$

$$\sum_{k=1}^{C} z_{jk} = 1 \qquad \forall j$$

$$\sum_{i=1}^{M} y_{ik} \leq M_{max} \qquad \forall k$$

3.1 Similarity Matrices

Incidence matrices machine-part can be broken into two square matrices for the formulation of manufacturing cells: One corresponds to a part-part matrix and the other one to a machine-machine matrix. Thus, the objective of the part-part matrix is to represent the number of machines that process p_j and p_i parts. Similarly, the machine-machine matrix represents the number of parts that are processed by the machines m_j and m_i [19]. We employ this machine-machine matrix to minimize the workload of the SFLA, sharing work in the generation of the initial population, in the exchange of information by the frogs, in every memeplex, and finally to

Fig. 1 Machine-part and machine-machine matrices

Part	1	2	3	4	5	6	7
M 1	1			1			
a 2	1						1
c 3			1	1		1	
h 4			1	1		1	
i 5	1						1
n 6			1	1		1	
e 7	1				1		

Machine	1	2	3	4	5	6	7
M 1	2						2
a 2		2			2		
c 3			3	3		3	
h 4			3	3		3	
i 5		2			2		
n 6			3	3		3	
e 7	2						2

support the generation of the part-cell matrix used to compute the fitness value of every frog. The similarity machine-machine matrix resulting from the incidence machine-part matrix can be seen on right side of Fig. 1.

4 Shuffled Frog Leaping Algorithm

The Shuffled Frog Leaping Algorithm (SFLA) is a modern metaheuristic, which was introduced in 2003 by Eusuff and Lansey [1]. SFLA was designed by the basic individual memetic evolution and global information exchange among a population of frogs, representing feasible solutions to attack the problem. The population is divided between virtual frogs memeplexes, where local searches are performed. In each memeplexes, frogs undergo an evolution through being infected by the "ideas" of other frogs. Thus, the memetic evolution improves the quality of solutions and movements of frogs toward a goal [20]. After a certain number of evolutions in each memeplex, ideas are passed in a process of rearrangement [21]. Local search and the reordering process continue until convergence criteria are met [1].

An initial population of P frogs is generated randomly within the search space. For problems of S-Dimensions, a frog i is represented as $X_i = (x_{i1}, x_{i2}, \ldots, x_{is})$. After the initial generation, frogs are sorted in a descending order according to their fitness value. Thus the population is divided into m memeplexes, each of which consisting of n frogs. The filling process of the groups is as follows:

- The first frog goes to the first memeplex
- The second frog goes to the second memeplex
- The m frog goes to the mth memeplex
- The $m + 1$ frog goes to the first memeplex, etc.

Within the memeplex, the frog with the best and worst fitness are identified as X_b and X_w, respectively. Similarly, the frog with the best overall fitness is identified as X_g. Then the process is applied to the frog with worse fitness based on the following leaping rule:

$$d^j_w = Rand \times (x^j_b - x^j_w) \qquad (1)$$

where d^j_w is the change of the jth position of the frog to a new position

$$x^j_{new} = (x^j_w - d^j_w) \qquad (2)$$

$$d^j_{min} < d^j_w < d^j_{max} \qquad (3)$$

where $Rand$ is a random number between 0 and 1 and $d_m ax^j$ is the maximum change allowed in the frogs position. If this produces a better solution, the worst frog is replaced. The opposite way by calculating Eqs. 2 and 4 are repeated but with the best overall frog (i.e. X_g replaces X_b). If there is no possible improvement, then a new solution is generated randomly to replace that frog. Finally, this phase continues a number of iterations for each memeplex [1].

4.1 Discretization of Decision Variables

As previously mentioned, SFLA operates on continuous spaces, so it is necessary to use transfer functions and discretization methods to map the decision variables to a binary space as follow:

When the worst frog X_w of each memeplex changes all decision variables x_w^j with the best frog X_b its calculated d_w^j using Eq. (1), followed by the transfer function. Thus, a transfer function calculates the probability $(T(d_w^j))$ of the jth bit of the vector representing a solution of the worst frog to change from 1 to 0 and vice versa. After obtaining the value resulting from the transfer function, we employ a discretization method, in order to assign [0, 1] to the jth bit of the vector of the worst frog.

- Transfer Function

$$T(d_w^j) = |tanh(d_x^j)|$$

- Discretization Method

$$x_{new}^j = \begin{cases} x_b^j & \text{if} \quad rand < T(d_w^j) \\ 0 & \text{otherwise}; \end{cases}$$

5 Experimental Result

The effectiveness of our proposed approach has been tested using the incidence matrices [13]. In this paper, these 10 problems were solved using the model presented in Sect. 2 with different sets of parameters. For the experimental evaluation, the parameters employed are defined as follows: 16 Machines, 30 Parts, and a combination between 2 Cells with 8, 9, 10, 11, 12 Mmax. Concerning the SFLA, the configuration uses 30 frogs as initial population size, 6 for the memeplexes grouping, 5 for the iterations within each memeplexes, and 50,000 iterations for the SFLA. The algorithm has been implemented using Java and launched on a 2.4 GHz Intel Core i7 with 8GB RAM running Windows 8. Tables 1 and 2 show detailed information of the results obtained by our approach. Here, we compare our results with the optimal values reported in [13] (Opt) as well as with [15] Particle Swarm Optimization (PSO) and Simulated Annealing (SA) [13]. Table 3 shows the relative percentage derivation (RPD), which is computed as follows:

$$RPD = \frac{(Z - Z_{opt})}{Z_{opt}} \times 100 \qquad (4)$$

Table 1 Results of SFLA using 2 cells

Pblm	Mmax = 8				Mmax = 9				Mmax = 10			
	Opt.	SA	PSO	SFLA	Opt.	SA	PSO	SFLA	Opt.	SA	PSO	SFLA
1	11	11	11	11	11	11	11	11	11	11	11	11
2	7	7	7	7	6	6	6	6	4	10	5	4
3	4	5	5	4	4	4	4	4	4	4	5	4
4	14	14	15	14	13	13	13	13	13	13	13	13
5	9	9	10	9	6	6	8	6	6	6	6	6
6	5	5	5	5	3	3	3	3	3	5	3	3
7	7	7	7	7	4	4	5	4	4	4	5	4
8	13	13	14	13	10	20	11	10	8	15	10	8
9	8	13	9	8	8	8	8	8	8	8	8	8
10	8	8	9	8	5	5	8	5	5	5	7	5

Table 2 Results of SFLA using 2 cells

Pblm	Mmax = 11				Mmax = 12			
	Opt.	SA	PSO	SFLA	Opt.	SA	PSO	SFLA
1	11	11	11	11	11	11	11	11
2	3	4	4	3	3	3	4	3
3	3	4	4	3	1	4	3	1
4	13	13	13	13	13	13	13	13
5	5	7	5	5	4	4	5	4
6	3	3	4	3	2	3	4	2
7	4	4	5	4	4	4	5	4
8	5	11	6	5	5	7	6	5
9	5	8	5	5	5	8	8	5
10	5	5	7	5	5	5	6	5

where Z_{opt} is the best known optimum value and Z is the best optimum value reached by SFLA. Finally, the average value of 10 executions is depicted.

The experimental results show that the proposed SFLA provides high quality solutions and good performance within 2 cells reaching $RPD = 0$ for all tested instances. The employment of the similarity matrix improved the generation of the solutions in early stages, sharing workload with the SFLA, and in final stages supporting the generation of the part-cell matrix for every best possible solution. This allows us to obtain better convergences rates (see example in Fig. 2) and provides robustness to the approach which is able to provide the global optimum in all the executions for all the instances as illustrate the average values in Table 3.

Table 3 Average and relative percentage derivation

Pblm	Mmax = 8		Mmax = 9		Mmax = 10		Mmax = 11		Mmax = 12	
	Avg.	RPD (%)	Avg.	RPD (%)	Avg.	RPD (%)	Avg.	RPD (%)	Avg.	RPD (%)
1	11	0	11	0	11	0	11	0	11	0
2	7	0	6	0	4	0	3	0	3	0
3	4	0	4	0	4	0	3	0	1	0
4	14	0	13	0	13	0	13	0	13	0
5	9	0	6	0	6	0	5	0	4	0
6	5	0	3	0	3	0	3	0	2	0
7	7	0	4	0	4	0	4	0	4	0
8	13	0	10	0	8	0	5	0	5	0
9	8	0	8	0	8	0	5	0	5	0
10	8	0	5	0	5	0	5	0	5	0

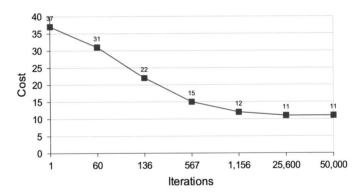

Fig. 2 Convergence chart for Problem 1 solved by SFLA with Mmax = 8

6 Conclusion and Future Work

In this work, we have proposed a Shuffled Frog Leaping Algorithm to solve Manufacturing Cell Design Problems. The experimental results show the capability of our
approach which is able to report optimum values for the whole set of tested instances.
The SFLA demonstrated excellent results when transfer function and discretization
method are employed. The method used is basically elitist allowing the frog to move
toward better solutions. Similarity is another interesting feature of our approach,
which allows to share workload in the initial and final computing process, generating optimal solutions with better convergence rates. Finally, concerning future work,
we aim at exploring other metaheuristics and evaluate their behaviour by integrating
discretization and similarity matrices for solving similar manufacturing problems.

Acknowledgments Ricardo Soto is supported by Grant CONICYT/FONDECYT/INICIACION/ 11130459, Broderick Crawford is supported by Grant CONICYT/FONDECYT/1140897, and Fernando Paredes is supported by Grant CONICYT/FONDECYT/1130455.

References

1. Eusuff, M., Lansey, K.: Optimization of water distribution network design using the shuffled frog leaping algorithm. J. Water Resour. Plan Manag. **129**, 210–225 (2003)
2. Burbidge, J.L.: Production flow analysis. Prod. Eng. **42**(12), 742–752 (1963)
3. Kusiak, A.: The part families problem in flexible manufacturing systems. Ann. Oper. Res. **3**, 279–300 (1985)
4. Xambre, A.R., Vilarinho, P.M.: A simulated annealing approach for manufacturing cell formation with multiple identical machines. Eur. J. Oper. Res. **151**, 434–446 (2003)
5. Shargal, M., Shekhar, S., Irani, S.A.: Evaluation of search algorithms and clustering efficiency measures for machine-part matrix clustering. IIE Trans. **27**(1), 43–59 (1995)
6. Seifoddini, H., Hsu, C.-P.: Comparative study of similarity coefficients and clustering algorithms in cellular manufacturing. J. Manuf. Syst. **13**(2), 119–127 (1994)
7. Srinivasan, G.: A clustering algorithm for machine cell formation in group technology using minimum spanning tree. Int. J. Prod. Res. **32**(9), 2149–2158 (1994)
8. Deutsch, S.J., Freeman, S.F., Helander, M.: Manufacturing cell formation using an improved p-median model. Comput. Ind. Eng. **34**(1), 135–146 (1998)
9. Atmani, A., Lashkari, R.S., Caron, R.J.: A mathematical programming approach to joint cell formation and operation allocation in cellular manufacturing. Int. J. Prod. Res. **33**(1), 1–15 (1995)
10. Adil, G.K., Rajamani, D., Strong, D.: A mathematical model for cell formation considering investment and operational costs. Eur. J. Oper. Res. **69**(3), 330–341 (1993)
11. Purcheck, G.: A linear-programming method for the combinatorial grouping of an incomplete set. J. Cybern. **5**, 51–58 (1975)
12. Olivia-Lopez, E., Purcheck, G.: Load balancing for group technology planning and control. Int. J. MTDR **19**, 259–268 (1979)
13. Boctor, F.F.: A linear formulation of the machine-part cell formation problem. Int. J. Prod. Res. **29**(2), 343–356 (1991)
14. Soto, R., Kjellerstrand, H.: Durn, O., Crawford, B., Monfroy, E., Paredes, F.: Cell formation in group technology using constraint programming and Boolean satisfiability. Expert Syst. Appl. **39**, 11423–11427 (2012)
15. Durán, O., Rodriguez, N., Consalter, L.: Collaborative particle swarm optimization with a data mining technique for manufacturing cell design. Expert Syst. Appl. **37**(2), 1563–1567 (2010)
16. Wu, T., Chang, C., Chung, S.: A simulated annealing algorithm for manufacturing cell formation problems. Expert Syst. Appl. **34**(3), 1609–1617 (2008)
17. Venugopal, V., Narendran, T.T.: A genetic algorithm approach to the machine-component grouping problem with multiple objectives. Comput. Indu. Eng. **22**(4), 469–480 (1992)
18. Gupta, Y., Gupta, M., Kumar, A., Sundaram, C.: A genetic algorithm-based approach to cell composition and layout design problems. Int. J. Prod. Res. **34**(2), 447–482 (1996)
19. Oliveira, S., Ribeiro, J.F.F., Seok, S.C.: A spectral clustering algorithm for manufacturing cell formation. Comput. Ind. Eng. **57**(3), 1008–1014 (2009)
20. Bhattacharjee, K.K., Sarmah, S.: Shuffled frog leaping algorithm and its application to 0/1 knapsack problem. Appl. Soft Comput. **19**, 252–263 (2014)
21. Lion, S., Atiquzzaman, M.: Optimal design of water distribution network using shuffled complex evolution. J. Instrum. Eng. **44**, 93–117 (2004)

A Hybrid Bat-Regularized Kaczmarz Algorithm to Solve Ill-Posed Geomagnetic Inverse Problem

M. Abdelazeem, E. Emary and Aboul Ella Hassanien

Abstract The aim of geophysical inverse problem is to determine the spatial distribution and depths to buried targets at a variety of scales; it ranges from few centimetres to many kilometres. To identify ore bodies, extension of archaeological targets, old mines, unexploded ordnance (UXO) and oil traps, the linear geomagnetic inverse problem resulted from the Fredholm integral equation of the first kind is solved using many strategies. The solution is usually affected by the condition of the kernel matrix of the linear system and the noise level in the data collected. In this paper, regularized Kaczmarz method is used to get a regularized solution. This solution is taken as an initial solution to bat swarm algorithm (BA) as a global swarm-based optimizer to refine the quality and reach a plausible model. To test efficiency, the proposed hybrid method is applied to different synthetic examples of different noise levels and different dimensions and proved an advance over using the Kaczmarz method.

Keywords Bat-regularized Kaczmarz Algorithm · Geomagnetic inverse problem · Swarm-based optimizer

1 Introduction

Beside the inherent ambiguity in the geophysical potential field inverse problem that make it difficult to interpret, the potential field inverse problem is generally an ill-posed one. This is mainly due to instability against error/noise levels in the input

M. Abdelazeem
National Research Institute of Astronomy and Geophysics (NRIAG), Helwan, Egypt

E. Emary (✉) · A.E. Hassanien
Faculty of Computers and Information, Cairo University, Giza, Egypt
e-mail: eidemary@yahoo.com
URL: http://www.egyptscience.net

E. Emary · A.E. Hassanien
Scientific Research Group in Egypt (SRGE), Cairo, Egypt

A.E. Hassanien
Faculty of Computers and Information, BeniSuef University, Bani Sweif, Egypt

© Springer International Publishing Switzerland 2016
T. Gaber et al. (eds.), *The 1st International Conference on Advanced Intelligent System and Informatics (AISI2015), November 28–30, 2015, Beni Suef, Egypt,*
Advances in Intelligent Systems and Computing 407, DOI 10.1007/978-3-319-26690-9_24

observed data. Before we use any numerical technique to solve such problems, one has to calculate the condition number of the kernel matrix of the linear system if the thought problem is linear. For higher condition numbers, the traditional methods of inversion will not be stable enough against different noise levels that usually accompany any field geophysical data and one have to apply regularizing strategies such as truncated singular value decomposition (TSVD), L-curve, trust region subproblem (TRS) and many others to reach a plausible solution. Similar gravity problems in geophysics was solved using TSVD by [1] and a comparative study for the 2-D geomagnetic problem was done using both TSVD and parameterized TRS in [2]. Tikhonov and Arsenin [3] was the first introduce to the concept of regularization.

In the meantime, a very large condition number of the coefficient matrix A in a linear system of equations $Ax = b$ implies that some/all of the equations are numerically linearly dependent. Numerical tools such as the Singular Value Decomposition SVD can help to improve the model and lead to a modified system with a better-conditioned matrix. The efficiency of such methods varies with the condition number, level of noise in data and the dimension of the kernel matrix. On the other hand, the evolutionary techniques including different swarm-based algorithms, neural networks and genetic algorithms proved to be less sensitive to either noise or ill-posedness. Such methods are used in geophysical potential field inverse problem by many authors. Sweilam et al. [4] used the PSO in solving spontaneous potential simple body problem, [5] used PSO with stretching in solving gravity simple body problem. Al-Garni [6, 7] solved the simple body magnetic problem using the Neural Networks, [8] developed the geothermal gradient map of the northern Western Desert in Egypt using the Neural Networks and [9] enhanced interpretation of magnetic survey data from archaeological sites using artificial neural networks. Those and many other publications used different artificial intelligent concepts to get the global minimum through steering the members, genes or weights by some controlling factors. This job is not always easy and may take a lot of time and iterations. The initial solution is an effective factor for convergence. In this paper, a hybrid combination between the global artificial minimizer, resembled by the bat algorithm, is used in conjunction with kaczmarz regularization strategy to improve the efficacy of inversion flow and overcome the ill-posedness of the geophysical potential field problems.

The paper is organized as follows: Sect. 2 presents the proposed hybrid Bat-regularized Kaczmarz Algorithm to Solve Ill-posed Geomagnetic Inverse Problem. Section 3 presents the synthetic examples and the results. Section 4 concludes and suggests the future work.

2 The Proposed Regularized Kaczmarz Model

The aim of the geomagnetic inverse problem is to detect the buried magnetic bodies from the measured geomagnetic field signal. In the 2-dimensional space, the model domain is subdivided into two-dimensional prisms of equal sides and each prism is extended to infinity in the third direction perpendicular to the plane. Each prism has an unknown magnetic susceptibility (physical property). The problem can be expressed as:

$$\min \| Ax - b \|^2 \tag{1}$$

where, A is the kernel matrix, which expresses the geometrical relation between each prism and each measuring point, x is the vector of unknown magnetic susceptibility of prisms and b is the vector of measured magnetic field. This problem has proved to be mostly ill-posed [2].

The classical Tikhonov regularization approach is one of the most popular regularization approaches for solving discrete forms of ill-conditioned linear algebraic system. Tikhonov regularization problem takes the form:

$$\min \| Ax - b \|^2 + \alpha \| x \|^2 \tag{2}$$

where, $A \in \Re^{n \times m}$, $b \in \Re^m$, $\alpha > 0$ and $\|\|$ is the Euclidean norm. This is always solved by recent iterative methods based on solving Euler equations (regularized normal equations):

$$(A^T A + \alpha^2 I)x_\alpha \tag{3}$$

where α is the regularization parameter and x_α is the regularized solution. Here, the condition number in 3 is equal about the square of condition number in 2. So, it is impossible to solve such problem 2 using popular iterative methods.

In this paper the method of [10] which is based on the famous Kaczmarz method [11] and its version [12] is used as an initial solution to the bat swarm optimizer.

The system in Eq. 3 is transformed to augmented system of linear equation [13] which is always consistent and determined for $\alpha > 0$. The randomized Kaczmarz algorithm [14] used by solving the following system:

$$\begin{bmatrix} \omega I_m & A \\ A^T & -\omega I_n \end{bmatrix} \begin{bmatrix} y \\ x \end{bmatrix} = \begin{bmatrix} b \\ 0 \end{bmatrix} \Leftrightarrow \tilde{A}_\omega z = b \tag{4}$$

where $\omega = \sqrt{\alpha}$, $\tilde{A}_\omega$ is non-singular for $\alpha > 0$. The spectral condition number of this system is equal to square root of the condition number of system (3) as proved in [13].

As the kernel function decays with increasing depth of the prism, the inverted model will mostly be concentrated near the surface. The depth weighting matrix is introduced to overcome such natural geometric decay. It was reasonable to approximate the decay with depth by a diagonal weighting matrix of the form:

$$W(m) = \frac{1}{(z_m^2 + \varepsilon)^\beta} \tag{5}$$

where $\beta = 0.9$ [15] and ε is a very small number to prevent singularity when z is close to 0.

Bat swarm optimization initialized with Kaczmarz solution and generates many other random solutions in the space and applies successive iterations to further enhance the solutions and find the optimal one.

Bats are fascinating animals for its advanced capability of echolocation. Bats emit a very loud sound pulse and listen for the echo that bounces back from the surrounding objects. The correlation between the bounced and the reflected pulses decides the obstacle type. Their pulses vary in properties and can be correlated with their hunting strategies, depending on the species. Most bats use short, frequency modulated signals to sweep through about an octave, while others more often use constant frequency signals for echolocation. Their signal bandwidth varies depends on the species, and often increased by using more harmonics [16].

When hunting for prey, the rate of pulse emission can be sped up when they near their prey. Such short sound bursts imply the fantastic ability of the signal processing power of bats. The loudness also varies from the loudest when searching for prey and to a quieter base when moving towards the prey.

All bats use echolocation to sense distance and to discriminate between background barriers and targets; food or prey. Bats randomly fly with velocity V_i at position X_i with varying wavelength λ and loudness A_0 to search for prey. They can adjust the wavelength of the emitted pulses and the emission rate depending on proximity to the target. Loudness of the emitted pulses is also changes over time but bounded by $[A_0; A_{min}]$. Virtual bats adjust their position according to the following equations:

$$F_i = F_{min} + (F_{max} - F_{min})\beta \tag{6}$$

$$V_i^t = V_i^{t-1} + (X_i^t - X_*)F_i \tag{7}$$

$$X_i^t = X_i^{t-1} + V_i^t \tag{8}$$

where β is a random vector in the range [0, 1] drawn from uniform distribution. X^* is the current global best location. F_{min} and F_{max} represent the minimum and maximum frequency need depending on the problem.

V_i represents the velocity vector.

Probabilistically a local search is to be performed using a random walk as in the following equation:

$$X_{new} = X_* + \epsilon A^t \tag{9}$$

where A^t is the average loudness of all bats at this time and ϵ is a random number uniformly drawn from [−1, 1]. The updating of the loudness is performed using the following equation:

$$A_i^{t+1} = \delta A^T \tag{10}$$

where δ is a constant selected experimentally. r_i controls the application of the local search as shown on the algorithm and is updated using the Eq. 11.

$$r_i^{t+1} = r_i^0[1 - exp(-\gamma t)] \qquad (11)$$

where r_i^0 is the initial pulse emission rate and is a constant greater than 0. Algorithm 1 describing the virtual Bat algorithm.

---

input : Q frequency band
$\quad\quad\quad f_i$ pulse frequency
$\quad\quad\quad r_i$ pulse emission rate
$\quad\quad\quad A_i$ Loudness
$\quad\quad\quad \epsilon$ sound amplitude control parameter
$\quad\quad\quad \gamma$ pulse rate control parameter
output: Optimal Bat position and the corresponding fitness

Initialize a population of n bat' positions at random
Find the best solution based on fitness; g^* ;
while *Stopping criteria not met* **do**
$\quad$ **foreach** *bat$_i$* **do**
$\quad\quad$ Generate new solution by adjusting frequency(x_i^{new}); see Eqs. 6, 7 and 8
$\quad\quad$ **if** *rand $> r_i$* **then**
$\quad\quad\quad$ | Perform local search around global best *best* (x_i^{new});
$\quad\quad$ **end**
$\quad\quad$ **if** *rand $< A_i$ and fitness of new solution is better than original one* **then**
$\quad\quad\quad$ Update the position of *bat$_i$* to bex_i^{new}
$\quad\quad\quad$ Increase emission rate r_i; 11
$\quad\quad\quad$ Decrease Loudness A_i; 10
$\quad\quad$ **end**
$\quad\quad$ Update the best solution
end

Algorithm 1: Algorithm for bat swarm optimization

---

3 Experimental Results and Discussion

The new hybrid method was applied to three synthetic examples using different noise levels and earth model dimension in each case to prove its efficiency. The Kacmarz method is adjusted to run on the same time as the hybrid Kacmarz-BA method.

Synthetic Sample-1 (earth model) The first arrangement suggests earth model consists of 640 prism arranged in 10 rows and 64 columns. The dimension of each prism is 1 m $\times$ 1 m and assumed to extend to infinity. The ambient field is assumed to be 40,000 nT and assumed to have the reduced to pole field. The Azimuth of the measured data is assumed to be zero. The data points are 640, i.e., our problem is an even-determined one. The condition number of the kernel matrix is found to be

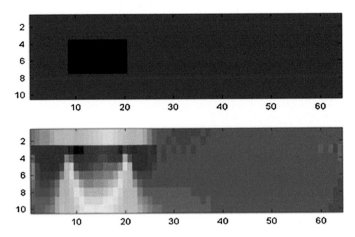

Fig. 1 Earth model of example (1) with 640 prisms and the inverted earth model using the proposed model

$9.1559e + 18$. Figure 1 shows the suggested magnetic body and its corresponding inverse solution.

The second arrangement consists of 160 prisms arranged in 5 rows and 32 columns with the same parameters and suggested earth model. The condition number of the kernel matrix is also $9.1559e + 18$.

The third arrangement consists of 80 prisms arranged in 5 rows and 16 columns and the condition number of its kernel is $9.1559e + 18$. The three different arrangements are reconstructed using the Kacmarz and the Kacmarz-BA method at different noise levels ranging from 100 to 60 dB.

The numerical results for this experiments is mentioned in Table 1. The obtained fitness value is tabulated and we can see that the BA initialized with Kacmarz outperforms the Kacmarz method running both algorithms for the same run time which can be interpreted by the super capability of BA to search the space for optimal solution even if many local minima exists. We can also remark that increasing the resolution for reduces the capability of the algorithms to fins optimal solution as the search space becomes very large and hence the error sum for the high resolution is much difficult than the low resolution reconstruction. We can also see that the main advance of the proposed hybrid system is much clear for high resolution reconstruction where the Kacmarz fails to converge to optimal solution.

Synthetic Sample-2 (earth model) This example has the same structure as the first one but with different suggested magnetic bodies as shown in Fig. 2. Table 2 presents the results for the same noise range used in example (1). We can remark from table that the hybrid proposed method is still outperforms Kaczmarz method on the different noise rates and the different sizes.

Table 1 Results of example (1) using Kaczmarz and the proposed hybrid Kaczmarz-bat algorithms

SNR	Dim	Misfit of field using Kaczmarz	Misfit of field using Kaczmarz/Bat	Misfit in model using Kaczmarz	Misfit of model using Kaczmarz/Bat
100	640	18.56654	15.70077	0.000004	0.000004
	160	4.364369	3.211048	0.000002	0.000002
	80	1.82478	2.315475	0.000002	0.000004
90	640	17.80673	15.93999	0.000004	0.000004
	160	4.254198	3.220406	0.000002	0.000002
	80	1.839903	2.346384	0.000002	0.000004
80	640	20.67425	17.63531	0.000005	0.000005
	160	5.898271	4.248078	0.000004	0.000004
	80	3.362344	5.026152	0.000004	0.000005
70	640	17.72511	17.26083	0.000004	0.000004
	160	4.397793	3.49102	0.000002	0.000002
	80	1.964255	2.575199	0.000002	0.000004
60	640	23.32973	21.53744	0.000004	0.000004
	160	5.45699	94.243028	0.000002	0.000002
	80	2.312926	3.843799	0.000002	0.000004

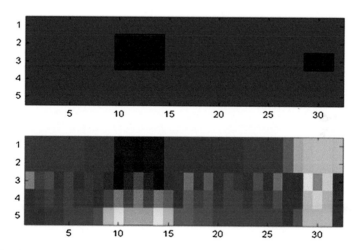

Fig. 2 Earth model of example (2) with 160 prisms and the inverted earth model using the proposed model

Synthetic Sample-3 (earth model) This example presents an earth model for three magnetic bodies in order to test the effect of complicated geological situation with ill-posed kernel on the method and its applicability, refer to Fig. 3. Results are summarized in Table 3. We can remark that the hybrid proposed system is still performing better than the Kaczmarz for the different noise levels and the different

Table 2 Results of example (2) using Kaczmarz and the proposed hybrid Kaczmarz-bat algorithms

SNR	Dim	Misfit of field using Kaczmarz	Misfit of field using Kaczmarz/Bat	Misfit in model using Kaczmarz	Misfit of model using Kaczmarz/Bat
100	640	20.14454	17.67157	0.000005	0.000005
	160	4.8768	4.057205	0.000004	0.000004
	80	3.007872	4.913887	0.000004	0.000005
90	640	19.98031	17.74035	0.000005	0.000005
	160	5.19905	4.045526	0.000004	0.000004
	80	3.207954	5.002001	0.000004	0.000005
80	640	20.67425	17.63531	0.000005	0.000005
	160	5.898271	4.248078	0.000004	0.000004
	80	3.362344	5.026152	0.000004	0.000005
70	640	21.02544	18.18722	0.000005	0.000005
	160	5.292449	4.127389	0.000004	0.000004
	80	3.3517	4.897991	0.000004	0.000005
60	640	21.67418	19.0649	0.000005	0.000005
	160	6.335895	5.229592	0.000004	0.000004
	80	4.417849	5.067848	0.000004	0.000005

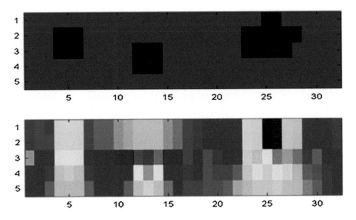

Fig. 3 The earth model with 160 prisms suggested in example (3) and the inverted one

resolutions regardless of the complications existing in the search space. The difference in the performance is minimum between the two used methods in case of low resolution data as the search space is respectively small and hence both methods can converge to the optimal solution.

Table 3 Results of example (3) using Kaczmarz and the proposed hybrid Kaczmarz-bat algorithms

SNR	Dim	Misfit of field using Kaczmarz	Misfit of field using Kaczmarz/Bat	Misfit in model using Kaczmarz	Misfit of model using Kaczmarz/Bat
100	640	34.29171	29.05363	0.000008	0.000009
	160	9.053332	7.218613	0.000006	0.000006
	80	5.408153	6.843847	0.000006	0.000008
90	640	32.52878	29.32763	0.000008	0.000009
	160	9.36744	7.284556	0.000006	0.000006
	80	5.210043	6.787388	0.000006	0.000008
80	640	33.22117	29.30881	0.000008	0.000009
	160	9.684629	7.226406	0.000006	0.000006
	80	5.115863	6.751183	0.000006	0.000008
70	640	34.45014	29.81535	0.000008	0.000009
	160	10.22134	7.522129	0.000006	0.000006
	80	5.157756	6.950278	0.000006	0.000008
60	640	38.6098	35.45318	0.000008	0.000009
	160	11.88874	9.294612	0.000006	0.000006
	80	6.157133	7.277374	0.000006	0.000008

4 Conclusion and Future Work

In this paper a scheme for geophysical inverse system using a hybrid bat swarm optimization initialized with regularized Kaczmarz solution is introduced to improve the solutions of ill-posed geomagnetic inverse problem. The proposed scheme is evaluated using different synthetic data samples with different noise levels to test for algorithm robustness against noise. Different earth model dimensions are applied to test model resolution. The proposed hybrid method outperforms the Kaczmarz regularizing method especially in high resolution data reconstruction thanks to the capability of the bat algorithm in adaptively searching the space for the optimal solution. This ability is more obvious during the inversion of noisy synthetic models.Future work will concentrate on applying the algorithm on different real data with different geological complexities and different resolutions.

References

1. El-Azeem, A.B.D.M., Sweilam, N.H., Gobashy, M., Nagy, A.M.: Two dimensions gravity inverse problem using adaptive pruning L-curve. Paper published in Journal of Mathematical and Physical Society of Egypt (2007)
2. Abdelazeem, M.: Regularizing ill-posed magnetic problems using a parameterized trust-region sub-problem. Contrib. Geophys. Geodesy **43**(2), 99–123 (2013)

3. Tikhonov, A.N., Arsenin, V.Y.: Solutions of Ill-Posed Problems. V. H. Winston & Sons, New York (1977)
4. Sweilam, N.H., El-Azeem, A.B.D.M., Metwally, K.: Self potential signals inversion to simple polarized bodies using the particle swarm optimization method: a visibility study. J. Appl. Geophys. ESAP (6) (2007)
5. Sweilam, N.H., Gobashy, M.M., Hashem, T.: Using particle swarm optimization with function stretching (SPSO) for inverting gravity data: a visibility study. Proc. Math. Phys. Soc. Egypt **86**(2), 259–281 (2008)
6. Al-Garni, M.A.: Interpretation of magnetic anomalies due to dipping dikes using neural network inversion. Arab. J. Geosci. (2015, in press)
7. Al-Garni, M.A.: Interpretation of some magnetic bodies using neural networks inversion. Arab. J. Geosci. **2**, 175–184 (2009)
8. Mohamed, H.S., Abdel Zaher, M., Senosy, M.M., Saibi, H., El Nouby, M. Fairhead, J.D.: Correlation of aerogravity and BHT data to develop a geothermal gradient map of the northern western desert of Egypt using an artificial neural network. Pure Appl. Geophys. (2014)
9. Bescoby, D.J., Cawley, G.C., Chroston, P.N.: Enhanced interpretation of magnetic survey data from archaeological sites using artificial neural networks. Geophysics **71**(5), H45–H53 (2006)
10. Ivanov, A.A., Zhdanov, A.I.: Kaczmarz algorithm for Tikhonov regularization problem. Appl. Math. E-Notes **13**, 270–276 (2013)
11. Kaczmarz, S.: Angenherte Auflsung von Systemen linearer Gleichungen. Bull. Internat. Acad. Polon. Sci. Lett. A **35**, 335–357 (1937)
12. Vasilchenko, G.P., Svetlakov, A.A.: A projection algorithm for solving systems of linear algebraic equations of high dimensionality. Comput. Math. Math. Phys. **20**, 1–8 (1980)
13. Zhdanov, A.I.: The method of augmented regularized normal equations. Comput. Math. Phys. **52**, 194–197 (2012)
14. Strohmer, T., Vershynin, R.: A randomized Kaczmarz algorithm for linear systems with exponential convergence. J. Fourier Anal. Appl. **15**, 262–278 (2009)
15. Boulanger, O., Chouteau, M.: Constraints in 3D gravity inversion. Geophys. Prospect. **49**, 265–280 (2001)
16. Yang, X.-S.: A New Metaheuristic Bat-Inspired Algorithm. In: Gonzalez, J.R., et al. (eds.) Nature In- spired Cooperative Strategies for Optimization (NISCO). Studies in Computational Intelligence, vol. 284, pp. 65–74. Springer, Berlin (2010)

A Fish Detection Approach Based on BAT Algorithm

Mohamed Mostafa Fouad, Hossam M. Zawbaa, Tarek Gaber,
Vaclav Snasel and Aboul Ella Hassanien

Abstract Fish detection and identification are important steps towards monitoring fish behavior. The importance of such monitoring step comes from the need for better understanding of the fish ecology and issuing conservative actions for keeping the safety of this vital food resource. The recent advances in machine learning approaches allow many applications to easily analyze and detect a number of fish species. The main competence between these approaches is based on two main detection parameters: the time and the accuracy measurements. Therefore, this paper proposes a fish detection approach based on BAT optimization algorithm (BA). This approach aims to reduce the classification time within the fish detection process. The performance of this system was evaluated by a number of well-known machine learning classifiers, KNN, ANN, and SVM. The approach was tested with 151 images to detect the Nile Tilapia fish species and the results showed that k-NN can achieve

M.M. Fouad (✉)
Arab Academy for Science, Technology, and Maritime Transport, Cairo, Egypt
e-mail: mohamed_mostafa@aast.edu
URL: http://www.egyptscience.net

M.M. Fouad · T. Gaber
IT4Innovations, VSB-Technical University of Ostrava, Ostrava, Czech Republic

T. Gaber
Faculty of Computers and Informatics, Suez Canal University, Ismailia, Egypt

H.M. Zawbaa
Faculty of Mathematics and Computer Science, Babes-Bolyai University,
Cluj-Napoca, Romania

H.M. Zawbaa
Faculty of Computers and Information, Beni-Suef University, Beni Suef, Egypt

V. Snasel
Faculty of Electrical Engineering and Computer Science,
VSB-Technical University of Ostrava, Ostrava, Czech Republic

A.E. Hassanien
Faculty of Computers and Information, Cairo University, Cairo, Egypt

M.M. Fouad · H.M. Zawbaa · T. Gaber · A.E. Hassanien
Scientific Research Group in Egypt (SRGE), Giza, Egypt

© Springer International Publishing Switzerland 2016
T. Gaber et al. (eds.), *The 1st International Conference on Advanced Intelligent System and Informatics (AISI2015), November 28–30, 2015, Beni Suef, Egypt,*
Advances in Intelligent Systems and Computing 407, DOI 10.1007/978-3-319-26690-9_25

273

high accuracy 90 %, with feature reduction ratio close to 61 % along with a notice-able decrease in the classification time.

Keywords BAT algorithm · SURF · Fish detection · Tilapia fish · k-NN · ANN · SVM

1 Introduction

A great number of fish species over the world suffers from pollution and overfishing practices. Moreover, the human construction through rivers such as dams that nega-tively affects the number of fishes and their diet sources. Monitoring of any diversity change in the quantities of any fish species is important to ensure the sustainability of one of the protein sources for the world populations. Usually the common methods used by marine biologists are human observations, caught and counted using casting nets, or using sonars [1]. Recently, computer vision [2–5], relying on the application of machine learning techniques [6, 7], are used to detect, count, classify and study fish behavior of different fish species.

Several studies have considered the fish detection and species recognition in water, such as the work in [6] where the authors used an underwater camera to detect swimming fish in an open sea. The proposed model is based on inter-class similarities to construct a hierarchical tree classification. The trajectory voting is used as a sec-ond phase to eliminate the unknown classes or less confident decisions. Although the model can classify 15 species of fish with a hight accuracy that reached 97.5 % using the flat Support Vector Machine (SVM) classifier, but the model may reject new fish species which are not been sampled within the model. These rejected classes may require further analysis.

Since the shape extraction is important for species recognition, a number of approaches use it as reference for fish classification. Examples include using the Principal Component Analysis (PCA) to detect fish through the shape knowledge [7]. The Haar-classifier is trained to locating snout and tail of fish in the underwater image sequences. The detection of both snout and tail points are used as references for constructing the shape model of the fish. to efficiently capture the main varia-tions of a training set, the PCA statistical procedure is used to remove poor contrast boundaries, background clutter and occlusions caused by overlapping with neigh-boring fish.

Accurate extraction of the shape is important for fish classification. However, the shape extraction may fail if it relies on outer boundary (edge) detection [8] as shape cannot be detected correctly if the scene is unclear or contains a large swarm of fish [9]. One of the solution is to use color information along with the shape as the work reported in [10]. This work used the neural networks to detect stationary species in deep sea.

To evaluate which feature extraction technique could be efficient for the classifica-tion accuracy, a comparison in [11] was conducted between a Scale Invariant Feature

Transform (SIFT), a Speed Up Robust Features (SURF) algorithms. The evaluation results showed the superiority of the Support Vector Machine (SVM) over Artificial Neural Networks (ANN) and k-Nearest Neighbor (k-NN) algorithms when they applied to the SURF-based and SIFT-based features. The a SVM classifier using the the linear function achieved at 94.44 % using the SURF-based features and 69.57 % using the SIFT-based features.

The rest of this paper is organized as follows: Sect. 2 gives an overview about the BAT algorithm and Sect. 3 presents the proposed system. The results and the discussion are reported in Sect. 4. The paper is concluded in Sect. 5.

2 The Bat Algorithm: An Overview

The bat bird has a distinguish strategy for tracking its preys and avoiding obstacles. This flaying animal uses the echolocation process for navigation and it has the ability to differentiate between the sending signals and their reflections. The reflected signals is used for measuring the distance between the bat and the prey. Yang [12] has studied the bat behavior and proposed a new meta-heuristic optimization algorithm known as The Bat Algorithm. This algorithm can be modeled using the following three general rules [13]:

- The bats use echolocation sensing to measure the distance and they can easily distinguish between the prey and obstacles.
- Each bat fix minimum frequency, varying wavelength, and loudness through their random searching for a prey. Based on the proximity of the prey, each bat can adjust its frequency of their emitted pulses.
- The variation of loudness should be limited through a constant range of values

As a meta-heuristic algorithm, the bat optimization generates a number of new solutions within the search space through adjusting frequency, and updating velocities till a single solution is ranked among the best found solutions [14], Algorithm 1 shows the general structure of the BA. This meta-heuristic algorithm has been applied widely in a number of structural optimization problems such as in [14–16].

3 The Proposed Approach

The proposed approach can be divided into two main procedures: the training procedure and the testing procedure. Each of these procedures consists from a number of phases as illustrated in Fig. 1. A description of these phases are give below.

In the training procedure, the approach starts by a set of n raw-images that pass through a pre-processing phase to unify their scale-invariance and to reduce the noise

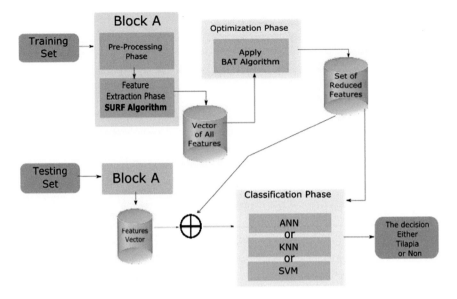

Fig. 1 The proposed detection approach

foreach *Input image* **do**

  1. Initialize the bat positions X_i and Velocity V_i and define the pulse frequency F_i, the pulse emission rate r_i and the loudness A_i per bat.
  2. While stopping criteria are not met

    a   For every bat

        i   Generate a new solution by calculating the updated frequency, velocity and position.

        ii If *rand* > r_i then

            –   Select a solution among the best solutions set and perform a random local search around it.

        iii End If

        iv Generate a new random solution

        v   If *rand* < A_i and $F(X_i) < F(X*)$

            –   Accept the new solution

            –   Increase r_i and decrease A_i

        vi End If

        vii Reselect the best solution $X*$

    b   End For

  3. End While

end

Algorithm 1: Bat Algorithm (BAT)

levels. Then, in order to focus on the intensity information within the image, the pre-porcessed images were converted from RGB image to gray-scale. In the feature extraction phase, the SURF algorithm was applied. The algorithm selects the Hessian matrix based detector to determine the region of interest for each image $I \in n$ at the scale σ as illustrated in Eq. 1.

$$H = \begin{bmatrix} I_{xx}(x, \sigma) \ I_{xy}(x, \sigma) \\ I_{xy}(x, \sigma) \ I_{yy}(x, \sigma) \end{bmatrix}. \tag{1}$$

where I_{xx} is the second order Gaussian smoothed image derivatives which detects signal changes in two orthogonal directions.

The SURF algorithm determines for each feature a 4?4 sub-window to give a description for each feature as weighted wavelet coefficients after calculating the *Haar–wavelet* response in horizontal and vertical directions. As a result, the final deliverable of the SURF is a 64-*dimensional vector(V)* of descriptors for each I image [17]. Finally, the features were labeled according to the corresponding classes. The following formula models the feature vector:

$$V = \left\{ x_I, y_I \right\}_{I=1}^{n} \tag{2}$$

where x_i represents the features of interest and y_i describes the class label (Tilapia, or Non-Tilapia Class).

The produced features vector was then given as an input to the optimization phase at which the BA was applied to reduce the number of features. Such process aims to minimize the classification time while keep the classification accuracy at high rate. Table 1 shows the values adjust for the BA within the optimization phase. The output of the optimization phase, i.e. the reduced feature vector, was then given to a number of machine learning classifiers to test the classification accuracy and to be build the model.

Table 1 The BAT adjusted parameters

Parameter	Value(s)
No. of bats	5
Num. of itterations	100
Loudness (A)	0.5
Pulse rate (r)	0.5
Frequency minimum ($Qmin$)	0
Frequency maximum ($Qmax$)	2
Problem dimension	SURF features vector = 64 dimensions
Search domain	[0, 1]

In the testing procedure, a set of unlabeled images n' was used. This set went through the same pre-processing and feature extraction phases which were done in the training procedure. As illustrated in Fig. 1, before the classification phase of the testing procedure, the produced features vector was passed to a logical gate function so this vector is filtered by the reduced vector obtained in the optimization phase done in the training procedure. This process will avoid running the BAT algorithm again in the testing procedure.

It is worth noting that, the BAT algorithm used for feature selection was considered the following points: (1) The solution space here represents all possible solutions of features and hence the bat positions represent a binary selection of feature sets, (2) Each feature is considered as an individual dimension with uniform distribution in range of [0, 1]. To decide if a feature will be selected or not, its position value will be threshold with a constant threshold (ε). (3) The selection parameter (ε) is very effective in the performance of the bat algorithm as it controls the diversity of the solution obtained at a given iteration. So, in this work we made adaptive tuning for the selection parameter (ε) throughout iteration numbers. In the beginning of the iterations, there was a need for large diversity, hence a large value of (ε) was adjusted for more exploration and discovering through the search space for an optimal solution. At the end of optimization approach, the bats should reach near the optimal solution, therefore, a less diversity is required (intensification the search near this solution), so a minimum value for (ε) was set. The initial value for the (ε) was set as half of the search range, to be inside the search space and it is only an initial value. The parameter (ε) is calculated in each iteration using Eq. (3).

$$\varepsilon = \varepsilon_0 - \frac{t * \varepsilon_0}{N_{gen}}, \tag{3}$$

where ε_0 is the initial value for ε, t is the iteration number, and N_{gen} is the maximum number of iterations for the BAT optimization.

4 Experimental Results and Discussion

The proposed fish classification approach was implemented using the MATLAB simulation tool. A self-collected database for fish was used. This is because there is no well-known data set for Nile Tilapia fish. The database was collected through capturing a number of fish photos with different transformations (scale change, rotation, illumination, image blur, viewpoint change, and compression). The dataset consists of 96 images of Tilapia fish and 55 images of Non-Tilapia fish, Fig. 2 shows samples of these images.

To evaluate that the aim of our proposed system, achieving high accuracy rate using the least features optimized by the BAT algorithm, two scenarios were designed. The first scenario was to the classification rate before and after applying the BA to the feature extracted by the SURF algorithm. Table 2 summarizes the reduc-

Fig. 2 Sample images of the Nile Tilapia data set

Table 2 BA features' reduction ratio classifier

Experiment No.	BA features' reduction ratio (%)
1	39.0
2	53.0
3	45.3
4	48.4
5	45.3
6	40.6
7	46.8
8	56.2
9	60.9
10	54.6
11	57.8

tion ratio obtained by running different experiments using BA and without using BA. So, in the second scenarios, we will see if a small set (optimal feature set) of these features may be enough to indicate the classification results within shorten time.

The secondly scenario was to test the compatibility of the approach with a number of classifiers such as the Artificial Neural Network (ANN), the k-Nearest Neighbors(k-NN), and the Support Vector Machine that based on the radial basis function kernel (SVM-rbf). The ratio between the number of correct and false matches among the reference images and the set of n' has been used as an evaluation criteria of the classification results. Moreover, the results of these classifiers were evaluated using two well-known statistical equations [18]; *the recall* (Eq. 4), and *the precision* (Eq. 5) along with the accuracy measurements.

$$Recall = \frac{no.\ of\ correct\ positives}{total\ no.\ of\ positives} \tag{4}$$

$$Precision = \frac{no.\ of\ correct\ positives}{no.\ of\ correct\ positives + no.\ of\ false\ positives} \tag{5}$$

The results of the second scenario are summarized in Tables 3, 4 and 5. These results reported that the redaction ratio positively affected the classification time

Table 3 Results of Tilapia detection based on k-NN classifier

Exp.	Precision (%)		Recall (%)		Accuracy (%)		Classification time (s)	
No.	No BA	With BA	No BA	With BA	No BA	With_BA	No BA	With BA
1	50	88.9	22	88.9	30.7	88.9	1.041248	0.94
2	66.5	80	22	44	33	57	1.127216	0.91
3	75	100	33	44	46	61.5	1.06	0.98
4	75	100	66.5	55.5	70.5	71	1.10	1.12
5	83	100	55.5	44	66.5	61.5	1.11	0.98
6	100	85	55.5	66.5	71	75	1.04	1.40
7	83	85	55.5	66.5	66.5	75	1.03	0.43
8	100	100	22	77.8	36	87.5	3.68	1.06
9	83	100	55.5	77.8	66.5	87.5	1.11	0.99
10	100	100	33	33	50	50	1.41	0.96
11	66.5	100	22	55.5	33	71.4	1.023	0.97

Table 4 Results of Tilapia detection based on ANN classifier

Exp.	Precision (%)		Recall (%)		Accuracy (%)		Classification time (s)	
No.	No BA	With BA	No BA	With BA	No BA	With_BA	No BA	With BA
1	58.3	66.5	77.8	88.9	66.5	76	79.09	103.18
2	54.5	66.5	66.5	88.9	60	76	105.95	72.15
3	57	63.6	88.9	77.8	69.5	70	58.83	82.21
4	53.8	55.5	77.8	55.5	63.6	55.5	88.63	97.26
5	71.4	85.7	55.5	66.5	62.5	75	87.32	78.58
6	71.4	77.8	55.5	77.8	62.5	77.8	78.72	40.77
7	71.4	60	55.5	66.5	62.5	63	97.01	70.36
8	55.5	85.7	55.5	66.5	55.5	75	84.05	134.38
9	63.6	50	77	55.5	70	52.6	87.22	75.34
10	85.7	80	66.5	88.9	75	84	71.76	77.39
11	60	60	66.5	33	63	43	108.64	103.03

(taken short time) while at the same time achieving high accuracy, reaching 100 %
in some experiments.

Since the BA is one of the stochastic-based global optimization algorithm, some
results may deviated (noisy) based on the quality of the selected features as it is
appear within the outputs of some experiments. Despite this fact, the BA reduction
showed superior classification precision, recall, and accuracy with the k-NN classifier. For example experiment number 9 where the accuracy increase from 66.6 to
87.5 % with a feature reduction ratio near to 61 % along with a decrease in classification time reached to 11 %. So, it can be claimed that the reduction of features
gained by BA improved the accuracy rate of most of the experiments conducted by

Table 5 Results of Tilapia detection based on SVM classifier

Exp.	Precision (%)		Recall (%)		Accuracy (%)		Classification time (s)	
No.	No BA	With BA	No BA	With BA	No BA	With BA	No BA	With BA
1	66.5	63.6	66.5	77.8	66.5	70	1.29	1.13
2	60	63.6	66.5	77.78	63	70	1.59	1.19
3	66.5	57	88.9	88.9	76	69.5	1.37	1.24
4	77.8	62.5	77.8	55.5	77.8	63	1.43	1.28
5	85.7	66.5	66.5	66.5	75	66.5	1.30	1.37
6	75	87.5	66.5	77.8	70.5	84	1.26	2.51
7	55.5	66.5	55.5	66.5	55.5	66.5	1.29	1.24
8	70	80	77.8	88.9	73.7	82	1.65	1.55
9	100	70	44	77.8	78	73.7	1.69	2.77
10	60	87.5	66.5	77.8	63	84	1.44	1.43
11	85.7	85.7	66.5	66.5	75	75	1.23	1.30

each of the used classifiers. The reason behind this is that the classifier performance improved as the reduction phase removed the noisy and duplicate features.

Unlike the results obtained by [11], the k-NN in our proposed system showed its superiority over the other tested classifiers in terms of accuracy and time reduction. Its accuracy reached near to 90 % in some experiments. Also, one of the findings of this paper is that the ANN classifier required a higher classification times than the others, this time is usually consumed through the establishment procedure of the neural network structure. Another important finding is that both the ANN and SVM reached quickly to a hyperplane decision for some experiments.

5 Conclusions and Future Directions

The current paper introduced an approach for fish detection and identification. The approach made use of the BAT algorithm to reduce the number of features extracted using the SURF algorithm. These features were used to classify the Nile Tilapia species (the used data set). The experimental results showed when the BAT algorithm has improved the classification rate (K-NN, ANN, and SVM were used) while at the same time classification time was minimized. It was proved that the k-NN is the best one among (ANN, and SVM) in terms of classification time and accuracy. The other classifiers (ANN, and SVM) may need more modifications for more enhancements for the detection accuracy and lowering the classification time. Further researches may provide these required enhancements.

Acknowledgments This paper has been elaborated in the framework of the project New creative teams in priorities of scientific research, reg. no. CZ.1.07/2.3.00/30.0055, supported by Operational Programme Education for Competitiveness and co-financed by the European Social Fund and the state budget of the Czech Republic and supported by the IT4Innovations Centre of Excellence project (CZ.1.05/1.1.00/02.0070), funded by the European Regional Development Fund and the national budget of the Czech Republic via the Research and Development for Innovations Operational Programme. This work was partially supported by the IPROCOM Marie Curie initial training network, funded through the People Programme (Marie Curie Actions) of the European Union's Seventh Framework Programme FP7/2007-2013/ under REA grant agreement No. 316555. This fund only apply for one author (Hossam M. Zawbaa). Also, we wish to acknowledge the efforts of Rehab Adly Shehabeldin who supports in the data set collection process.

References

1. Wolff, L.M., Sabah, B.-H.: Imaging sonar-based fish detection in shallow waters. In: Oceans-St. John's, IEEE, pp. 1–6 (2014)
2. Tharwat, A., Gaber, T., Hassanien, A.E.: Cattle identification based on muzzle images using gabor features and SVM classifier. In: Advanced Machine Learning Technologies and Applications, pp. 236–247. Springer International Publishing, Heidelberg (2014)
3. Tharwat, A., Gaber, T., Hassanien, A.Ella., Shahin, M.K., Refaat, B.: SIFT-based arabic sign language recognition system. In: Afro-European Conference for Industrial Advancement, pp. 359–370. Springer International Publishing, Heidelberg (2015)
4. Gaber, T., Tharwat, A., Snasel, V., Hassanien, A.E.: Plant identification: two dimensional-based vesus one dimensional-based feature extraction methods. In: 10th International Conference on Soft Computing Models in Industrial and Environmental Applications, pp. 375–385. Springer International Publishing (2015)
5. Semary, N.A., Tharwat, A., Elhariri, E., Hassanien, A.E., Fruit-based tomato grading system using features fusion and support vector machine. In: Intelligent Systems' 2014, pp. 401–410. Springer International Publishing (2014)
6. Huang, P.X., Boom, B.J., Fisher, R.B.: Hierarchical classification with reject option for live fish recognition. Mach. Vis. Appl. **26**(1), 89–102 (2014)
7. Ravanbakhsh, M., Shortis, M.R., Shafait, F., Mian, A., Harvey, E.S., Seager, J.W.: Automated fish detection in underwater images using shapebased level sets. Photogram. Rec. **30**(149), 46–62 (2015)
8. Ravanbakhsh, M.M., Shortis, F., Shafait, Ajmal, M., Euan, H.,J. Seager.: An application of shape-based level sets to fish detection in underwater images. In: Geospatial Science Research 3 Symposium (GSR_3), Rheinisch-Westfaelische Technische Hochschule Aachen Lehrstuhl Informatik V, vol. 1307, pp. 1–9 (2014)
9. Shortis, M.R., Mehdi, R., Faisal, S., Euan, S.H., Ajmal, M., James, W.S, Philip, F.C., Danelle, E.C., Duane, R.E.: A review of techniques for the identification and measurement of fish in underwater stereo-video image sequences. In: SPIE Optical Metrology, International Society for Optics and Photonics, pp. 87910G–87910G (2013)
10. Mehrnejad, M., Alexandra, B.A., David, C., Maia, H.: Detection of stationary animals in deep-sea video. Oceans-San Diego 1–5 (2013)
11. Fouad, M.M.M., Hossam, M., Zawbaa, N.El.-B., Aboul, E.H.: Automatic nile tilapia fish classification approach using machine learning techniques. In: Hybrid Intelligent Systems (HIS), 2013 13th International Conference on, IEEE, pp. 173–178 (2013)
12. Yang, X.-S.: Bat algorithm for multi-objective optimisation. Int. J. Bio-Inspired Comput. **3**(5), 267–274 (2011)
13. Yang, X.-S.: A new metaheuristic bat-inspired algorithm. In: Nature inspired cooperative strategies for optimization (NICSO 2010), pp. 65–74. Springer, Berlin (2010)

14. Talatahari, S., Kaveh, A.: Improved bat algorithm for optimum design of large-scale truss structures. Int. J. Optim. Civ. Eng5 **2**, 241–254 (2015)
15. Aguirre, P.E.R., Pedro, F.P., de Guadalupe Cota Ortiz, M.: Multi-objective optimization using bat algorithm to solve multiprocessor scheduling and workload allocation problem. Comput. Sci. 2 **2**, 41–51 (2015)
16. Hasancebi, O., Carbas, S.: Bat inspired algorithm for discrete size optimization of steel frames. Adv. Eng. Softw. **67**, 173–185 (2014)
17. Zhu, L., Ying, W., Bo, Z., Xiaozheng. Z.: A fast image stitching algorithm based on improved SURF. In: Tenth International Conference on Computational Intelligence and Security (CIS), pp. 171–175. IEEE (2014)
18. Tao, Y., Skubic, M., Han, T.Y., Xia, Y., Chi, X.: Performance evaluation of SIFT-Based descriptors for object recognition. In: Proceedings of the International Multiconference of Engineers and Computer Scientists (IMECS), vol. II. Hong Kong (2010)

Part IV
Hybrid Intelligent Systems

Hybrid Differential Evolution and Simulated Annealing Algorithm for Minimizing Molecular Potential Energy Function

Ahmed Fouad Ali, Nashwa Nageh Ahmed,
Nagwa Abd el Moneam Sherif and Samira Mersal

Abstract In this paper, we present a new hybrid differential evolution algorithm with simulated annealing algorithm to minimize a molecular potential energy function. The proposed algorithm is called Hybrid Differential Evolution and Simulated Annealing Algorithm (HDESA). The problem of minimizing the molecular potential energy function is very difficult, since the number of local minima grows exponentially with the molecular size. The proposed HDESA is tested on a simplified model of a molecular potential energy function with up to 100° of freedom and it is compared against 9 algorithms. The experimental results show that the proposed algorithm is a promising algorithm and can obtain the global or near global minimum of the molecular potential energy function in reasonable time.

Keywords Differential evolution algorithm · Simulated annealing algorithm · Globe optimization problems · Potential energy function

1 Introduction

The steady state or the most stable state of a molecule can be formulated as a global optimization problem. The minimization of the potential energy function is a very hard problem and in most cases it is a non-convex problem, which has many local minimizers. Many researchers apply their works to solve this problems see for example [1, 3, 6, 8–10, 12, 14]. In this paper, a new hybrid differential evolution algo-

A.F. Ali (✉)
Faculty of Computers and Informatics, Department of Computer Science,
Suez Canal University, Ismailia, Egypt
e-mail: ahmed_fouad@ci.suez.edu.eg

A.F. Ali
Member of Scientific Research Group in Egypt, Cairo, Egypt

N.N. Ahmed · N.A. el Moneam Sherif · S. Mersal
Faculty of Science, Department of Mathematic, Suez Canal University,
Ismailia, Egypt

© Springer International Publishing Switzerland 2016
T. Gaber et al. (eds.), *The 1st International Conference on Advanced Intelligent System and Informatics (AISI2015), November 28–30, 2015, Beni Suef, Egypt,*
Advances in Intelligent Systems and Computing 407, DOI 10.1007/978-3-319-26690-9_26

287

rithm and simulated annealing algorithm is presented in order to minimize the potential energy function with up to 100° of freedom. The proposed algorithm is called Hybrid Differential Evolution and Simulated Annealing Algorithm (HDESA). The proposed algorithm has a powerful capability of performing wide exploration and deep exploitation processes. These two processes have been invoked in HDESA through two strategies as follows. The first strategy is a dimension reduction by dividing the population into small partitions. The differential evolution (DE) operators are applied in each partition. The second strategy is applying the simulated annealing on the best obtained solution from the DE algorithm in order to avoid trapping in local minima. The simulated annealing algorithm can increase the diversity of the search. Invoking these two strategies in HDESA can improve the performance of the proposed algorithm and obtains optimal or near optimal solution in a reasonable time. The rest of the paper is organized as follows. Section 2 describes the mathematical form of the molecular potential energy function. Section 3 gives the details of the proposed HDESA algorithm. The numerical experiments are presented in Sect. 4. The conclusion of this paper is summarized in Sect. 5.

2 The Problem of Minimizing Molecular Potential Energy Function

The potential energy of a molecule is derived from molecular mechanics, which describes molecular interactions based on the principles of Newtonian physics. An empirically derived set of potential energy contributions is used for approximating these molecular interactions. The molecular model considered here consists of a chain of m atoms, where $m = 4$ centered at $x_1, \ldots, x_m$, in a 3-dimensional space as shown in Fig. 1. For every pair of consecutive atoms x_i and x_{i+1}, let $r_{i,i+1}$ be the bond length which is the Euclidean distance between them. For every three consecutive atoms x_i, x_{i+1}, x_{i+2}, let $\theta_{i,i+2}$ be the bond angle corresponding to the relative position of the third atom with respect to the line containing the previous two. Likewise, for every four consecutive atoms $x_i, x_{i+1}, x_{i+2}, x_{i+3}$, let $\omega_{i,i+3}$ be the angle, called the torsion angle, between the normal through the planes determined by the atoms x_i, x_{i+1}, x_{i+2} and $x_{i+1}, x_{i+2}, x_{i+3}$. The force field potentials corresponding to bond lengths, bond angles, and torsion angles are defined respectively [6] as follow.

Fig. 1 Coordinate set of atomic chain

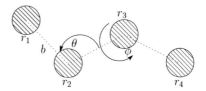

$$E_1 = \sum_{(i,j)\in M_1} c_{ij}^1 (r_{ij} - r_{ij}^0)^2, \tag{1}$$

$$E_2 = \sum_{(i,j)\in M_2} c_{ij}^2 (\theta_{ij} - \theta_{ij}^0)^2, \tag{2}$$

$$E_3 = \sum_{(i,j)\in M_3} c_{ij}^3 (1 + \cos(3\omega_{ij} - \omega_{ij}^0)), \tag{3}$$

where c_{ij}^1 is the bond stretching force constant, c_{ij}^2 is the angle bending force constant, and c_{ij}^3 is the torsion force constant. The constant r_{ij}^0 and θ_{ij}^0 represent the "preferred" bond length and bond angle, respectively, and the constant ω_{ij}^0 is the phase angle that defines the position of the minima. The set of pairs of atoms separated by k covalent bond is denoted by M_k for $k = 1, 2, 3$.

In addition to the above, there is a potential E_4 which characterizes the 2-body interaction between every pair of atoms separated by more than two covalent bonds along the chain. We use the following function to represent E_4:

$$E_4 = \sum_{(i,j)\in M_3} \left(\frac{(-1)^i}{r_{ij}} \right), \tag{4}$$

where r_{ij} is the Euclidean distance between atoms x_i and x_j.

The general problem is the minimization of the total molecular potential energy function, $E_1 + E_2 + E_3 + E_4$, leading to the optimal spatial positions of the atoms. To reduce the number of parameters involved in the potentials above, we simplify the problem considering a chain of carbon atoms as follow.

$$E = \sum_{(i,j)\in M_3} (1 + \cos(3\omega_{ij})) + \sum_{(i,j)\in M_3} \left(\frac{(-1)^i}{r_{ij}} \right), \tag{5}$$

By calculating the value of r_{ij}, the total potential energy can be calculated as follow.

$$E = \sum_{(i,j)\in M_3} \left(1 + \cos(3\omega_{ij}) + \frac{(-1)^i}{\sqrt{10.60099896 - 4.141720682(\cos(\omega_{ij}))}} \right), \tag{6}$$

Finally, the function $f(x)$ can be defined as follow.

$$f(x) = \sum_{i=1}^{n} \left(1 + \cos(3x_i) + \frac{(-1)^i}{\sqrt{10.60099896 - 4.141720682(\cos(x_i))}} \right), \tag{7}$$

and $0 \leq x_i \leq 5, \quad i = 1, \dots, n.$

3 Hybrid Differential Evolution Algorithm for Minimizing Molecular Potential Energy Function

In this section, we highlight the main components of the proposed algorithm. In the HDESA algorithm, we tried to combine the DE algorithm [13] and the SA algorithm [11]. The main steps of the HDESA algorithm is summarized as follow.

3.1 The Proposed HDESA Algorithm

In this subsection, the structure of the proposed algorithm and its flowchart are presented in Algorithm 1 and Fig. 2 as follow. The main steps of the proposed algorithm are described as follow.

- **Step 1** The algorithm starts by setting the initial counter G, the amplification factor F and crossover factor CR. **Lines 1–2**.
- **Step 2** The initial population is generated randomly and each solution in the population is evaluated by calculating its function value. **Lines 3–4**

Algorithm 1 The proposed HDESA algorithm

1: Set the generation counter $G := 0$.
2: Set the initial value of F and CR.
3: Generate an initial population P^0 randomly.
4: Evaluate the fitness function of all individuals in P^0.
5: **repeat**
6:      Set $G = G + 1$. {**Generation counter increasing**}.
7:      The population $P^{(G)}$ is partitioning into small partitions with size $P \times v$
8:      **for** $i = 1; i \leq part_{no}; i + +$ **do**
9:          Select random indexes r_1, r_2, r_3, where $r_1 \neq r_2 \neq r_3 \neq i$.
10:         $v_i^{(G)} = x_{r_1}^{(G)} + F \times (x_{r_2}^{(G)} - x_{r_3}^{(G)})$. {**Mutation operator**}.
11:         $j = rand(1, D)$
12:         **for** $(k = 0; k < D; k + +)$ **do**
13:             **if** $(rand(0, 1) \leq CR$ or $k = j$ **then**
14:                 $u_{ik}^{(G)} = v_{ik}^{(G)}$ {**Crossover operator**}
15:             **else**
16:                 $u_{ik}^{(G)} = x_{ik}^{(G)}$
17:             **end if**
18:         **end for**
19:         **if** $(f(u_i^{(G)}) \leq f(x_i^{(G)}))$ **then**
20:             $x_i^{(G+1)} = u_i^{(G)}$ {**Greedy selection**}.
21:         **else**
22:             $x_i^{(G+1)} = x_i^{(G)}$
23:         **end if**
24:     **end for**
25:     Apply the standard simulated annealing on the overall best obtain solution from DE algorithm.
26: **until** Termination criteria satisfied.
27: Present the best solution.

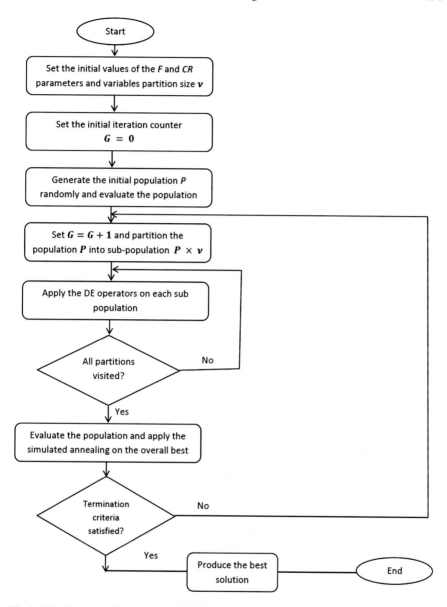

Fig. 2 The flowchart of the proposed HDESA algorithm

- **Step 3** The following steps are repeated until termination criteria satisfied
- **Step 3.1** The population is partitioning into small partitions with size $P \times v$ **Line 7**
- **Step 3.1.1** The mutation and crossover operators are applied on the each partition. **Lines 8–18**.
- **Step 3.1.2** The best solutions are selected by applying the greedy selection operator and the overall solution is assigned. **Lines 19–24**.

- **Step 3.2** The simulated annealing algorithm is applied on the overall solution in order to ovoid trapping in local minima. **Line 25**
- **Step 4** The best solution is presented. **Line 27**

4 Numerical Experiments

The general performance of the proposed algorithm and the efficiency of it are investigated in the following subsections. HDESA algorithm was programmed in MAT-LAB. HDESA is compared against 9 benchmark algorithm. Before discussing the results, we summarize the parameter setting of the HDESA algorithm as follow.

4.1 Parameter Setting

HDESA parameters are summarized with their assigned values as shown in Table 1. These values are based on the common setting in the literature or determined through our preliminary numerical experiments.

4.2 The General Performance of the Proposed HDESA with the Minimization of Molecular Potential Energy Function Problem

The general performance of the proposed HDESA algorithm is shown in Fig. 3, by plotting the function values (mean errors) versus the number of iterations (function evaluations) for molecules size 20, 40, 60, 80 and 100. Figure 3 shows that the function values are rapidly decreases while the number of iterations are slightly increases (few number of iterations).

Table 1 Parameter setting

Parameters	Definitions	Values
P	population size	25
v	number of variables in each partition	5
F	amplification factor	0.5
CR	crossover rate	0.9
T_{max}	initial temperature	d
T_{min}	final temperature	0
β	temperature reduction value	1
max_{itr}	maximum iteration number	20d

Fig. 3 The general performance of the proposed HDESA with the minimization of molecular potential energy function problem

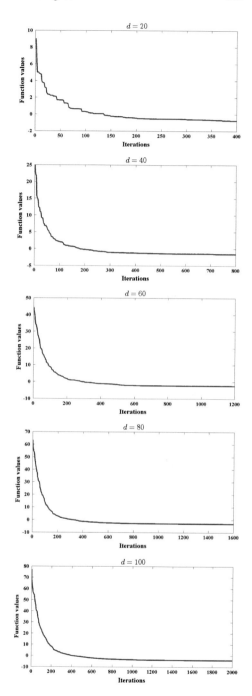

4.3 Comparison Between the Proposed HDESA Algorithm and Other Algorithms

In order to test the efficiency of the proposed HDESA algorithm, we compare it against two sets of benchmark algorithms. The first set of algorithms contains four different real coded genetic algorithms (RCGAs), WX-PM, WX-LLM, LX-LLM [7] and LX-PM [4]. These four algorithms are based on two real coded crossover operators Weibull crossover WX and LX [5] and two mutation operators LLM and PM [4]. The second set of algorithms contains 5 benchmark algorithms, variable neighborhood search based algorithm (VNS), (VNS-123), (VNS-3) algorithms [6]. In [6], four variable neighborhood search algorithms, VNS-1, VNS-2, VNS-3, VNS-123 were developed. They differ in the choice of random distribution used in the shaking step for minimization of a continuous function subject to box constraints. Genetic algorithm (GA) [3], (rHYB) method [3] denotes the staged hybrid GA with a reduced simplex and a fixed limit for simplex iterations and (qPSO) method [2] is a quadratic hybrid particle swarm optimization (PSO) in which quadratic approximation operator is hybridized with PSO. The termination criterion of HDESA is to run $20d$ generations, all results have been averaged over 20 independent runs. The results of the comparative methods are taken from their original papers and reported in Tables 2 and 3. The results in Tables 2 and 3 represent the problem size d and the mean number of function evaluation. The best mean values of function evaluations are marked in bold face. The—sing in Table 3 means that the results of the corresponding method are not reported in the original paper. We can conclude from Tables 2 and 3 that the proposed HDESA is a promising algorithm and it is faster than other algorithms.

Table 2 Mean number of function evaluations of HDESA and other algorithms with 20–100 dimensions

d	WX-PM	LX-PM	WX-LLM	LX-LLM	HDESA
20	15,574	23,257	28,969	14,586	**10,350**
40	59,999	71,336	89,478	39,366	**21,135**
60	175,865	280,131	225,008	105,892	**35,395**
80	302,011	326,287	372,836	237,621	**48,795**
100	369,376	379,998	443,786	320,146	**56,500**

Table 3 Mean number of function evaluations of HDESA and other methods with 20–100 dimensions

d	VNS-123	VNS-3	GA	qPSO	rHYB	HDESA
20	23,381	9887	36,626	–	35,836	**10,350**
40	57,681	25,723	133,581	–	129,611	**21,135**
60	142,882	39,315	263,266	–	249,963	**35,395**
80	180,999	74,328	413,948	–	387,787	**48,795**
100	254,899	79,263	588,827	–	554,026	**56,500**

5 Conclusion

Differential evolution algorithm (DE) has a promising performance when it applied to solve global optimization problems, however it suffers from the premature convergence when all solutions trapped in local minima and the algorithm becomes unable to escape from stagnation. In this paper, we tried to minimize the potential energy function and avoid trapping in local minima by proposing a new hybrid differential evolution algorithm and a simulated annealing algorithm. The proposed algorithm is called Hybrid Differential Evolution and Simulated Annealing Algorithm (HDESA). The simulated annealing algorithm can help the proposed algorithm to escape from trapping in local minima and increases the diversity of the search. The proposed HDESA algorithm is applied to minimize the molecular potential energy function with up to $100°$ and compared against 9 different algorithms. The numerical results show that the proposed HDESA algorithm is a promising algorithm and more precise than compared algorithms.

References

1. Ali, A.F., Hassanien, A.E.: Minimizing molecular potential energy function using genetic Nelder-Mead algorithm. In: 8th International Conference on Computer Engineering & Systems (ICCES), pp. 177–183 (2013)
2. Bansal, J.C., Deep, S.K., Katiyar, V.K.: Minimization of molecular potential energy function using particle swarm optimization. Int. J. Appl. Math. Mech. **6**(9), 1–9 (2010)
3. Barbosa, H.J.C., Lavor, C., Raupp, F.M.: A GA-simplex hybrid algorithm for global minimization of molecular potential energy function. Ann. Oper. Res. **138**, 189–202 (2005)
4. Deep, K., Thakur, M.: A new mutation operator for real coded genetic algorithms. Appl. Math. Comput. **193**(1), 211–230 (2007)
5. Deep, K., Thakur, M.: A new crossover operator for real coded genetic algorithms. Appl. Math. Comput. **188**(1), 895–912 (2007)
6. Dražić, M., Lavor, C., Maculan, N., Mladenović, N.: A continuous variable neighborhood search heuristic for finding the three-dimensional structure of a molecule. Eur. J. Oper. Res. **185**, 1265–1273 (2008)
7. Grünewald, S., Laranjeira, F., Walraven, J., Aguado, A., Molins, C.: Improved tensile performance with fiber reinforced self-compacting concrete. In: Parra-Montesinos, G.J., Reinhardt, H.W., Naaman, A.E. (eds.) HPFRCC 6. RILEM Bookseries, vol. 2, pp. 51–58. Springer, Heidelberg (2012)
8. Hedar, A., Ali, A.F., Hassan, T.: Genetic algorithem and tabu search based methods for molecular 3D-structure prediction. Int. J. Numer. Algebr. Control Optim. (NACO) **1**(1), 187–205 (2011)
9. Hedar, A., Ali, A.F., Hassan, T.: Finding the 3D-structure of a molecule using genetic algorithm and tabu search methods, in: Proceeding of the 10th International Conference on Intelligent Systems Design and Applications (ISDA2010), Cairo, Egypt (2010)
10. Kovačević-Vujčić, V., čangalović, M., Dražić, M., Mladenović, N.: VNS-based heuristics for continuous global optimization, In: Hoai An, L.T., Tao, P.D. (Eds), Modelling. Computation and Optimization in Information Systems and Management Sciences, Hermes Science Publishing Ltd, pp. 215–222 (2004)
11. Kirkpatrick, S., Gelatt, C.D., Vecchi, M.P.: Optimization by simulated annealing. Science **220**(4598), 671–680 (1983)

12. Pardalos, P.M., Shalloway, D., Xue, G.L.: Optimization methods for computing global minima of nonconvex potential energy function. J. Glob. Optim. **4**, 117–133 (1994)
13. Storn, R., Price, K.: Differential evolution-a simple and efficient heuristic for global optimization over continuous spaces. J. Glob. Optim. **11**, 341–359 (1997)
14. Wales, D.J., Scheraga, H.A.: Global optimization of clusters, crystals and biomolecules. Science **285**, 1368–1372 (1999)

A Hybrid Classification Model for EMG Signals Using Grey Wolf Optimizer and SVMs

Esraa Elhariri, Nashwa El-Bendary and Aboul Ella Hassanien

Abstract Electromyography (EMG) signal is an electrical indicator for neuromuscular activation. It provides direct access to physiological processes enabling the muscle to generate force and produce movement in order to accomplish countless functions. As a successful classification of the EMG signal is basically dependent on the selection of the best parameters carefully, this paper proposes a hybrid optimized classification model for EMG signals classification. The proposed system implements grey wolf optimizer (GWO) combined with support vector machines (SVMs) classification algorithm in order to improve the classification accuracy via selecting the optimal settings of SVMs parameters. The proposed approach consists of three phases; namely pre-processing, feature extraction, and GWO-SVMs classification phases. The obtained experimental results obviously indicate that significant enhancements in terms of classification accuracy have been achieved by the proposed GWO-SVMs classification system. It has outperformed the typical SVMs classification algorithm via achieving an accuracy of over 90 % using the radial basis function (RBF) kernel function.

Keywords Grey wolf optimization (GWO) · Features extraction · Electromyography (EMG) signal · Support vector machines (SVMS)

E. Elhariri
Faculty of Computers and Information, Fayoum University, Fayoum, Egypt

N. El-Bendary (✉)
Arab Academy for Science, Technology, and Maritime Transport, Cairo, Egypt
e-mail: nashwa.elbendary@ieee.org

A.E. Hassanien
Faculty of Computers and Information, Cairo University, Cairo, Egypt

A.E. Hassanien
Faculty of Computers and Information, Beni-Suef University, Beni-suef, Egypt

E. Elhariri · N. El-Bendary · A.E. Hassanien
Scientific Research Group in Egypt (SRGE), Giza, Egypt
URL: http://www.egyptscience.net

© Springer International Publishing Switzerland 2016
T. Gaber et al. (eds.), *The 1st International Conference on Advanced Intelligent System and Informatics (AISI2015), November 28–30, 2015, Beni Suef, Egypt,*
Advances in Intelligent Systems and Computing 407, DOI 10.1007/978-3-319-26690-9_27

297

1 Introduction

Electromyography (EMG) signal, which is a measurement of the electrical activity in muscles as a result of contraction, is one of the electrophysiological signals that has been expansively studied. EMG signals have been applied both clinically as well as in the engineering domain for assistive technologies and rehabilitation engineering. The EMG signal consists of discrete waveforms; namely, motor unit action potentials (MUPs) that result from the recurring discharges of motor units (MUs), which are groups of muscle fibers [1, 2]. By placing certain electrodes on the skin, the summation of action potentials from the muscle fibers can be recorded to represent the EMG signal. Moreover, for recording the EMG signals in varying degrees of voluntary muscle activity, a typical needle EMG treatment is performed using a concentric needle electrode.

Feature extraction and function classification represent a significant challenge in processing and analyzing the EMG signals. Many machine learning (ML) techniques are used for the classification problems. The success of any classification system is dependent on selecting its parameters carefully. This research focuses on using support vector machines (SVMs) for solving the problem of EMG signals classification. SVMs proved its efficiency as a classification method. But, SVMs face some challenges, when adopted in real practical applications. Setting the optimal parameters of SVMs is one of these challenges. This research presents a method for selecting the best parameters for SVMs via applying an optimization algorithm [3–5]. Selecting these parameters correctly guarantees to obtain the best classification accuracy [4]. SVMs have two types of parameters (penalty constant C parameter and kernel functions parameters), and the values of these parameters affect the performance of SVMs [3].

This paper presents a hybrid optimized classification system for EMG signals classification. The proposed system implements grey wolf optimizer (GWO) combined with support vector machines (SVMs) classification algorithm in order to achieve enhanced EMG classification accuracy via selecting the optimal settings of SVMs parameters to be used later for assistive technologies and rehabilitation engineering. Grey Wolf Optimization (GWO) algorithm is a new meta-heuristic method, which is inspired by grey wolves, to mimic the hierarchy of leadership and grey wolves hunting mechanism in nature. This research chooses GWO algorithm. The reason of this choice is that comparing with some of the most well-known evolutionary algorithms such as Particle Swarm Optimization (PSO), Genetic Algorithm (GA), Ant Colony Optimization (ACO), Evolution Strategy (ES), and Population-based Incremental Learning (PBIL), The results showed that the GWO algorithm has the ability to provide very competitive results in terms of improved local optima avoidance. The proposed approach consists of three phases; namely pre-processing, feature extraction, and GWO-SVMs classification phases. The obtained experimental results obviously indicate significant enhancements in terms of classification accuracy achieved by the proposed GWO-SVMs classification system compared to classification accuracy achieved by the typical SVMs classification algorithm.

The rest of this paper is organized as follows. Section 2 presents research work considering processing and analyzing the EMG signals. Section 3 introduces the proposed optimized classification system and describes its different phases; namely pre-processing, feature extraction, and classification phases. Section 4 presents the tested EMG dataset and discusses the obtained experimental results. Finally, Sect. 5 presents conclusions and discusses future work.

2 Related Work

Generally, there is a number of current researches that tackle the problem of EMG signals classification problem. Authors in [6] presented a new rehabilitation robotics control design based on multilayer perceptron neural network and the analysis of real-time EMG. The proposed system consists of three main phases namely; signal preprocessing, feature extraction, and classification phases. For signal preprocessing, signal rectification, removing DC offset (mean value of waveform) from EMG signal and creation of envelope curve were applied. Then for feature extraction phase, time domain, frequency domain, and dynamic system features were calculated to be used as a feature vector. Finally, a multilayer perceptron neural network is used as a binary classifier for human actions. Experimental results showed that the proposed system achieved an accuracy of 90 % for clapping and handshaking, 91 % for kneeing and pulling, 75 % for hammering and headering, 98 % for running and hugging and 88 % for elbowing and slapping. In [7], authors proposed a new method based on back propagation neural network classifier for classification of myopathy patient's and healthy subjects using EMG signal. The proposed method has four basic steps, which are preprocessing, singular value decomposition, feature extraction, and classification. In this research, authors used The extracted singular values as a feature vector. Finally, they applied back propagation neural network as a classifier. Experimental results showed that the proposed method achieved an accuracy of 96.75 %.

Moreover, in [8], authors present a comparison of different algorithms for EMG signal classification in order to find an effective machine learning algorithm for EMG signals classification. The presented framework for classification used multi-scale principal component analysis (MSPCA) for de-noising, discrete wavelet transform (DWT) for feature extraction, and different machine learning algorithms for classification. CART, C4.5 and random forests (RF) decision tree classification algorithms have been applied to classify EMG signal into myopathic, ALS (Amyotrophic lateral sclerosis) or normal. Using different performance measures, the obtained results showed that the best accuracy of 96.67 % is achieved using a combination of DWT and RF algorithms.

In this paper, a hybrid optimized classification system for EMG signals is presented. The proposed system utilizes grey wolf optimization algorithm along with SVMs classification algorithm to improve classification accuracy via selecting the best parameters of SVMs.

3 The Proposed Hybrid GWO-SVMs
Classification Approach

The proposed classification approach consists of three phases; namely, pre-processing, feature extraction, and classification phases. Figure 1 describes the general
structure of the proposed approach.

3.1 Pre-processing Phase

During pre-processing phase, the proposed approach applies the following steps:

1. Apply EMG signal rectification.
2. Remove any DC offset of the signal.
3. Create an envelope curve of EMG signal via applying a 4th degree low pass but-
   terworth filter at 10 Hz.
4. Segment each time series to 15 windows.

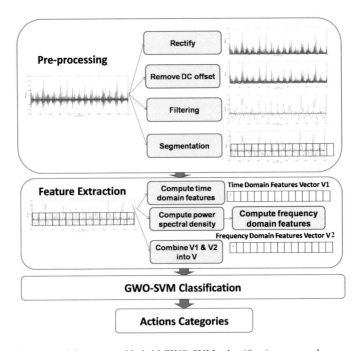

Fig. 1 Architecture of the proposed hybrid GWO-SVMs classification approach

3.2 Feature Extraction Phase

Based on the literature, features in analysis of the EMG signal can be generally divided into three main groups; namely *time domain*, *frequency domain*, and *time-frequency or time-scale* features [9–11]. Time-scale features cannot be directly used by themselves and features extracted from time-frequency or time-scale methods should be reduced their high dimensions before being sent to a classifier. Hence, only two feature sets, *time domain features* and *frequency domain features*, with twenty-three features, have been considered for experiments conducted in this research [6, 12].

Time Domain Features: The time domain features considered in this paper are integrated EMG (IEMG), mean absolute value (MAV), modified mean absolute value type 1 (MAV), modified mean absolute value type 2 (MAV), simple square integral (SSI), variance of EMG (VAR), absolute value of the 3rd, 4th, and 5th temporal moment (TM3, TM4 and TM5), root mean square (RMS), V-order, log detector (LOG), waveform length (WL), average amplitude change (AAC) and difference absolute standard deviation value (DASDV) [6, 12].

Frequency Domain Features: The frequency domain features considered in this paper are total power (TTP), mean frequency (MNF), median frequency (MDF), peak frequency (PKF), mean power (MNP), The 1st, 2nd, and 3rd spectral moments (SM1, SM2 and SM3) [9].

During feature extraction phase, the proposed approach implements steps shown in Algorithm 1.

```
1: for Each action (Normal/Aggressive) do
2:    Compute time domain features for each channel of the eight channels.
3: end for
4: for Each action (Normal/Aggressive) do
5:    Compute the Power spectral density (PSD) and frequency for each channel of the eight
      channels.
6:    Compute frequency domain features.
7: end for
8: Form a 1D feature vector via combining all features.
```
Algorithm 1: GWO-SVMs feature extraction phase

3.3 GWO-SVMs Classification Phase

Finally, for classification phase, the proposed approach applied a hybrid GWO-SVMs model that employs grey wolf optimization (GWO) combined with support vector machines (SVMs) algorithm to improve the classification accuracy via selecting the optimal SVMs parameters setting for One-against-One multi-class SVMs.

Support Vector Machines (SVMs): One of the most widely used machine learning techniques for classification and regression of high-dimensional datasets with excellent results is support vector machines (SVMs) [13–15]. SVMs can solve any classification problem via attempting to find an optimal separating hyperplane between two classes. Maximizing the margin around the separating hyperplane between a positive and negative classes is the main aim of SVMs [13–15].

Let $S = \{(x_1, y_1), (x_2, y_2), \ldots, (x_n, y_n)\}$ is a training dataset with n samples, where x_i is a n-dimensional feature vector and $y_i \in -1, 1$ is the class label (classes C_1 and C_2). Geometrically, finding an optimal hyperplane with the maximal margin to separate two classes requires to solve the optimization problem, as shown in Eqs. (1) and (2).

$$maximize \sum_{i=1}^{n} \alpha_i - \frac{1}{2} \sum_{i,j=1}^{n} \alpha_i \alpha_j y_i y_j . K(x_i, x_j) \tag{1}$$

$$Subject-to : \sum_{i=1}^{n} \alpha_i y_i, 0 \le \alpha_i \le C \tag{2}$$

where, α_i is the weight assigned to each training sample x_i. If $\alpha_i > 0$, x_i is called a support vector. C is a regulation parameter used to trade-off the training accuracy and the model complexity. K is a kernel function, which is used to measure the similarity between two samples.

Grey Wolf Optimization Grey wolf optimizer (GWO) is a new meta-heuristic technique. It can be applied for solving optimized problems and achieves excellent results [16, 17]. In fact, the GWO mimics the grey wolves' leadership hierarchy and hunting mechanism. To simulate the leadership hierarchy, there are four types of grey wolves which are alpha (α), beta (β), delta (δ) and omega (ω). Those four types can be used for simulating the leadership hierarchy. The hunting (optimization) is guided by three wolves(α, β and δ). The ω wolves follow them [17, 18]. During the hunt process, it is known that grey wolves surround their prey. Mathematically, this is modeled by Eqs. (3), (4) [16, 17]:

$$D = |C.X_p(t) - X(t)| \tag{3}$$

$$X(t+1) = X_p(t) - A.D \tag{4}$$

where t is the current iteration, A and C are coefficient vectors, X_p is the vector of the prey position, and X indicates the vector of the grey wolf position. A and C vectors can be calculated as shown in Eqs. (5), (6).

$$A = 2a.r_1 - a \tag{5}$$

$$C = 2.r_2 \tag{6}$$

where components of a are linearly decreased from 2 to 0, over the course of iterations and r_1, r_2 are random vectors in [0, 1]. To mimic the hunting process of grey wolves, assume that the α (the best candidate solution), β and δ have a superior knowledge about the possible position of prey. So, the best obtained three solutions are saved so far and force other search agents (including ω) to update their positions according to the position of the best search agents. To update the grey wolves positions, Eqs. (7–9) are being applied [17, 18].

$$D_\alpha = |C_1.X_\alpha - X|, D_\beta = |C_2.X_\beta - X|, D_\delta = |C_3.X_\delta - X| \qquad (7)$$

$$X_1 = X_\alpha - A_1.(D_\alpha), X_2 = X_\beta - A_2.(D_\beta), X_3 = X_\delta - A_3.(D_\delta) \qquad (8)$$

$$X(t+1) = \frac{X_1 + X_2 + X_3}{3} \qquad (9)$$

GWO can be summarized by steps shown in Algorithm 2.

1: Initialize the grey wolf population

$$X_i, (i = 1, 2, …, n)$$

2: Initialize a, A, and C
3: Calculate the fitness of each search agent, as: X_α = the best search agent, X_β = the second best search agent, X_δ = the third best search agent
4:
5: **while** $t < Maxnumberofiterations$ **do**
6:    **for** <each agent> **do**
7:       Update the position of the current search agent by equations (7-9).
8:    **end for**
9:    Update a, A, and C
10:   Calculate the fitness of all search agents
11:   Update X_α, X_β, and X_δ
12:   $t = t + 1$
13: **end while**
14: return X_α

Algorithm 2: Grey Wolf Optimization

Details of the Proposed GWO-SVMs Hybrid Approach The input are training dataset feature vectors and their corresponding classes, GWO initialized parameters, whereas the output is the optimal SVMs parameters and the action name of each sample in the testing dataset. SVMs was trained and tested using different kernel functions (multilayer perceptron (MLP), radial basis function (RBF), and polynomial) and a 3 folds cross-validation. Details of the proposed approach is shown at (Fig. 2).

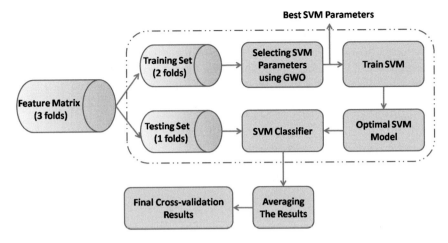

Fig. 2 Proposed GWO-SVMs optimized classification approach

4 Experimental Results

Simulation experiments in this article are done considering the (EMG Physical Action Dataset) that was downloaded from UCI-Machine Learning Repository [19]. This dataset includes 10 normal and 10 aggressive physical actions that measure the human activity. The data have been collected using the Delsys EMG wireless apparatus. In this research, a dataset of only 3 subjects is used. They used a total of 8 electrodes, which corresponds to 8 input time series one for a muscle channel (ch1−8). Also, simulation experiments in this research used a dataset of total 900 (3 subjects * 20 experiments * 15 actions per experiments) instances of actions for both training and testing datasets with 3-fold cross-validation. Training dataset is divided into 20 classes representing the different normal and aggressive actions. Normal actions are bowing, clapping, handshaking, hugging, jumping, running, seating, standing, walking, waving, while aggressive actions are elbowing, frontkicking, hamering, headering, kneeing, pulling, punching, pushing, sidekicking, slapping. The proposed GWO-SVMs approach has been implemented considering the One-against-One multi-class SVMs system to select the best parameters for SVMs penalty cost parameter, which varied between (1 and 1000) and kernel functions parameters. Different kernel functions have been tested. The most popular kernel functions are:

- **Gaussian radial basis function (RBF):** σ parameter varied between (1 and 100).

$$k(x, y) = exp^{(-\frac{||x-y||^2}{2\sigma^2})} \tag{10}$$

- **Polynomial function:** d degree paramete varied between (1 and 10).

$$k(x, y) = (x^T y + 1)^d \tag{11}$$

- **Multi layer perceptron (MLP) function**: α varied between (1 and 100), while c_1 varied between (-50 and -1).

$$tanh(\alpha x^T y + c_1) \tag{12}$$

The accuracy of the proposed system can be calculated using this equation:

$$Accuracy = \frac{Number\ of\ correctly\ classified\ samples}{Total\ number\ of\ testing\ samples} * 100 \tag{13}$$

Figure 3 shows classification accuracies obtained via applying the proposed hybrid GWO-SVMs against SVMs classification approaches, using one-against-one multi-class approach and 3-fold cross-validation.

It is noticed from (Fig. 3) that accuracy achieved by SVMs MLP kernel function without GWO optimization is very low and equals to 48.33 %, while it achieved an accuracy of 64.1 % using GWO-SVMs, which means that the accuracy increased by $\approx$15.77 %. For RBF kernel function, it is noticed from (Fig. 3) that by trying different values for σ parameter, the accuracy change significantly. Accuracy changed from 40.44 to 79.56 % without optimization, while it has increased to 90.78 % using GWO-SVMs proposed approach. Finally, for polynomial kernel function, an accuracy achieved without optimization changes from 65.78 to 78.44 %, while it has increased to 81.44 % using GWO-SVMs.

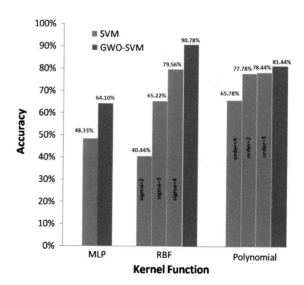

Fig. 3 Results of GWO-SVMs and SVMs kernel functions using one-against-one multi-class approach and 3-fold cross-validation

5 Conclusions and Future Work

This paper proposes a hybrid optimized classification model for EMG signals classification. The proposed system implements grey wolf optimizer (GWO) combined with support vector machines (SVMs) classification algorithm in order to improve the classification accuracy via selecting the optimal settings of SVMs parameters. Obtained experimental results showed that the proposed GWO-SVMs classification system has outperformed the typical SVMs classification algorithm via achieving an accuracy of over 90 % using the radial basis function (RBF) kernel function. Also, experimental results showed that trying different values for the σ parameter of the RBF kernel function positively changed the accuracy in a significant way. That is achieved accuracy changed from 40.44 to 79.56 % without applying GWO optimization algorithm, then increased to 90.78 % using GWO-SVMs proposed approach. For future work, it is planned to enhance the system proposed in this paper in order to be widely considered in many clinical and engineering applications. Performance enhancements will be achieved via considering other bio-inspiring optimization algorithms and more effective features. Features selection techniques will be considered, in order to improve system performance.

References

1. Fuglsang-Frederiksen, A.: The utility of interference pattern analysis. J. Muscle Nerve **23**, 18–36 (2000)
2. Subasi, A.: Classification of EMG signals using combined features and soft computing techniques. J. Appl. Soft Comput. **12**(8), 2188–2198 (2012). Elsevier
3. Garšva, G., Danenas, P.: Particle swarm optimization for linear support vector machines based classifier selection. J. Nonlinear Anal. Model. Control **19**(1), 26–42 (2014)
4. Gaspar, P., Carbonell, J., Oliveira, J.L.: n the parameter optimization of Support Vector Machines for binary classification. J. Integr. Bioinform. **9**(3), 201 (2012)
5. Huang, C.-L., Wang, C.-J.: A GA-based feature selection and parameters optimizationfor support vector machines. J. Expert Syst. Appl. **31**(2), 231–240 (2006). Elsevier
6. Elbagoury, B.M., Schrader, T., Altaf, M.M., Banawan, S.A., Roushdy, M.: Design of neural network for rehabilitation robotics. In: Proceedings of the 2014 International Conference on Circuits, Systems and Control, Interlaken, Switzerland, pp. 89–95. 22–24 Feb 2014
7. Patidar, Mukesh, Jain, Nitin, Parikh, Ashish: Classification of normal and myopathy EMG signals using BP neural network. Int. J. Comput. Appl. **69**(8), 12–16 (2013)
8. Gokgoz, Ercan, Subasi, Abdulhamit: Comparison of decision tree algorithms for EMG signal classification using DWT. J. Biomed. Signal Process. Control **18**, 138–144 (2015)
9. Oskoei, M.A., Hu, H.: Myoelectric control systems—A survey. J. Biomed. Signal Process. Control **2**(4), 275–294 (2007)
10. Oskoei, M.A., Hu, H.: Support vector machine based classification scheme for myoelectric control applied to upper limb. IEEE Trans. Biomed. Eng. **55**(8), 1956–1965 (2008)
11. Zecca, M., Micera, S., Carrozza, M.C., Dario, P.: Control of multifunctional prosthetic hands by processing the electromyographic signal. J. Crit. Rev. Biomed. Eng. **30**(4–6), 459–485 (2002)
12. Phinyomark, A., Phukpattaranont, P., Limsakul, C.: Feature reduction and selection for EMG signal classification. J. Expert Syst.Appl. **39**(8), 7420–7431 (2012)

13. Wu, Q., Zhou, D.-X.: Analysis of support vector machine classification. J. Comput. Anal. Appl. **8**(2), 99–119 (2006). Springer
14. Zawbaa, H.M., El-Bendary, N., Abraham, A, Hassanien, A.E.: SVM-based soccer video summarization system. In: Proceeding of The Third IEEE World Congress on Nature and Biologically Inspired Computing (NaBIC2011), Salamanca, Spain, pp. 7–11. 19–21 Oct 2011
15. Zawbaa, H.M., El-Bendary, N., Hassanien, A.E., Kim, T.-H.: Machine learning-based soccer video summarization system. In: Proceeding of Multimedia, Computer Graphics and BroadcastingFGIT-MulGraB (2), Jeju Island, Korea, vol. 263, pp.19-28, 8–10 Dec 2011. Springer
16. Mirjalili, S., Mirjalili, S.M., Lewis, A.: Grey wolf optimizer. J. Adv. Eng. Softw. **69**, 46–61 (2014). Elsevier
17. El-Gaafary, A.A.M., Mohamed, Y.S., Hemeida, A.M., Mohamed, A.-A.A.: rey wolf optimization for multi input multi output system. Univers. J. Commun. Netw. **1**(1), 1–6 (2015)
18. Emary, E., Zawbaa, H.M., Grosan, C., Hassenian, A.E.: Feature subset selection approach by gray-wolf optimization. Afro-European Conference for Industrial Advancement, vol. 334, pp. 1–14. Springer International Publishing, Switzerland (2015)
19. Machine Learning Repository: EMG Physical Action Data Set. Accessed 18 Aug 2015. https://archive.ics.uci.edu/ml/datasets/EMG

New Fuzzy Decision Tree Model for Text Classification

Ben Abdessalam Wahiba and Ben El Fadhl Ahmed

Abstract In this paper, a supervised automatic text documents classification using the fuzzy decision trees technique is proposed. Whatever the algorithm used in the fuzzy decision trees, there must be a criterion for the choice of discriminating attribute at the nodes to partition. For fuzzy decision trees two heuristics are usually used to select the discriminating attribute at the node to partition. In the field of text documents classification there is a heuristic that has not yet been tested. This paper tested this heuristic.

1 Introduction

Supervised classification is performed to assign automatically and independently one or more documents to one or more predefined categories [1]. There are various techniques for supervised classification, among the most known: the Bayesian networks, support vector machines, k-nearest neighbors, decision trees, etc. Among these techniques, only the decision trees easily generate a set of rules justifying the generated classification decisions. Other techniques generate in a more difficult and complicated way such set of rules.

Despite the wide spread of decision trees, this technique suffers from a problem that may affect its effectiveness: the problem of continuous value attributes. Let us take the example of a tree that will classify two men according to their sizes.

B.A. Wahiba
Kingdom of Saudi Arabia, Taif University, Taif, Saudi Arabia

B.E.F. Ahmed (✉)
Higher Institute of Management, Tunis University, Tunis, Tunisia
e-mail: benelfadhl.ahmed@yahoo.fr

© Springer International Publishing Switzerland 2016
T. Gaber et al. (eds.), *The 1st International Conference on Advanced Intelligent System and Informatics (AISI2015), November 28–30, 2015, Beni Suef, Egypt,*
Advances in Intelligent Systems and Computing 407, DOI 10.1007/978-3-319-26690-9_28

The first has a height of 181 cm; the second has a height of 180. The tree classifies a man as tall, if he has a height strictly larger than 180. In this example the tree will classify the first man as tall, but not the second, despite the invisible difference between the two sizes in the real world. One of the solutions used to solve this type of problem is the integration of fuzzy set theory with decision trees. This theory describes the phenomena of the real world in a graduated way closer to the reality [2].

A fuzzy decision tree is a good choice to use in the field of text classification to manage a big problem which is the uncertainty and the ambiguity necessarily related to the use of human language terms in the documents to be classified.

Different models have been developed in the literature to construct fuzzy decision trees. Most of these models are based on the fuzzy ID3 algorithm [3], which is an extension of the ID3 algorithm [4].

For the discrimination attribute selection in the fuzzy decision tree, two heuristics have been used in literature: The first is based on the minimization of the fuzzy entropy; the second is based on minimizing the classification ambiguity [5]. In the area of text classification with fuzzy decision tree, in our literature search, only the first heuristic has been implemented and tested [6]. Concerning the second heuristic based on the minimization of the classification ambiguity, it has not been yet implemented, nor tested.

Minimizing ambiguity has been used by [7] in their model with sample classification on sport to practice according to the state of the climate described by four attributes. We will study and apply this heuristic for text classification.

The rest of this paper is organized as follows: the second section presents the calculation of the classification ambiguity in a document classification context. The third section details our fuzzy decision tree model. In the fourth section we present the results of our model experiments. We end this paper with an analysis and discussion of the results found in the experimental section, and some perspectives.

2   Calculation of the Classification Ambiguity

Before starting the details of the classification ambiguity calculus, we have to prepare Table 1 as an example of the learning space that we will use in our model. For attributes, symbols L, M and H respectively designate the following linguistic terms "Low", "Medium", and "High". These terms are the definition domain of the fuzzy variable "weight of the attribute in a document."

In Table 1, the lines present the learning space documents. The columns present the attributes describing documents and the classes of these documents. We have transformed each attribute weight with continuous values into a fuzzy variable with values as one of the three linguistic terms: low, medium or high.

Table 1 An example of the learning space

	Attributes												Classes of documents			
	A1			A2			A3			A4						
	L	M	H	L	M	H	L	M	H	L	M	H	C1	C2	C3	C4
D1	0.9	0.1	0.0	1.0	0.0	0.0	0.4	0.5	0.1	0.0	0.4	0.6	0.1	0.3	0.2	0.6
D2	0.8	0.2	0.0	0.0	0.6	0.4	0.6	0.3	0.1	1.0	0.0	0.0	0.5	0.7	0.2	0.1
D3	0.0	0.7	0.3	0.5	0.2	0.3	0.1	0.7	0.2	0.0	0.5	0.5	0.3	0.2	0.4	0.2
D4	0.2	0.7	0.1	0.3	0.7	0.0	0.0	0.2	0.8	0.6	0.3	0.1	0.0	0.5	0.1	0.3
D5	0.0	1.0	0.0	1.0	0.0	0.0	0.3	0.4	0.3	0.0	0.0	1.0	0.7	0.1	0.4	0.5
D6	0.2	0.6	0.2	0.0	0.3	0.7	1.0	0.0	0.0	0.1	0.7	0.2	0.5	0.3	0.4	0.1
D7	0.9	0.1	0.0	0.5	0.5	0.0	0.0	0.3	0.7	0.3	0.4	0.3	0.1	0.8	0.5	0.4

The numerical values crossing documents with the attributes are calculated through membership functions which convert the continuous value weight of each attribute into a set of membership degrees. The details of these membership functions are presented in Sect. 3.2.

The method used to calculate the weight of each attribute is TF × IDF method. The numerical values crossing documents with classes present the membership degrees calculated for each document to each class. In Sect. 2.4 we present our proposed method to calculate these degrees. According to [7], to calculate the classification ambiguity, we have to calculate some pre values:

- The truth degree of a fuzzy rule.
- The classification possibility with fuzzy evidence.
- The ambiguity of the evidence and fuzzy classification with fuzzy partitioning.

2.1 The Truth Degree of a Fuzzy Rule

A fuzzy rule is generally of the form "if A then B", which defines a fuzzy relation between the two fuzzy sets A and B. According to Yuan and Shaw [7] a fuzzy rule is true, means that A implies B (A → B). They define the implication of A and B based on the principle of "subsethood", that is A is a subset of B. In fuzzy logic, A is a subset of B, if for every element u belonging to the universe of discourse U the membership degree of u to A is less than the membership degree of u to B ($\mu_A(u) \leq \mu_B(u)$). Often this condition is not satisfied for all u belonging to U, in this case we speak about partial "subsethood" or "the truth degree of the rule". To calculate the truth degree of a rule of the form "if A then B", denoted S (A, B), we use the following formula:

$$S(A, B) = \frac{M(A \cap B)}{M(A)} = \frac{\sum_{u \in U} \min(\mu A(u), \mu B(u))}{\sum_{u \in U} \mu A(u)} \tag{1}$$

With, M is a function measuring the cardinality of a set.

As an example, let us calculate the truth degree of the following rule: "if A1 is L (low) then C = C3" denoted S (A1, C3).

$$S\ (A1,\ C3)\ =\ (0.2\ +\ 0.2\ +\ 0.0\ +\ 0.1\ +\ 0.0\ +\ 0.2\ +\ 0.5)\ /$$
$$(0.9 + 0.8 + 0.0 + 0.2 + 0.0 + 0.2 + 0.9) = 0.4$$

2.2 The Classification Possibility with Fuzzy Evidence

Yuan and Shaw define the fuzzy evidence as a condition (simple or compound) using fuzzy variables (variables with linguistic terms as values). Example, "A3 is M" is a fuzzy evidence.

The possibility Π of the classification of an object in a class C_i, knowing a fuzzy evidence E is calculated using the following formula:

$$\Pi(C_i|E) = S(E, C_i) / \max_J S(E, C_J) \qquad (2)$$

With, the index i is used to define a particular class C_i, j is a counter used to browse the various classes. Let us consider the fuzzy evidence E1 = "A3 is M". The calculation of Π (Class|E1), requires the calculation of the 4 following values: S (E1, C1), S (E1, C2), S (E1, C3), S (E1, C4). S (E1, C1) = 0.5, S (E1, C2) = 0.58, S (E1, C3) = 0.66, S (E1, C4) = 0.7. Π (Class|E1) = {0.5, 0.58, 0.66, 0.7}, so after applying normalization (by dividing all the values by the biggest one) and sorting we will have: Π (Class|E1) = {1, 0.94, 0.82, 0.71}.

2.3 Classification Ambiguity with Fuzzy Evidence and Fuzzy Partitioning

In their model [7], Yuan and Shaw provide the formulas for calculating the classification ambiguity, first, with the verification of one fuzzy evidence, second, with the verification of several fuzzy evidences.

In the First Case: Knowing a fuzzy evidence E, the classification ambiguity noted G (Class|E) can be defined as follows:

$$G(\text{Classe}|E) = \sum_{i=1}^{n} (\Pi(C_i|E) - (\Pi C_{i+1}|E)) * \ln(i) \qquad (3)$$

With, $\Pi(C_i|E) > \Pi(C_{i+1}|E)$ and n presents the number of the classes. As an example let us calculate the classification ambiguity with the fuzzy evidence E1 = "A3 is M" noted G (Class|E1):

G (Class|E1)) = (1 − 0.94) * ln (1) + (0.94 − 0.82) * ln (2) + (0.82 − 0.71) * ln (3) + (0.71 − 0) * ln (4) = 1.18

How to calculate the classification ambiguity if then we will partition by the attribute A2?

In the Second Case: Let F be a fuzzy evidence, P a set of k fuzzy evidences {E1, ..., Ek} set in the universe of discourse U. Yuan and Shaw de ne the fuzzy partition of P in F as: P|F = {E1 ∩ F,..., Ek ∩ F} To calculate G (P | F), we use the following formula:

$$G\ (P|\ F)\ =\ \sum_{i=1}^{k}\ w\ (E_i|F) * G\ (E_i \cap F) \tag{4}$$

With, $w\ (E_i|\ F) = M\ (E_i \cap F)\ /\sum_{j=1}^{k}\ (E_j \cap F)$.

In the first case of this section we calculated G (Π (class|E1)), we will continue the partition with the attribute A2. Consider the following fuzzy evidences: E2: "A2 is L", E3: "A2 is M" and E4: "A2 is H". To calculate the value of the classification ambiguity after partitioning by the attribute A2, we have to calculate G (A2|E1).

G (A2|E1) = w (E2|E1) * G(E2 ∩ E1) + w(E3|E1) * G(E3 ∩ E1) + w(E4|E1) * G (E4 ∩ E1).

To calculate w (E2|E1), we use the following formula: w (E2|E1) = M (E2 ∩ E1) /(M (E2 ∩ E1) + M (E3 ∩ E1) + M (E4 ∩ E1)). To calculate M (E2 ∩ E1) it is sufficient to calculate the sum of the minimum of the two degrees belonging to E1 and E2 (which are the columns M of the attribute A3 and L of the attribute A2) for all documents in learning space. The same method is used to calculate w (E3|E1) and w (E4|E1).

To calculate G (E2 ∩ E1), we have to calculate G (Π (C_i|E2 ∩ E1)).

Π (C_i|E2 ∩ E1) = S(E2 ∩ E1, C_i)/max_j S(E2 ∩ E1, CJ). S(E2 ∩ E1, C_i) = M(E2 ∩ E1 ∩ C_i)/M (E2 ∩ E1)). We have to calculate the sum of the minimum of the three membership degrees of each document for the two evidences E2, E1, and the class C_i, divided after by the sum of the minimum of the two membership degrees for both E1 and E2 evidences. The same calculation is applied to calculate the entire formula G (A2|E1).

2.4 The Calculation of Document Membership Degrees to the Different Classes

To calculate the membership degree of each document to each class in Table 1 we relied on Bayesian classifiers. The calculation of membership degree of a document Di to a class Cj can be seen as the calculation of the value of $P(C_j|D_i)$. According to Bayes' theorem and after some simplifications we have [8]:

$$P\big(C_j|D_i\big) = \big(P\big(D_i|C_j\big) * P\big(C_j\big)\big) \tag{5}$$

To calculate P(D_i|C_j), it is sufficient to calculate P(Ai1|Cj) * P(Ai2|Cj)* ... P (Ain|Cj), with Aie, is an attribute belonging to the document D_i, e varies between 1 and n. n presents the number of attributes without repetition in the document D_i. To calculate P(Ai1|C_j) it is sufficient to calculate the number of documents belonging to the class C_j and containing Ai1, after, divide the result by the total number of documents [9]. Having a single attribute Aie whose probability P(Aie|C_j) equals to zero will affect the probability of the entire document. To solve this problem, simply [10] add the value one for each numerator and n in the denominator. So, $P(Ai1|C_j) = (q + 1)/(n_i + \text{total number of document})$. With q presents the number of documents belonging to the class C_j and containing the attribute Ai1. n_i is the total number of attributes without repetition in the document D_i. The calculation of $P(C_j)$ is done by dividing the number of documents of class C_j by the total number of documents in all the classes, so:

$$P\left(C_j\right) = \left|C_j\right| / \sum_{i=1}^{m} \left|C_i\right| \tag{6}$$

m is the number of classes in the learning space.

3 The Proposed Model

Figure 1 shows the various stages of our fuzzy decision tree model for the text documents classification. Underlined steps are ones we have modified to fit the context of text documents classification with fuzzy decision tree using the minimization of the classification ambiguity.

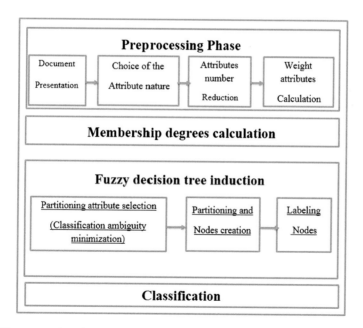

Fig. 1 The proposed model

3.1 Preprocessing Phase

For the document presentation we have chosen the vector format and the bag-of-words model. We chose the words as attributes to present the documents. For the attributes number reduction we first eliminated the stop words. Secondly we used a digital processing which orders the attributes based on a given measure then keeps only the first m attributes. The value of m depends on the experiments. We used two measures: the document frequency and CHI2 measure.

3.2 Membership Degrees Calculation

We used the triangular membership functions presented by [6]. (In the example we used in Sect. 2, we set the number of membership functions at 3 to simplify calculations). These functions are defined as follows:

$$
T_1(x) = \begin{cases} 1 & \text{if } x < C_1 \\ (C_2 - x)/(C_2 - C_1) & \text{if } x \in [C_1, C_2] \\ 0 & \text{if } C_2 < x \end{cases}
$$

$$
T_i(x) = \begin{cases} 1 & \text{if } x > C_{i+1} \\ (C_{i+1} - x)/(C_{i+1} - C_i) & \text{if } x \in [C_i, C_{i+1}] \\ (x - C_i)/(C_i - C_{i-1}) & \text{if } x \in [C_{i-1}, C_i] \\ 0 & \text{if } x < C_{i-1} \end{cases}
$$

$$
T_k(x) = \begin{cases} 1 & \text{if } x > C_k \\ (x - C_k)/(C_k - C_{k-1}) & \text{if } x \in [C_{k-1}, C_k] \\ 0 & \text{if } x < C_{k-1} \end{cases}
$$

To determine the number of the membership functions and their centers, [6] used an iterative algorithm that is based on a density distribution function and a measure called the F-statistic.

3.3 Induction of Fuzzy Decision Tree

Generating the tree is based on the algorithm proposed by [7]. This algorithm is detailed below. Before starting the details of the algorithm, a parameter must be set: β: This setting allows managing the evolution of the size of the tree. (In our experiments we set β to 0.7 as did Yuan and Shaw)

Below the general algorithm for the induction of fuzzy decision tree.

Begin Algo_Fuzzy_Tree
1. Search among the list of the attributes the one having the smallest classification ambiguity.
2. Use the found attribute to partition the current node.
3. For each child node Do
   i. Calculate the truth degree T_i for each class C_i.
   ii. **If** there is $T_i > \beta$
       The node is a leaf, its class is that with the highest truth degree.
   Else
       a. Calculate initial classification ambiguity G_i of the child node.
       b. Search among the remaining attributes if there is one with an ambiguity G after partition $< G_i$.
       c. **If** there is one
           Partition the current node and go to step 3.
       Else
           Consider the node as a leaf and assign it the class with the highest truth degree.
End Alg_Fuzzy_Tree

3.4 The Classification with Fuzzy Decision Tree

The classification of a document passes through two sub-steps: the conversion of the tree into a set of rules and the use of these rules for the classification.

For the first sub-step, each branch path leading from the root to a leaf is a rule. The set of attributes visited from the root to the leaf form the condition part of the rule. In the leaf, the class with the highest truth degree is the conclusion part of the rule.

After rules generation, we can classify the new documents. For classical decision trees, a new document undergoes one rule. With fuzzy decision trees, a document to classify may undergo several rules at once, so a document can be classified into different classes with different degrees. Classification in this case is obtained by the following steps:

1. For each rule, calculate the membership degree of the condition part of the rule. Calculating this membership degree is done by multiplying the membership degrees of the different attributes used in the condition part, the membership degree of the conclusion is the same as the membership degree of the condition.
2. If two rules are used to classify a document in the same class with two different degrees, retain the rule with the greatest degree.
3. If the rules give different classes for the same document, retain the class of the rule with the highest membership degree.

Table 2 Comparison of three different algorithms using the document frequency

Number of used attributes	50	100	150	200	228
Original model of Yuan and Shaw	11.78	11.78	11.78	11.78	11.78
Proposed model	23.27	28.36	33.73	41.95	45.18
Fuzzy ID3	26.59	30.84	35.62	44.29	48.67

Table 3 Comparison of three different algorithms using the measurement CHI2

Number of used attributes	50	100	150	200	228
Original model of Yuan et Shaw	11.78	11.78	11.78	11.78	11.78
Proposed model	22.34	26.12	30.88	39.54	43.24
Fuzzy ID3	25.27	28.52	34.57	43.12	45.52

4 Experimentation and Results

To evaluate our model we used a known measure: the macro F1.

The used data set to test our fuzzy decision tree system is the set "Reuters". Initially this set contains 21578 documents [11, 12]. Different versions (or subsets) were generated later from the original version to adapt the needs of researchers.

In our experimentation, we used the subset R8 of the set "Reuters". It consists of the 8 most frequent categories counting in total 7674 documents, divided in 5485 learning documents and 2189 test documents. The performed tests concern three models: the original model of Yuan and Shaw, the model we have proposed, and the fuzzy ID3 algorithm with 5 different values for the number of used attributes. The tests are operated following two scenarios. The first uses the frequency of the attribute in each document to reduce the number of attributes (Table 2). The second uses the index CHI 2 (Table 3). We used these two indices to reduce the number of attributes because it has been shown that these two indices offer the best classification results [13].

In both scenarios we chose the following values for the attributes number to consider in the learning phase: 50, 100, 150, 200, and 228. Over than 228 attributes the system launches a RAM saturation exception.

5 Analysis and Discussion

The application of the Yuan and Shaw model in its original version (Tables 2 and 3) gives very poor results (a rate of 11, 78 % for the measurement F1). In their model Yuan and Shaw gave directly the membership degrees of documents to different classes without specifying how to calculate these degrees. Applying the model of Yuan and Shaw directly for the documents classification, is assigning binary degrees of belonging to different classes. If the document in the training set belongs to a class i, the membership degree to this class will be equal to one, zero if else. With such a distribution of membership degrees we get poor results because these membership degrees interfere in the calculation of the truth degree, the criterion that controls stopping the tree development. Applying our model significantly improves the classification results but these results are still not as good as those achieved by the fuzzy ID3 algorithm. We note that the maximum value of the macro-F1 in the different used models did not exceed 48 %. We recall here that the purpose of our experiments is not to measure the maximum efficiency for our model, but to compare its effectiveness with the famous fuzzy ID3 algorithm. We note that our fuzzy tree is not optimized, and the results of the classifications of our model can be further improved by optimizing the decision tree. This optimization can be done by testing other types of attributes (n-grams, the lemmas, the stems, etc.), by changing the membership function, optimizing the pruning of the tree (in our experiments we only used the pre pruning without implementing a post-pruning technique), increasing the number of used attributes, testing different values for β parameter used in the tree induction algorithm, etc. For the index used to reduce the number of attributes, we note that the frequency of the attribute offer better results than those found with CHI2 index. This is explained by the nature of each index. Indeed, the frequency of the attribute, promotes the most common attributes in the training set. An attribute with a non-zero frequency will have a greater opportunity to have a non-zero degree later in the fuzzification phase. The membership degrees having a zero value affect the performance of the model of Yuan and Shaw because they present absorbing elements in some formulas used in this model. Specifically the formula that calculates the truth degree of each classification rule and the formula that calculates the degree of total belonging to a rule. These two formulas are based on the two operators "min" and "product".

6 Conclusion

In this paper, we proposed a new fuzzy decision tree model for the text documents classification. We tested a new heuristic based on the minimization of the classification ambiguity for the choice of discriminating attribute in the tree nodes partition step. Also we have proposed a method based on Bayes theorem to calculate the membership degrees of the training set documents for the different classes. In the fuzzification step, we used a method proposed by [6] to calculate the membership degrees of the attributes used in the tree induction phase. As it has been verified by its authors, this method has the advantage to autonomously find the appropriate number of linguistic terms for each attribute. In the experimentation results, our system gives better results than those presented by the original model of Yuan and Shaw. But compared with the famous fuzzy ID3 algorithm, our system cannot exceed the results of this algorithm, at the same time it presents close results. In future works we will try to surpass even ID3 fuzzy algorithm, by testing other values for the parameter β, testing other operators different from the used "product" and "min", and trying various other interpretations for the involvement operation.

References

1. Sebastiani, F.: Machine learning in automated text categorization. ACM computing survey. **34** (1), 1–47 (2002)
2. Janikow, C.Z., Kawa, K.: Fuzzy decision tree FID. In: Proceedings of the Annual Meeting of the North American Fuzzy Information Processing Society, pp. 379–384 (2005)
3. Matiasko, K., Bohacik, J., Levashenko, V., Kovalik, S.: Learning fuzzy rules from fuzzy decision tree. J. Inf. Control Manage. Syst. 4(2), 143–154 (2006)
4. Quinlan, J.R.: Induction of decision trees. Mach. Learn. **1**, 81–106 (1986)
5. Wang, X., Chen, B., Qian, G., Ye, F.: On the optimization of fuzzy decision trees. Fuzzy Sets Syst. **112**, 117–125 (2000)
6. Wang, Y., Wang, Z.O.: Text categorization rule extraction based on fuzzy decision tree. In: Proceedings of International Conference on Machine Learning and Cybernetics, vol. 4, pp. 2122–2127 (2005)
7. Yuan, Y., Shaw, M.J.: Induction of fuzzy decision trees. Fuzzy Sets Syst. **69**, 125–139 (1995)
8. Raheel, S.: L'apprentissage arti ciel pour la fouille de donnees multilingues: application pour la classification automatique des documents arabes. Ph.D. thesis defended on October 22, 2010. Higher National School of Information and Communication Sciences, University of Lyon 2 (2010)
9. Rehel, S.: Categorisation automatique de textes et cooccurrence de mots provenant de documents non etiquetes. Faculty of Science and Engineering, University LAVAL, QUEBEC (2005)
10. Witten, I.H., Frank, E.: Data mining: Practical Machine Learning Tools and Techniques, 2nd edn, p. 2005. Morgan Kaufmann, San Francisco, CA (2005)
11. Reuters: http://www.cs.umb.edu/smimarog/textmining/datasets (2007)
12. Cardoso-Cachopo, A., Oliviera, A.L.: Semi supervised single label text categorization using centroid-based classi ers. In: SAC'07 11–15 March 2007, Seoul, Korea (2007)
13. Yang, Y., Pederson, J.: A comparative study on feature selection in text categorization. In: Fisher, J.D.H. (ed.) The Proceedings of the Fourteenth International Conference on Machine Learning (ICML' 97), pp. 412–420 (1997)

An E-mail Filtering Approach Using Classification Techniques

Eman M. Bahgat, Sherine Rady and Walaa Gad

Abstract E-mail is one of the most popular ways of communication due to its accessibility, low sending cost and fast message transfer. However, Spam emails appear as a severe problem affecting this application of today's Internet. Filtering is an important approach to isolate those spam emails. In this paper, an approach for filtering spam email is proposed, which is based on classification techniques. The approach analyses the body of Email messages and assigns weights to terms (features) that can help identifying spam and clean (ham) emails. An adaptation is proposed that tries to reduce the dimensionality of the extracted features, in which only determined (meaningful) terms are regarded by consulting a dictionary. A thorough comparative study has been studied among different classification algorithms that prove the efficiency of the filtering approach proposed. The approach has been evaluated using Enron dataset.

Keywords Email filtering · Spam email · Classification · Enron

1 Introduction

Electronic email has become one of the most important applications for computer users. The average email messages sent daily have reached 3.4 billion in 2012 [1]. Such explosive growth leads to several problems, such as unsolicited commercial email, heavy network traffic, and computer worms which are frequently spread via

E.M. Bahgat (✉) · S. Rady · W. Gad
Faculty of Computer and Information Sciences, Ain Shams University, Cairo, Egypt
e-mail: eman.bahgat4@cis.asu.edu.eg

S. Rady
e-mail: srady@cis.asu.edu.eg

W. Gad
e-mail: walaagad@cis.asu.edu.eg

© Springer International Publishing Switzerland 2016
T. Gaber et al. (eds.), *The 1st International Conference on Advanced Intelligent System and Informatics (AISI2015), November 28–30, 2015, Beni Suef, Egypt,*
Advances in Intelligent Systems and Computing 407, DOI 10.1007/978-3-319-26690-9_29

emails. Therefore, a big challenge is to manage the huge number of emails efficiently. To solve this problem, an Email Filtering approach is significantly required.

Common email filters should filter incoming email automatically. The filtering process results in a set of categories or classifications, such as spam and ham. The filtering process can be decided and executed based on the email origin or header [2] (i.e. source) or based on the email content [3].

Origin-based filtering monitors the source of the e-mail, which is stored in the domain name and address of the sender device. Such filtering preserves two types email; white-list and black-list. Usually, the new email source is compared with a database to know how it is classified (i.e. spam history). In such technique, however, spammers regularly change the email source, address and IP. Therefore, content-based filtering is the second way to review the email content depending on a proposed analysis technique. It is a type of filtering that recommends items for a user based on the description of previously evaluated information available from the content [3].

Generally, email spam will cause severe problems for Internet users. The spam email is usually unsolicited and is sent in bulk. The sender of spam email also does not target the recipient personally, but the spam invades users without their consent and fills their email box. Besides the time consumed in checking and deleting spam emails, they overload the network bandwidth by useless data packages. All these factors will increase the operating costs, impact work productivity and privacy, and can harm the network infrastructure and the recipient's device if the email is pernicious.

In this paper, a classification-based email filtering approach is proposed. The approach is content (body)-based, in which a detailed comparative study among many classification algorithms have been studied for filtering emails. Five classification algorithms have been experimentally tested; these are Naïve Bayes, Support Vector Machine, Bayesian logistic regression, J48 and Random Forest. In addition, an adaptation is proposed, in which the dimensionality of the extracted features is reduced by considering the meaningfulness of the terms. This allowed for keeping a reduced feature set that contains determined (meaningful) terms only. The proposed methods have been evaluated using the Enron dataset and the performance have been recorded using precision, recall, F-score and accuracy measures.

2 Related Work

Email filtering has been conducted in the literature based on several classification methods. Some of the methods used for filtering spam email are Naive Bayes (NB) [4–6], k-Nearest Neighbor (KNN) [5], logistic regression [6], Artificial Neural Networks (ANN) [7], C4.5 classifier [7, 8], Multi-Layer Perceptron (MLP) [7], AdaBoost [9, 10], Support Vector Machine (SVM) [4, 5, 10] and Random Forest (RF) [10].

These classification methods have been applied based on the content of the email. In [8], a classification method based on ensemble learning and decision tree has been presented to classify spam email. Some researchers have shown that ensemble learning is efficient for the email classification. An email dataset has been used named SPAM-Email, The machine learning algorithms used are C4.5, Naive Bayes, SVM and kNN. The authors claimed accuracy of the proposed work of 94.2 %. In [11] a filter approach has been applied based on the content of the email, two behavioral features (the URL and the time the email was dispatched) and eight keywords known as bag-of-spam-words to discriminate the spam and legitimate emails. The dataset used is the SpamAssassin corpus. The work was applied using Random Forest classifier. The authors claimed that the accuracy was 99.2 % with spam rate 42 %.

The high dimensionality of features has also been studied in the email filtering problem. In [12], a feature selection criteria based on some similarity coefficients is used to increase the accuracy of spam filtering and detection rate. The Spambase dataset was used. Authors in [13] proposed an improvement in mutual information method with word frequency and average word frequency to calculate the relation between a feature and a class. The experiments are tested based on the English corpus (PU1's) and Chinese corpus email dataset. Other methods are also proposed in [14].

Some comparative analysis for the different classifiers has been done, such as in [15, 16]. In [15], five algorithms were used for email filtering: RF, Bagging, AdaBoost, SVM, and NB. The experiments were conducted based on four datasets: CSDMC2010, SpamAssassin, LingSpam, and Enron-Spam. A spam classification method was applied that uses features based on email content-language and some readability features related to the message (e.g. document length and word length). More et al. [16] presented another comparative analysis for Logistic Regression, Neural Network, NB and RF. They used Enron dataset. Their result shows that the RF classifier gives the highest accuracy. Moreover, a comparative analysis has been presented for the different classifiers MLP, NB, and J48 in [7]. The dataset was collected from UCI repository. The results show that the NB takes the least time to build the model and J48 has the best prediction accuracy.

3 Proposed Work

In this work, a model for email filtering is proposed, as shown in Fig. 1. It consists of three main processing steps: (1) Email Pre-processing, (2) Email Representation and (3) Classification.

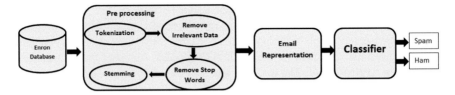

Fig. 1 Architecture of proposed model

3.1 Pre-processing Step

In the pre-processing step, tokens are extracted separately from email subject and body. This is followed by the removing of the irrelevant tokens such as symbols and numbers. Next the punctuation and the stop words (e.g. "a", "for", "to") are eliminated. Finally, a stemming procedure is applied to bring all similar words to the root or the base of the word (e.g. "helps", "helped" and "helping" to "help"). The stemming is done using Porter algorithm [17]. The purpose of the pre-processing step is to eliminate irrelevant tokens and reduce the size of extracted data for the classifier.

3.2 Email Representation

In this step, each word is assigned a weight proportional to the number of occurrences of this word in each email. We use term frequency and inverse term frequency, which are commonly used for indexing web documents.

Term Frequency (TF): measures how frequently a term occurs in a document. Since every document is different in length, it is possible that a term would appear much more times in long documents than shorter ones. Thus, the term frequency is often divided by the document length as a way of normalization.

$$tf(t,d) = \frac{f_d(t)}{\max[f_d(t)]} \tag{1}$$

where $f_d(t)$ is frequency of term t in document d.

Inverse Document Frequency (IDF): measures how important a given term is. While computing TF, it is known that certain terms may appear a lot of times but have little importance. Thus, we need to weigh the frequent terms while scale up the rare ones, by computing the following:

$$IDF(t) = \log\left(\frac{N}{df_t}\right) \tag{2}$$

where df_t is the no. of documents with term t, and N is the total no. of documents.

Finally, we compose the TF − IDF as the multiplication of Eqs. (1) and (2):

$$TF - IDF = tf(t,d) \times IDF(t) \tag{3}$$

In the context of the extracted tokens or terms, we propose two features set representation for evaluation:

- First: by considering all the email extracted terms as features after the previously mentioned preprocessing step.
- Second: by considering the meaningfulness of the terms. This is performed by consulting an English dictionary in order to identify determined or meaningful words. The set of the non-determined (non-meaningful words) were not ignored, but rather their frequency occurrence is stored as the single feature representing these tokens. This proposed representation is considered a method of reducing the high dimensionality of features.

3.3 Classification

In this step, a classifier is used which tries to analyze the email data to build a predicting model. Different types of classifiers are studied (e.g. probabilistic methods, tree-based). Five classifiers were used for the evaluation of the accuracy of the proposed model: Naïve Bayes, SVM, Bayesian Logistic Regression, J48 and Random Forest.

The Naive Bayes algorithm is a simple probabilistic classifier that calculates a group of probabilities by counting the combinations and frequency of terms in a given training dataset. The Bayesian classifier makes a conditional independence assumption between the attributes and this reduces the number of parameters to be estimated dramatically. Hence, the characterization is Naive yet the algorithm tends to perform well and learn rapidly in various supervised classification problems. It is based on the principle that most events are dependent and that the probability of an event occurring in the future can be detected from the previous occurrence of that event [5, 6].

Naïve Bayesian classifiers use the Bayes rule [4]:

$$P(c_j|d) = \frac{P(d|c_j)P(c_j)}{P(d)} \tag{4}$$

where $P(c_j|d)$ is the probability of instance d belong to class c_j, $(j = 1, 2, \ldots, n)$ and n is the number of classes. In our case, the spam filtering n equals to 2.

$P(d|c_j)$ is the probability of generating instance d given class c_j,
$P(c_j)$ is the probability of occurrence of class c_j,
$P(d)$ is the probability of occurrence of instance d.

The symbol d represents the set of features $(t_1, t_2, \ldots t_m)$, where m is the number of features. The Eq. (4) can then be changed to:

$$P(c_j|d) = \frac{P(c_j) \prod_{i=1}^{m} P(t_i|c_j)}{P(d)} \qquad (5)$$

Support vector machine (SVM) is a supervised learning method used in classification and regression. It tries to construct an N-dimensional hyper plane that separates the data into two classes. It can also be extended for more than 2 class classification. SVM can handle high dimensional feature space effectively [4, 10].

C4.5 decision tree (DT) and J48 algorithm are rule-based algorithms based on a group of rules which take advantage of the sequential structure of decision tree branches [7, 8]. The idea is to split the data into smaller subsets based on the attribute values for that item found in the training data. Using the concept of information entropy, J48 builds decision tree from a group of training data. At each node of the tree, J48 selects one attribute that divides its set of instances into subsets effectively. The J48 algorithm then visits each node recursively until no splits are available. The built trees are methods of induction that can be used for prediction effectively.

Logistic regression is seen to be similar to ordinary regression. It is used to analyze the relationship between a dependent variable and one or more independent variables, and helps to discover how suitable is the model and the significance of the relationships as well (between the dependent and independent variables) [6]. It combines the independent variables to estimate the probability that a particular event will occur. For a given case, logistic regression calculates the probability that a case with a particular set of values of the independent variable is a member of the modeled category.

Random Forest (RF) is an ensemble classifier that consists of a combination of many tree predictors. A randomized selection of features is used to split each tree independently and with the same distribution for all trees in the forest. RF runs efficiently on large datasets and its learning is fast [10].

4 Experimentation

Enron-Spam is a large public email database collection which consists of the personal emails of approximately 150 Enron employees. It focuses on six Enron employees with large mail-boxes. The Enron dataset is divided into six different subsets [18].

For our work, we extract a subset of the Enron Corpus "Enron1" yielding 300 emails (32 % spam, 68 % ham). The dataset was divided randomly into two parts: the first part is used for training the classifier, while the second part is used for the testing. Testing is done by using 10-fold cross validation method.

4.1 Measuring Performance

The performance of the classifiers is measured using recall, precision, F-score and accuracy. Recall represents the percentage of correctly identified positive cases and defined as:

$$Recall = \frac{TP}{TP + FN} \tag{6}$$

Precision reflects the number of real predicted examples and defined as:

$$Precision = \frac{TP}{TP + FP} \tag{7}$$

F-score, simply, is the harmonic mean of precision and recall, defined by:

$$F - score = \frac{2 \times Precision \times Recall}{Precision + Recall} \tag{8}$$

The overall accuracy has been also defined by:

$$Accuracy = \frac{TP + TN}{TP + TN + FP + FN} \tag{9}$$

In the previous equations, TP (True Positive) is the number of instances correctly classified to that class. TN (True Negative) is the number of instances correctly rejected from that class. FP (False Positive) is the number of instances incorrectly rejected from that class. FN (False Negative) is the number of instances incorrectly classified to that class.

4.2 Results and Discussion

The selected Enron dataset has been processed as explained in Sect. 3 by applying firstly the Email Pre-processing and Email Representation steps. Next, the selected set of classifiers has been studied using two different feature sets:

- First: by considering all the email extracted terms as features. After the pre-processing step, we get a total of 6424 features in this case study.
- Second: by considering only the meaningful terms performed by consulting our constructed dictionary (172,821 entries). Hereby, determined or meaningful words are identified. In this case study, the set of non-determined (non-meaningful words) were not ignored, but rather their frequency occurrence is stored as the single feature representing this category. This proposed

Table 1 Performance of first experimental study

Evaluation criteria	Classifiers				
	Naïve Bayesian	SVM	Bayesian logistic regression	J48	Random forest
Precision	0.944	0.953	0.977	0.803	0.888
Recall	0.943	0.953	0.977	0.807	0.883
F-measure	0.943	0.953	0.977	0.804	0.878
Execution time (s)	0.17	0.66	0.35	0.9	0.09

method reduces the total number of features to 3636 (43.5 % reduction than first case study).

Two experimental studies have been conducted that match the above mentioned feature set extraction, in which the TF-IDF (Eq. 3) has been applied for email representation. In each study, the classifiers performance is measured using the precision, recall, F-score, and accuracy.

Table 1 shows the different classifiers performance for the first study. The Bayesian logistic regression recorded the best precision with a normalized mean value of 0.977. This is followed by the Naïve Bayes and SVM recording almost similar results of 0.95. The J48 classifier has the least mean precision result of value 0.807. The other performance measures, Recall and F-Measure, almost take similar performance like the precision.

The results for the second experimental study are recorded in Table 2. The best classifiers are SVM and Bayesian logistic regression as recorded by their mean precision measure (0.96, having similar results). J48 has the least precision result of (0.76). In this study, it is clear that most of the classifiers performances are slightly affected by the feature set reduction, except for the SVM which is increased. It is also noticed that the execution time decreased than the previous study due to feature reduction.

Figure 2a, b summarizes the accuracy performance measure for the first and second experiment studies respectively, as measured by Eq. (9). Similarly, it is noticed that the Naïve Bayes, SVM and Logistic Regression outperforms the J48 and Random Forest, which complies with the performance obtained in Tables 1 and 2.

The proposed approach has also been compared to the related work in [15, 16], which have presented classification methods for the Enron dataset. This comparison is shown in Fig. 3a, b. The figures show the common classifiers used in the related work (in black) and our proposed work (in grey).

Figure 3a shows that the accuracy of Naïve Bayes for [15] is 0.733, while in our results it is 0.943, which is significantly better. In addition, SVM classifier has 0.783 in terms of accuracy, while in our performance it is 0.953 which is also higher. Figure 3b shows a comparison of different classifiers for our proposed model against the work done in [16]. In [16], it was concluded that Random Forest classifier outperforms the rest although not enough experiments were presented.

Table 2 Performance of second experimental study (reduced feature set)

Evaluation criteria	Classifiers				
	Naïve Bayesian	SVM	Bayesian logistic regression	J48	Random forest
Precision	0.923	0.961	0.96	0.76	0.867
Recall	0.923	0.96	0.96	0.76	0.867
F-measure	0.923	0.96	0.96	0.76	0.867
Execution time (s)	0.09	0.35	0.2	0.57	0.07

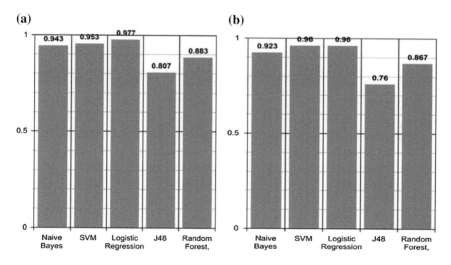

Fig. 2 **a** Accuracy for different classifiers of the first experimental study. **b** Accuracy for different classifiers of the second experimental study

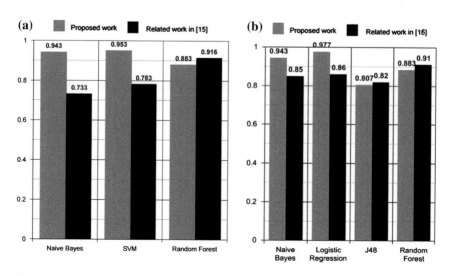

Fig. 3 **a** Comparing accuracy value of proposed work against related work in [15]. **b** Comparing accuracy value of proposed work against the related work in [16]

Comparing both results, it is clear that our work has highly significant accuracy for the Naïve Bayes and Logistic Regression, which outperforms the related work. Nevertheless, for the Decision Tree and Random Forest, it is slightly lower than the related work in [16].

5 Conclusion and Future Work

In this paper, an email filtering approach using classification techniques is proposed and studied. In the approach, two ways of presenting features are suggested. In the first, features are extracted from body content based on web document analysis methods. In the second way, we tried to reduce the dimensionality of these extracted features by selecting the determined (meaningful) terms only using a constructed dictionary. Experimental studies have been conducted using several classifiers and compared to existing related work using the same dataset. The recorded results prove the efficiency of the proposal filtering approach. The dictionary based filtering had an acceptable performance with faster filtering execution.

In the future work, we will try to enhance our model by considering some more methods for reducing the dimensionality of features.

References

1. Radicati, S., Hoang, Q.: Email statistics report. The Radicati Group Inc., London (2012)
2. Lai, C.C., Tsai, M.C.: An empirical performance comparison of machine learning methods for spam e-mail categorization. In: Fourth International Conference on Hybrid Intelligent Systems HIS'04, pp. 44–48. IEEE (2004)
3. del Castillo, M.D., Serrano, J.I.L.: An interactive hybrid system for identifying and filtering unsolicited e-mail. In: Intelligent Data Engineering and Automated Learning–IDEAL. Lecture Notes in Computer Science, vol. 4224, pp. 779–788. Springer, Berlin (2006)
4. Islam, M.S., Al Mahmud, A., Islam, M.R.: Machine learning approaches for modeling spammer behavior. In: Proceedings of the Information Retrieval Technology: The 6th Asia Information Retrieval Societies Conference, AIRS, Taipei, Taiwan, vol. 6458, pp. 251–260. Springer, Berlin (2010)
5. Blanzieri, E., Bryl, A.: A survey of learning-based techniques of email spam filtering. Technical Report DIT-06-056, University of Trento, Information Engineering and Computer Science Department, 2008
6. Mitchell, T.: Generative and discriminative classifiers: naive Bayes and logistic regression. http://www.cs.cm.edu/~tom/NewChapters.html. (2005)
7. Renuka, D.K., Hamsapriya, T., Chakkaravarthi, M. R., Surya, P.L.: Spam classification based on supervised learning using machine learning techniques. In: International Conference on Process Automation, Control and Computing (PACC), pp. 1–7. IEEE (2011)
8. Shi, L., Wang, Q., Ma, X., Weng, M., Qiao, H.: Spam email classification using decision tree ensemble. J. Comput. Inf. Syst. **8**(3), 949–956 (2012)
9. Islam, M., Zhou, W.: Architecture of adaptive spam filtering based on machine learning algorithms. In: ICA3PP, LNCS, vol. 4494, pp. 458–469. Springer, Berlin (2007)

10. Islam, R., Xiang, Y.: Email classification using data reduction method. In: Proceedings of the 5th International ICST Conference on Communications and Networking in China, pp. 1–5. IEEE (2010)
11. Bhat, V.H., Malkani, V.R., Shenoy, P.D., Venugopal, K.R., Patnaik, L.M.: Classification of email using beaks: behavior and keyword stemming. In: TENCON IEEE Region 10 Conference, pp. 1139–1143. IEEE (2011)
12. Abdelrahim, A.A., Elhadi, A.A.E., Ibrahim, H., Elmisbah, N.: Feature selection and similarity coefficient based method for email spam filtering. In: International Conference on Computing, Electrical and Electronics Engineering (ICCEEE). IEEE (2013)
13. Ting, L., Yu, Q.: Spam feature selection based on the improved mutual information algorithm. In: Fourth International Conference on Multimedia Information Networking and Security (MINES), IEEE (2012)
14. Wang, R., Youssef, A.M., Elhakeem, A.K.: On some feature selection strategies for spam filter design. In: Canadian Conference on Electrical and Computer Engineering (CCECE'06) pp. 2186–2189. IEEE (2006)
15. Shams, R., Mercer, R.E.: Classifying spam emails using text and readability features. In: 13th International Conference on Data Mining (ICDM). IEEE (2013)
16. More, S., Kulkarni, S.: Data mining with machine learning applied for email deception. In: International Conference on Optical Imaging Sensor and Security. IEEE (2013)
17. Porter, M.F. An algorithm for suffix stripping. Program **14.3**, 130–137 (1980)
18. http://csmining.org/index.php/enron-spam-datasets.html

Part V
Cloud Computing and Big Data Mining

Exploring Big Data Environment for Conversation Data Analysis and Mining on Microblogs

Rami Belkaroui, Dhouha Jemal and Rim Faiz

Abstract Today, social media services and multiplatform applications such as microblogs, forums and social networks gives people the ability to communicate, interact and generate content which establish social and collaborative backgrounds. These services now embodies the leading and biggest repository containing millions of Big social Data that can be useful for many applications such as measure public sentiment, trends monitoring, reputation management and marketing campaigns. But social media data are essentially unstructured that's what makes it so interesting and so hard to analyze. Making sense of it and understanding what it means will require all new technologies and techniques, including the emerging field of big data. In addition, social media is a key model of the velocity and variety which are main characteristics of Big Data. In this paper, we propose a new approach to retrieve conversation on microblogging sites that combine Big Data environment and social media analytics solutions. The goal of our approach is to present a more informatives result and solve the information overload problem within Big Data environment. The proposed approach has been implemented and evaluated by comparing it with Google and Twitter Search engines and we obtained very promising results.

Keywords Social media · Big data analysis · Data mining · Conversation retrieval · Social networks analytics

R. Belkaroui (✉) · D. Jemal
LARODEC, ISG Tunis, University of Tunis, Bardo, Tunisia
e-mail: rami.belkaroui@gmail.com

D. Jemal
e-mail: Dh.jemal@gmail.com

R. Faiz
LARODEC, IHEC Carthage, University of Carthage, Carthage Presidency, Tunisia
e-mail: rim.faiz@ihec.rnu.tn

© Springer International Publishing Switzerland 2016
T. Gaber et al. (eds.), *The 1st International Conference on Advanced Intelligent System and Informatics (AISI2015), November 28–30, 2015, Beni Suef, Egypt,*
Advances in Intelligent Systems and Computing 407, DOI 10.1007/978-3-319-26690-9_30

1 Introduction

Social media is considered as a part of the Web 2.0 movement, which gave birth to a huge volume of data produced by users called User Generated Content (UGC). This massive explosion of volume and data types' diversity implies the imposition of the Big Date in our technological landscape. Furthermore, many of the most important sources of Big Data are relatively new. The huge amount of information from social networks, are only as old as the networks themselves; for example Facebook was launched in 2004, Twitter in 2006. In addition, the term Big data [24] has been mainly used in two contexts, firstly as a technological challenge when dealing with data-intensive domains such as high energy physics, astronomy or internet search, and secondly as a sociological problem when data about us is collected and mined by companies such as Facebook, Google, mobile phone companies, retail chains and governments.

In the current era, social media services attract more and more users and tend to become a solid media for simplified collaborative communication due to the ease and the speed of information sharing especially in real time. Thus, social media services have completely changed the manner in which people communicate and share information. They [13] give people the ability not only to communicate, interact and collaborate with each other, reply to messages from others and create conversations. In addition, social media such as Twitter and Facebook represents an enormous public archive of human thought that captures the ideas, opinions and debates taking place around the world on almost any topic at any moment in time. The analysis of those communications can be useful in various ways such as measuring public sentiment, trend monitoring, reputation management, marketing campaigns, customer behavior and news analysis. Additionally, the analysis of Big social Data has become substantially to help businesses gain a better understanding of their consumers by incorporating social feedback, which should improve decision-making process. Although, according to Salesforce research, 89 % of business leaders believe that Big Data will revolutionize business operations in the same way the Internet did, and 83 % have pursued Big Data projects in order to seize a competitive edge.[1] As Big Data is becoming a hot topic in many areas where datasets are so large that they can no longer be handled effectively or even completely, it presents an opportunity to analyze expressions and behavioral traces available on social platforms. The question that arises is how to benefit the e-identity and user's opinions shared on social medias.

In this paper, we propose a new approach to retrieve conversation on microblogging sites that combine Big Data environment and social media analytics solutions. The proposed approach can be used to extract conversation from social media in order to provide relevant results correspondig to user's information needs based on their interactions analysis. In particular, the novelty of our approach is the ability to provide an informative relevant data to satisfy any information need in a large social

---

[1] http://www.forbes.com/sites/louiscolumbus/2015/05/25/roundup-of-analytics-big-data-business-intelligence-forecasts-and-market-estimates-2015/.

Big Data. The remainder of this paper is organized as follows. In Sect. 2, we give an overview of related work addressing the domains used in our work: Big Data analysis and social media analytics solutions. In Sect. 3, we present our proposed approach for conversation retrieval, then in Sect. 4 we discuss the obtained results. Finally, Sect. 5 concludes this paper and outlines our future work.

2 Related Work

Basically, our work lies at the intersection of two main domains: Big Data analysis and social media retrieval. This section is devoted to presenting these latter.

2.1 Big Data Analysis

Currently, a strong interest towards the term Big Data is arising in the literature. Several research works have focused on this actual research trend in this field. In this context, Sagiroglu et al. [22] presented an overview of big data's content, scope, samples, methods, advantages and challenges, detail Big Data its main components and discuss privacy concern. In [24], the authors have performed analysis on Flickr, Facebook and Google+ social media sites. Based on this analysis, they have discussed the privacy implications and also geo-tagged social media; an emerging trend in social media sites. The proposed concept in this paper helps users to get informed about the data relevant to them in such large social Big Data. In [9], open problems and actual research trends are highlighted with the aim of providing an overview of state of the art research issues and achievements in the eld of analytics over Big Data, and extend the discussion to analytics over big multidimensional data. This work presented several novel research directions a rising in this field, which plays a leading role in next-generation Data analysis.

According to [21], the expressions and behavioral traces from large numbers of individuals or groups on social platforms constitute Big Data. The authors proposed to analyze this data as an opportunity to provide valuable insights into the arrays of meaning and practice that emerge and manifest social platforms users online.

In [4], authors presented the problem of misleading claims to objectivity and accuracy for Big Data. They supposed that in the case of social media data, there is a 'data cleaning' process: making decisions about what attributes and variables will be counted, and which will be ignored. They described this process as inherently subjective. In addition to this question, there is the issue of data errors. Authors considered the large data sets from Internet sources as unreliable, prone to outages and losses, and these errors and gaps are magnified when multiple data sets are used together. For this purpose, [3] explains that as a large mass of raw information, Big Data is not self-explanatory. The authors wonder if the data can represent an 'objective truth' or is any interpretation necessarily biased by some subjective filter or the way that data is 'cleaned?'

2.2 Social Media Retrieval and Analysis

Extracting information from Social media Sites is one of the critical problems which have already been addressed over the past years due to its theoretical and practical signifiance. Social media have become a digital place where users can discuss public issues, share critical information during natural disasters and it will be possible to comment TV shows they are following. Users examine social media sites for answers, timely information (e.g., news, events), people information and topical information [10, 26]. However, search functionality provided by those sites is limited to keyword based retrieval to return the most recent posts. However, users are able neither to explore the results nor retrieve more relevant information based on the content [1], and may get lost or become frustrated by the information overload [2].

Recently, various researches have focused on these phenomena with a closer perspective [5, 18, 20]. In [19] the authors proposed a user-based tree model for retrieving conversations from microblogs. They only considered tweets that directly respond to other tweets by using the "@username" as a marker of addressivity. The advantage of this method is having a coherent conversation based on the direct links between users. The downside of this method is the fact of neglecting the tweet that does not contain the "@sign". Nevertheless, we can have tweets that are related to the conversation even though they are not directly linked with other tweets. In [8], the authors proposed a method to build conversation graphs. The authors focused on the particular case of a conversation formed by users replying to tweets. In this case, a tweet can only reply to one other tweet and a retweet are ignored, but users can get involved in conversations with other by commenting, liking and sharing other user's posts. Other related works concentrated on different aspects of microblogging conversations are [12, 25], the deal respectively with the tagging of conversations and the identification of topics. Kumar et al. [15], in their work has proposed a simple mathematical model that produces basic conversation structures taking into account the identities of each member of the conversation.

Besides, some recent studies aimed to identify how information flow depends on the structural properties of user interactions [6, 16]. Moreover, modeling the interaction among users plays a crucial role in order to understand how the information is disseminated in the network [11, 27] and how to maximize its spread [7, 14]. In [16], the authors proposed the first large scale analysis of both topological and temporal structures of retweets in order to catch users' popularity of, trending topics and temporal patterns describing the propagation of information in the network. The temporal dimension of the information spread plays an important role in characterizing users' retweets and identifying influential as well [17, 28]. Eventually, we have to mention that in dynamic interactions, social cascades have also been used to reveal that, different from static properties, the user's geographic locality is a key factor to characterize the way how the information circulates in the network [23].

In summary, retrieving information from social media sites represented always a challenge given its volume, inconsistent writing and noise. Most existing systems focus on term-based approach, but retrieving the relevant information is often neglected, leading to less satisfactory results while searching information.

3 Conversations Extraction Approach on Big Twitter Data

In recent years, conversations generated by Big Data environments (Twitter, Facebook, blogs, etc.) have flooded the web. Big Social Data have transformed the scale of exploratory analysis on the web and offered new means of performing tasks that were not feasible before. This highlighted how Big Data research has become central to this growth of knowledge. Conversations have become an essential feature of contents generated by most social media platforms. Particularly, on Twitter, many users post tweets freely in order to express what they are thinking about any event, topic, followed by some comments, retweet or favorite. People uses conversations to express their interests, feelings, and experiences about virtually any topic. Thus, the users' interactions essentially reflect the importance of different topics and can be used to improve the conversation retrieval task quality.

In this paper, we explore the Big Data environment to propose an information retrieval approach for microblogging sites. Particularly, we focused on Twitter based conversation retrieval task (Fig. 1). Our approach combines direct and indirect conversation aspects in order to extract extensive posts beyond conventional conversation. In addition, we defined a conversation as a set of short text messages posted by a user at specific timestamp on the same topic. These messages can be directly replied to other users by using "@username" or indirectly by liking, retweeting, commenting and other possible interactions.

For this purpose, we opted for a Big Data solution. Due to the huge amount of data, structure diversity and specially the task complexity, a Big Data solution can provide best performance and efficiency. Thus, MapReduce is presented as one of the most efficient Big Data solutions. The MapReduce programming model has been successfully used for many different purposes. Hence, the remedy of the user defined function, which is one of the main strength feature of the MapReduce model. While conducting our exploratory detection and analysis of the tweeter's conversation, we came across large conversations' features. Our approach is bases on the direct and indirect conversations and consist in 2 steps:

- Step 1: Constructing the direct reply tree using all tweets in reply directly to other tweets.
- Step 2: Detecting the relevant tweets related indirectly to a same reply tree which might be retweets, comments or other possible interactions in order to extract extensive posts beyond conventional conversation.

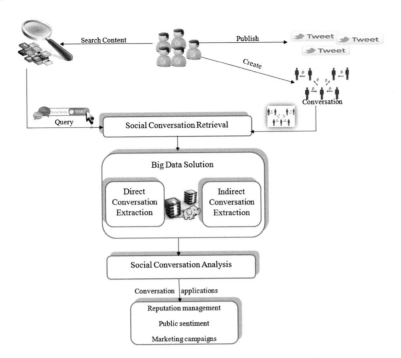

Fig. 1 Exploring Big Data environment for Twitter conversations extraction approach

3.1 Twitter Conversations Construction Using MapReduce

In this phase, we aim to collect all tweets in reply directly to other tweets. Obviously, a reply to a user will always begin with "@username". Our goal in this step is to create reply tree. The reply tree construction process consists of two algorithms: recursive root finder and iterative search algorithm.

For this purpose, we implement the two MapReduce phases: The map phase is implemented with a recursive algorithm to find T0 the root (first tweet published) of the conversation and H0 the hashtag used in the selected conversation. Thus, while the type of a given tweet is a reply or a retweet, the recursive algorithm boots until identifying the root. The iterative algorithm runs in parallel to seek the reminder of the conversation and extracting the used hashtag. The established conversations and the used hashtags, will be the input of the reduce phase of the MapReduce process. The goal of this phase is to classify conversations based on identical hashtags, using a grouping conversations algorithm.

3.2 *Indirect Reply Structure Using Conversational Features*

To the best of our knowledge, there has not been previous work on the structure of reply-based on indirectly conversation. In this step, we also used the MapReduce model to enrich conversations established during the first step. The goal of this step is to extract tweets that may be relevant to the conversation without the use of the @symbol. The conversations classification based on hashtags will facilitate the selection of the candidate tweets for each conversations cluster. Therefore, we define new features that may help to detect tweets related indirectly to a same conversation. The features are used during the MapReduce process. The Map phase is designed to calculate a score for each tweet relative to each conversation. The conversations are the output of the first step of our proposed approach. The features we used are:

- *URLs-based selection*:

Twitter allows users to include URLs as supplement information to their tweets. URLs that are contained in tweets can be considered as indicators for news related tweets. By sharing an URLs, an author would enrichment the information published in his tweet. In particular, if a tweet contains a URL that points to an external news resource, there is a very high possibility that this tweet is closely related to the linked resource. This feature is applied to collect tweets that share the same URL.

- *Hashtag-based selection*:

The # symbol, called hashtag, is used to mark a topic in a tweet or to follow conversation. Hashtag is meant to be identifier for conversations that rotate around the same topic. By including hashtag in a message(tweet), users indicate to which conversations their message is related to. We used this feature to collect tweets that share the same hashtags.

- *Time Difference and Publication dates*:

The time difference is highly important feature for detecting tweets linked indirectly to the conversation. We use the time attribute to efficiently remove tweets having a large distance in terms of time compared to conversation root. Date attribute are highly important for detecting conversations. Users tend to post tweets about conversational topic within a short time period. The euclidean distance has been used to calculate how similar two posts publication dates are.

- *Content-based Similarity*:

The criterion content refers to the thematic relevance traditionally calculated by IR systems standards. We compute the textual similarity between each element in t_i, t_j taking the maximum value as the similarity measure between two messages. The similarity between two elements is calculated using the well-known tf-idf cosine similarity, $sim(t_i, t_j)$.

After calculating the tweets' scores, a tweet sort relative to each conversation will be established during the reduce phase. Then, for each conversation, we suggest to select the five ranked tweet. The final output will be a conversation set, with an identified root for each one, and enriched by five extra tweets.

4 Experimentations and Results

The following experiment has been designed to gather some knowledge on the impact of our results on end-users. For this experiment, we have selected three events and queried our dataset using Google, Twitter search engine and our approach. Then, we have asked a set of assessors to rate the top-10 results of every search task, to compare these approaches.

4.1 Evaluation Metrics

In order to measure the results quality, we use the Normalized Discounted Cumulative Gain (NDCG) at 10 for all the judged events. In addition, we used a second metric which is the Precision at top 10. In the following, we first describe the experimental setting, then we present the results and we provide an interpretation of the data.

4.2 Experimental Settings

The dataset has been obtained by monitoring microblogging system Twitter posts over the period of March-May 2015. In particular, we used a sample of about 313 000 posts consisted of 1TB of data containing trending topic keywords using Twitter's streaming API. Trending topics (the most famous topic on twitter) have been determined directly by Twitter, and we have selected the most frequent ones during the monitoring period. To evaluate our search tasks' results, we have used a set of 100 assessors with three relevance levels, namely highly relevant (value equal to 2), relevant (value equal to 1) or irrelevant (value equal to 0). Every user was informed of three events happened during the sampling period. For each event we performed three searches using:

1. Google.
2. Twitter Search.
3. Our approach.

The evaluators were not aware from which system the results were retrieved. Every user for each search task was presented with three conversations selections, one for each of the previous options with the corresponding top-10 results.

4.3 Results Interpretation

We compare our conversation retrieval approach with the results returned by Google and by Twitter search engine using two metrics namely the P@10 and the NDCG@10. From this comparison, we obtained the values presented in Figs. 2 and 3 where we notice that our approach overcomes the results given by both of Google and Twitter.

The reason of these promising values is the fact that we combine a set of social conversational features and big data environment characteristics to retrieve conversation may have a significant impact on the users' evaluation. Focusing on the three messages selections, we observe that all conversations obtained with our approach receive higher scores with compared to Google and Twitter's selection. According to the free comments of some users and following the analysis of the posts in the three selections we can see that Google and twitter received lower scores not because they contained posts judged as less interesting, but because some posts were considered not relevant with regard to the searched topic.

Concentration on the three messages selections we observe that all conversations selections obtained with twitter search has higher scores with respect to Google's

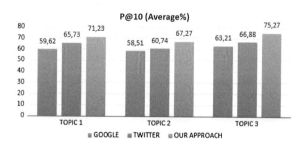

Fig. 2 P@10 Average scores of human evaluation for each approach

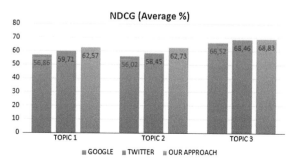

Fig. 3 NDCG Average scores of human evaluation for each approach

selection. These results lead us toward a more general interpretation of the collected data. It appears that the social metrics usage have a significant impact on the users' degree interest in the retrieved posts. In addition, the retrieving conversations process from Social Network differs from traditional Web information retrieval; it involves human communication aspects, like the degree interest in the conversation explicitly or implicitly expressed by the interacting people.

5 Conclusion

Only Big Data applications can enable users to manage social conversations. In addition, Big Data is usually associated to having volume, velocity, variety, variability and complexity which are also features of Social Media Analytics. That is why social media should be explored to make sense of what all the available data means to satisfy users' information needs.

This work proposed a new approach to retrieve conversation on microblogging sites that combine Big Data environment and social media analytics solutions. Our experimental results have highlighted many interesting points. First, combining social media features and the concept of direct conversation on Big data environment improves the relevance and informativeness of information search task and also provides results that are considered more satisfaction with respect to a traditional information retrieval approach. Future work will further research the conversational aspects by including human communication aspects, like the degree of interest in the conversation and their influence/popularity by gathering data from multiple sources from Social Networks.

References

1. Belkaroui, R., Faiz, R.: Towards events tweet contextualization using social influence model and users conversations. In: Proceedings of the 5th International Conference on Web Intelligence, Mining and Semantics, WIMS 2015, Larnaca, Cyprus, p. 3, July 13–15, 2015. http://doi.acm.org/10.1145/2797115.2797134
2. Bernstein, M., Hong, L., Kairam, S., Chi, H., Suh, B.: A torrent of tweets: managing information overload in online social streams. In. In Workshop on Microblogging: What and How Can We Learn From It? (CHI'10 (2010)
3. Bollier, D., Firestone, C.M.: The Promise and Peril of Big Data. Aspen Institute, Communications and Society Program Washington, DC, USA (2010)
4. Boyd, d., Crawford, K.: Six provocations for Big Data. In: A Decade in Internet Time: Symposium on the Dynamics of the Internet and Society (2011)
5. Bruns, A., Burgess, J.E.: #Ausvotes: how twitter covered the 2010 australian federal election. Commun. Polit. Cult. 44(2), 37–56 (2011), http://search.informit.com.au/documentSummary;dn=627330171744964;res=IELHSS
6. Cha, M., Mislove, A., Gummadi, K.P.: A measurement-driven analysis of information propagation in the flickr social network. In: Proceedings of the 18th International Conference on

World Wide Web, WWW'09, pp. 721–730. ACM, New York, NY, USA (2009). http://doi.acm.org/10.1145/1526709.1526806

7. Chen, W., Wang, Y., Yang, S.: Efficient influence maximization in social networks. In: Proceedings of the 15th ACM SIGKDD International Conference on Knowledge Discovery and Data Mining, KDD'09, pp. 199–208. ACM, New York, NY, USA (2009). http://doi.acm.org/10.1145/1557019.1557047

8. Cogan, P., Andrews, M., Bradonjic, M., Kennedy, W.S., Sala, A., Tucci, G.: Reconstruction and analysis of twitter conversation graphs. In: Proceedings of the First ACM International Workshop on Hot Topics on Interdisciplinary Social Networks Research, HotSocial'12, pp. 25–31. ACM, New York, NY, USA (2012). http://doi.acm.org/10.1145/2392622.2392626

9. Cuzzocrea, A., Song, I.Y., Davis, K.C.: Analytics over large-scale multidimensional data: the big data revolution! In: Proceedings of the ACM 14th International Workshop on Data Warehousing and OLAP, DOLAP'11, pp. 101–104. ACM, New York, NY, USA (2011). http://doi.acm.org/10.1145/2064676.2064695

10. Efron, M., Winget, M.: Questions are content: a taxonomy of questions in a microblogging environment. In: Proceedings of the 73rd ASIS&T Annual Meeting on Navigating Streams in an Information Ecosystem, ASIS&T '10, vol. 47, pp. 27:1–27:10, American Society for Information Science, Silver Springs, MD, USA (2010). http://dl.acm.org/citation.cfm?id=1920331.1920371

11. Gómez, V., Kappen, H.J., Kaltenbrunner, A.: Modeling the structure and evolution of discussion cascades. In: Proceedings of the 22Nd ACM Conference on Hypertext and Hypermedia, HT'11, pp. 181–190. ACM, New York, NY, USA (2011). http://doi.acm.org/10.1145/1995966.1995992

12. Huang, J., Thornton, K.M., Efthimiadis, E.N.: Conversational tagging in twitter. In: Proceedings of the 21st ACM Conference on Hypertext and Hypermedia, HT'10, pp. 173–178. ACM, New York, NY, USA (2010). http://doi.acm.org/10.1145/1810617.1810647

13. Jabeur, L.B., Tamine, L., Boughanem, M.: Uprising microblogs: A bayesian network retrieval model for tweet search. In: Proceedings of the 27th Annual ACM Symposium on Applied Computing, SAC'12, pp. 943–948. ACM, New York, NY, USA (2012). http://doi.acm.org/10.1145/2245276.2245459

14. Kempe, D., Kleinberg, J., Tardos, E.: Maximizing the spread of influence through a social network. In: Proceedings of the Ninth ACM SIGKDD International Conference on Knowledge Discovery and Data Mining, KDD'03, pp. 137–146. ACM, New York, NY, USA (2003). http://doi.acm.org/10.1145/956750.956769

15. Kumar, R., Mahdian, M., McGlohon, M.: Dynamics of conversations. In: Proceedings of the 16th ACM SIGKDD International Conference on Knowledge Discovery and Data Mining, KDD'10, pp. 553–562. ACM, New York, NY, USA (2010). http://doi.acm.org/10.1145/1835804.1835875

16. Kwak, H., Lee, C., Park, H., Moon, S.: What is twitter, a social network or a news media? In: Proceedings of the 19th International Conference on World Wide Web, WWW'10, pp. 591–600. ACM, New York, NY, USA (2010). http://doi.acm.org/10.1145/1772690.1772751

17. Lee, C., Kwak, H., Park, H., Moon, S.: Finding influentials based on the temporal order of information adoption in twitter. In: Proceedings of the 19th International Conference on World Wide Web, WWW'10, pp. 1137–1138. ACM, New York, NY, USA (2010). http://doi.acm.org/10.1145/1772690.1772842

18. Magnani, M., Montesi, D., Rossi, L.: Information propagation analysis in a social network site. In: International Conference on Advances in Social Networks Analysis and Mining (ASONAM), pp. 296–300, Aug 2010

19. Magnani, M., Montesi, D., Nunziante, G., Rossi, L.: Conversation retrieval from Twitter. In: Clough, P., Foley, C., Gurrin, C., Jones, G., Kraaij, W., Lee, H., Mudoch, V. (eds.) ECIR 2011. LNCS, vol. 6611, pp. 780–783. Springer, Heidelberg (2011)

20. Magnani, M., Montesi, D., Rossi, L.: Conversation Retrieval for Microblogging Sites, vol. 15, pp. 354–372. Springer, Netherlands (2012). http://dx.doi.org/10.1007/s10791-012-9189-9

21. Manovich, L.: Trending: The Promises and the Challenges of Big Social Data. Debates in the Digital Humanities, pp. 460–475 (2011)
22. Sagiroglu, S., Sinanc, D.: Big data: a review. In: International Conference on Collaboration Technologies and Systems (CTS), pp. 42–47 (2013)
23. Scellato, S., Mascolo, C., Musolesi, M., Crowcroft, J.: Track globally, deliver locally: improving content delivery networks by tracking geographic social cascades. In: Proceedings of the 20th International Conference on World Wide Web, WWW'11, pp. 457–466. ACM, New York, NY, USA (2011). http://doi.acm.org/10.1145/1963405.1963471
24. Smith, M., Szongott, C., Henne, B., von Voigt, G.: Big data privacy issues in public social media. In: 6th IEEE International Conference on Digital Ecosystems Technologies (DEST), pp. 1–6, June 2012
25. Song, S., Li, Q., Zheng, N.: A Spatio-temporal Framework for related topic search in microblogging. In: An, A., Lingras, P., Petty, S., Huang, R. (eds.) AMT 2010, LNCS, vol. 6335, pp. 63–73. Springer, Heidelberg (2010)
26. Teevan, J., Ramage, D., Morris, M.R.: #twittersearch: a comparison of microblog search and web search. In: Proceedings of the Fourth ACM International Conference on Web Search and Data Mining, WSDM'11, pp. 35–44. ACM, New York, NY, USA (2011) http://doi.acm.org/10.1145/1935826.1935842
27. Wang, D., Wen, Z., Tong, H., Lin, C.Y., Song, C., Barabási, A.L.: Information spreading in context. In: Proceedings of the 20th International Conference on World Wide Web, WWW'11, pp. 735–744. ACM, New York, NY, USA (2011). http://doi.acm.org/10.1145/1963405.1963508
28. Yang, J., Leskovec, J.: Patterns of temporal variation in online media. In: Proceedings of the Fourth ACM International Conference on Web Search and Data Mining, WSDM'11, pp. 177–186. ACM, New York, NY, USA (2011) http://doi.acm.org/10.1145/1935826.1935863

Enhancing Query Optimization Technique by Conditional Merging Over Cloud Computing

Eman A. Maghawry, Rasha M. Ismail, Nagwa L. Badr
and M.F. Tolba

Abstract Cloud computing has become a powerful distributed computing mode. A Cloud system has a characteristics strength such as scalability and heterogeneity against the traditional distributed paradigm. These characteristics lead to increased numbers of clients needs to access and process data from multiple distributed resources over a cloud environment with widely differing expectations. Therefore, query processing on such an environment needs to be adaptive to handling the concurrent queries. The aim of this paper is to improve the overall performance of the query execution. We focused on enhancing a query merging approach within a query processing architecture. This is done by considering different waiting times of submitted queries, therefore the queries are merged in case they have a positive impact on the query execution performance. The results show that our enhancement can improve the queries execution time over the original technique by 15 % and over the existing merging technique by 60 %.

Keywords Cloud computing · Query processing · Query optimization

E.A. Maghawry (✉) · R.M. Ismail · N.L. Badr · M.F. Tolba
Faculty of Computer and Information Sciences, Ain Shams University, Cairo, Egypt
e-mail: e_maghawry@yahoo.com

R.M. Ismail
e-mail: rashaismail@yahoo.com

N.L. Badr
e-mail: nagwabadr@cis.asu.edu.eg

M.F. Tolba
e-mail: fahmytolba@gmail.com

© Springer International Publishing Switzerland 2016
T. Gaber et al. (eds.), *The 1st International Conference on Advanced Intelligent System and Informatics (AISI2015), November 28–30, 2015, Beni Suef, Egypt,*
Advances in Intelligent Systems and Computing 407, DOI 10.1007/978-3-319-26690-9_31

1 Introduction

At present, the increasing amount of data has lead the users to store their own data on the cloud service provider's servers instead on their servers. Cloud service provides flexibility and scalability for data storage without any concerns regarding storage and maintainability issues. Cloud services are a fundamental component in cloud computing; the services that are offer range from software applications to virtualized platforms and infrastructures over cloud systems. It provides the users with the benefit of paying only for the amount of data they need to store for a specific period of time. Furthermore, the users can easily access their stored data from any geographical region through the Cloud Service Provider [1, 2]. Some of these cloud service providers are Amazon [3], Google [4] and Microsoft [5]. Users can deploy their databases on these clouds, which have virtual machines with pre-installed and pre-configured database systems [6]. The increasing the number of users accessing a cloud storage has resulted in an increasing demand to handle the high amount of concurrent queries accessing those resources. This may cause degradation in the queries evaluation performance in environments with rapidly changing computational properties, such as loads. Thus the movement of query processing into these heterogeneous environments means that it is necessary to consider unstable conditions. Therefore, query processing needs to be with a revising evaluation mechanism on such environments [7]. The query optimization is an essential module in query processing which generates and chooses the most efficient plan of how to execute the submitted queries [8].

The main challenge is how to provide efficient access to data storage during the queries evaluation over the distributed resources. As Clouds are generally built over wide area networks, query execution performance can be improved by minimizing communication overheads to avoid a slow query response time [9]. The main aim of this paper is to present an enhancement on the query optimization technique of our previous query processing architecture that was presented in [9]. Our proposed architecture overcomes the challenge of slow query response time by optimizing queries, then it schedules and assigns the queries to suitable resources. Furthermore, it monitors the queries evaluation in order to respond to any failure that may occur during the queries execution as presented in our previous works [10, 11]. Our query optimization technique was proposed to exploit and optimize the shared data among the submitted queries to improve the query processing efficiency. This technique is enhanced by conditional merging by considering different delay times of submitted queries in order to merge the queries that have a positive impact on the query execution performance. The evaluation of our query optimization technique is conducted over a real world by using the Amazon Elastic Compute Cloud (EC2) infrastructure provisioning service [12]. The results show that proposing our enhancement on the query optimization technique over a cloud environment offers significant benefits with regards to the overall query response time and the communication overheads.

The paper is organized as follows. Section 2: Reviews related works. Section 3: Describes the implementation of our enhancement on query merging techniques. Section 4: Presents the experimental environment. Section 5: Presents the results. Section 6: The conclusion is presented and discussed.

2 Related Work

In the query processing research area, past research has dealt with issues of query optimization. The Merge-Partition query reconstruction algorithm is presented in [13]. This algorithm can detect and employ data sharing opportunities across concurrent distributed queries and is done by combining multiple similar data requests dispatched to the same distributed data source to a common data request, with the benefits of reduced the average communication overheads. The work is related to the mechanism in [14] which proposed a request window mechanism that allows concurrent query executing to share the common result data. A resource selection module is proposed in [15] as another query optimization technique for query processing. They evaluate the query through selecting a sub-set of resources based on a ranking function. On the other hand, a modeling approach is presented in [16] to estimate the impact of concurrency on query performance for analytical workloads. Their solution depends on the analysis of query behavior and query interactions through sampling techniques to predict the resources contention. In addition, predicting the evaluation behavior of a time varying query workload through query interaction timelines, i.e., a fine grained estimation of the time segments during which discrete mixes are executed at the same time. In the same context, authors in [17] presented an approach relying on the fact that multiple requests that are executed at the same time, may have a positive behavior on the total execution time, because of caching or complementary resource consumption. They proposed a monitoring approach to derive those impacts between request types at runtime from measured execution times. Several techniques also are presented in [18] to dynamically re-distributing processor load assignments through considering the varying resource capabilities. They presented the Flux (Fault-tolerant Load-balancing eXchange) approach that was put forward in [19] which is placed between the consumer and producer phases in a dataflow pipeline to repartition stateful operators while the pipeline is still evaluated under loads. Authors in [20] presented DITN (Data In Network) method which relied on an alternate intra-fragment parallelism. In their method, each instance executes an independent select-project-join, with no tuple exchange between running instances. Their method handles also heterogeneous instances with different capabilities. Several techniques relied on machine learning to estimate the performance metrics of database queries as in [21–23]; their main aim was to create a prediction tool to predict the performance of new queries. Also a statistical approach for resource estimation was proposed in [24]. Their approach relied on predication accuracy in distribution or query plans. Also authors in [25, 26] used a queuing network to define performance metrics for response times and system throughputs. Their approach continually monitored the speed of query operators and used this information to modify the query plan by changing the way rows are routed. They also proposed several practical routing policies for a distributed stream management system. Many previous techniques didn't consider other queries running at the same time and load imbalance that may occur during the queries execution.

The query optimization technique was proposed in [9] which focused on exploiting the shared data between the submitted queries through merging the related queries to minimize the communication overheads. The proposed merging technique in [9] didn't consider the different waiting times of the queries which may have a negative impact on the query execution performance. Therefore, in this paper, we focused on enhancing the work regarding query optimization in the query processing architecture in [9] by taking into account the impact of merging on the query processing performance to improve the query response time and reduce the communication overheads.

3 The Implementation of the Enhanced Merging Technique

Our previous query processing architecture that is presented in [9] has the Query Optimization Sub-system. The main purpose of this sub-system is preparing the queries for execution through determining the most efficient way to dispatch them to the suitable resources. This is done by applying query optimization and scheduling. The scheduling technique determines the efficient ordering of the queries execution as presented in our previous work in [11]. The output of this sub-system is the list of the resources that are responsible for the queries execution.

The Query Optimization Sub-system has many processes as shown in Fig. 1. After accepting the queries from the remote users in our system, each query is transformed to a query tree. The tree contains two types of nodes; internal nodes that contain the operations involved in the query such as: select operation and the relations accessed by the query put into the leaves of the tree. Each node is traversed by the Optimize Queries process to detect the queries which satisfy three conditions to exploit the shared data among them. The group of optimized queries are dispatched to the Merge Queries process in order to construct new merged queries to eliminate data redundancy between the original queries.

Example: For a given submitted query with the form of q: $\Pi_L (\sigma_p (R))$ where L is the list of output attributes, P is the selection predicate and R are queried relations.

Let us assume the following assumptions:

- $L_o(qi)$ is the set of output attributes.
- $P(qi)$ is the selection predicates of qi.
- $L_c(qi)$ is the set of attributes that appear in $P(qi)$.

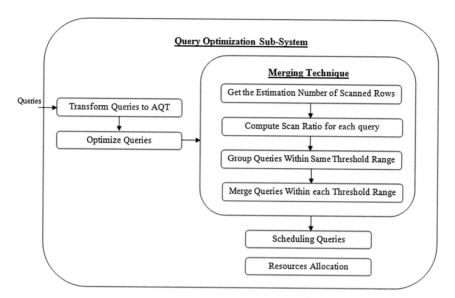

Fig. 1 The proposed enhanced merging technique architecture

For the two queries qi and qj, let R(qi ∩ qj) be their common answer. The two queries can be merged in the case of the three conditions presented in [13] that can all be satisfied:

- $L_c(qi) \subseteq L_o(qi)$.
- $L_c(qj) \subseteq L_o(qj)$.
- $L_o(qj) \subseteq L_o(qj)$.

The query reconstruction mechanism that was presented in [9] didn't consider if the merging process can bring a negative impact on the original queries execution time before merging them. For example, a short running query and a long running query can be merged if they satisfy the conditions in [9]; but in this case, the waiting time of the short running query will increase because of the merging step with the long running query, therefore it can lead to performance degradation. So there is a need before applying the merging process to ensure that it will improve the performance.

Therefore, in this paper we enhanced the merging process through adding another condition to overcome this challenge and improve the overall query processing performance. Our enhanced merging technique contains the following processes as shown in Fig. 1:

- *Get the Estimation Number of Scanned Rows*: For each submitted query, the estimated number of scanned rows are retrieved from the query optimizer of the database management system.

- *Compute Scan Ratio for each query*: For each query, if the estimated number of scanned rows is n and the relation size is r, the Scan Ratio can be computed as follows:

$$\text{Scan Ratio}\,(\%) = (n *100)\backslash r \qquad (1)$$

- *Group Queries Within The Same Threshold Range*: The queries are classified into groups based on the ratio of the estimated number of rows that will be scanned by each query within a specific relation. The queries that will scan the relation by a specific range of thresholds will be grouped with each other for merging.
- *Merge Queries Within each Threshold Range*: Queries can be merged as in [9] if their scan ratio is within the same specific range of thresholds. These ranges of thresholds are specified experimentally as presented in the evaluation section. By applying this enhancement in the merging process, we can ensure that the merged queries will have a positive impact on their execution performance.

4   Experimental Environment

The performance of the proposed work is evaluated on a real environment on Amazon EC2 [12] using standard small instances in the US West region. Amazon EC2 has gained so much popularity among enterprises and researchers that requiring instant and scalable computing power. It provides resizable compute capacity in a computational cloud. This platform allows users to pay only for the capacity that their applications actually need (pay-as-you-go model). We used small instances in our performance evaluation, because they are the default instance size and frequently demanded by users [27]. A small instance (t2.micro) has the following hardware configuration: 1 EC2 Compute Unit (i.e. 1 virtual core with 1 EC2 Compute Unit), 1 GB of main memory, 30 GB of local instance storage, and a 64-bit platform.

Our experimental environment consists of five machines—one master node and four worker nodes. For each worker node instance type, we used Microsoft Windows Server 2008 R2 as the operating system configured with a Microsoft SQL Server 2008 R2 as a database server. The master machine runs on a Inter Core 7, 6 GB of main memory, 2.60 GHz CPU and 1 TB hard disk, ten queries are sent to the system for processing. Example of the used queries are shown in Table 1. Each run was repeated five times and average values reported. After each execution, the database systems was restarted to clean up the bufferpool and to bring the database system to a steady state.

TPC-H database [28] is used as a dataset (scale factor 1) and deployed to test the proposed work. The TPC-H database has eight relations: Region, Nation, Customer, Supplier, Part, Partsupp, Orders, and Lineitem.

Table 1 Example of the used queries	Queries
	select o_orderkey, o_orderstatus, o_totalprice, o_orderdate from orders where o_orderdate > '11-01-1998'
	select o_orderkey, o_orderstatus, o_totalprice, o_orderdate from orders where o_orderdate > '11-01-1996'
	select o_orderkey, o_orderstatus, o_totalprice, o_orderdate from orders where o_totalprice < 50000

5 Evaluations

In order to specify the scan ratio threshold ranges, we executed a number of queries with different numbers of scanned rows. The scan ratio for each query is computed and the response time is recorded with small and big table sizes as shown in Figs. 2 and 3. As observed from the results, the change in the query response time between

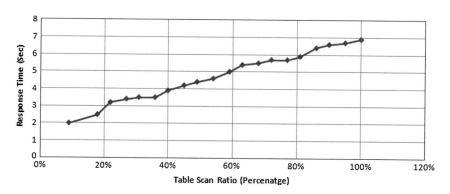

Fig. 2 Specifying the threshold ranges (small table size)

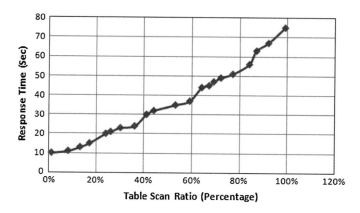

Fig. 3 Specifying the threshold ranges (big table size)

Fig. 4 Measuring the queries execution time

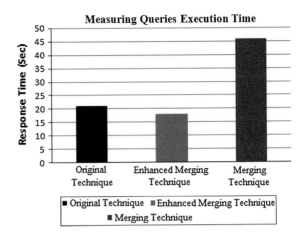

every increasing 20 % of a scanning table is almost to small so we set the boundaries of the threshold ranges as (0–20, 20–40, 40–60, 60–80, 80–100 %). Therefore for each submitted query, we get the estimated number of scanned rows from the query optimizer of the database management system, then compute the query scan ratio and it can be merged with another query has a scanned ratio that falls within the same 20 % ratio range threshold as observed in our experiments.

Figure 4 shows that the execution time of the queries by applying our enhancement can reduce the queries execution time over the original technique by 15 % and over the merging technique in [9] by 60 %. As the submitted queries were mixed between short and long running queries so using the merging technique in [9] resulted in increasing the waiting time of the short running query.

In order to evaluate the performance of our enhancement, the communication overheads of sending bytes from our system is computed and compared with applying the original technique without merging [15] and the merging technique as presented in [9].

Figure 5 shows that the merging technique in [9] has the minimum communication overheads because of increasing the number of merged queries without

Fig. 5 Measuring communication overheads

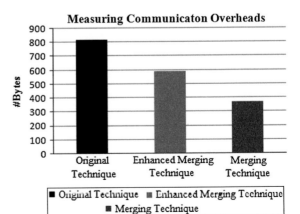

checking the impact of merging which resulted in decreasing the total number of submitted queries. Our enhancement can reduce the communication overheads over the original technique [15] by 27 %.

6 Conclusion

In this paper, an enhancement on query processing architecture is proposed to improve the overall query performance on the cloud computing environment. The architecture overcomes the challenge of a slow query response time. This paper is focused on enhancing the queries merging process by considering the impact of merging on the query execution performance. Our enhancement technique groups the related queries based on the number of rows that will be scanned. The queries can be merged only if their scan ratio is fallen within the same range of specific thresholds.

The results show that applying the enhancement on merging techniques improve the query execution performance over the original technique [15] by 15 % and with or without the merging technique in [9] by 60 %. In addition, Our enhancement can reduce the communication overheads over the original technique [15] by 27 %.

References

1. Samatha, N., Vijay, C.K., Raja, S.R.P.: Query optimization issues for data retrieval in cloud computing. Int. J. Comput. Eng. Res. **2**(1361–1364), 22 (2012)
2. Lang, W., Nehme, R. V. and Rae, I.: Database optimization for the cloud: where costs, partial results, and consumer choice meet. In: Biennial Conference on Innovative Data Systems Research (CIDR'15), California, USA (2015)
3. Amazon Web Services.: http://aws.amazon.com/
4. Google Apps.: www.google.com/Apps/Work
5. Microsoft Azure.: http://azure.microsoft.com/en-us/
6. Aboulnaga, A., Salem, K., Soror, A. A., Minha, U. F.: Deploying DATABASE APPLIANCES IN THE Cloud. In: IEEE Computer Society Technical Committee on Data Engineering (2009)
7. Gounaris, A., Smith, J., Paton, N.W., Sakellariou, R., Fernandes, A.A., Watson, P.: Adaptive workload allocation in query processing in autonomous heterogeneous environments. J. Distrib. Parallel Databases **25**(3), 125–164 (2009)
8. Lua, X., Guan, J.: A new approach to building histogram for selectivity estimation in query processing optimization. J. Comput. Math. Appl. **57**, 1037–1047 (2009)
9. Maghawry, E.A., Ismail, R.M., Badr, N.L., Tolba, M.F.: An enhanced resource allocation approach for optimizing a sub-query on cloud. In: Hassanien, A.E., Salem, A.-B.M., Ramadan, R., Kim, T.-h. (eds.) AMLTA 2012. CCIS, vol. 322, pp. 413–422. Springer, Heidelberg (2012)
10. Maghawry, E.A., Ismail, R.M., Badr, N.L., Tolba, M.F.: Queries based workload management system for the cloud environment. In: AMLTA 2014, vol. 488, pp. 77–86. Springer, Heidelberg (2014)

11. Maghawry, E.A., Ismail, R.M., Badr, N.L., Tolba, M.F.: An enhanced queries scheduler for query processing over a cloud environment. In: ICCES, pp. 409–414. IEEE (2014)

12. Amazon Elastic Compute Cloud (EC2).: http://aws.amazon.com/ec2/

13. Chen, G., Wu, Y., Liu, J., Yang, G., Zheng, W.: Optimization of sub-query processing in distributed data integration systems. J. Netw. Comput. Appl. **34**, 1035–1042 (2011)

14. Lee, R., Zhou, M., Liao, H.: Request window: an approach to improve throughput of RDBMS-based data integration system by utilizing data sharing across concurrent distributed queries. In: 33rd International Conference on Very Large Data Bases, pp. 1219–1230 (2007)

15. Liu, S., Karimi, A.H.: Grid query optimizer to improve query processing in grids. Future Gener. Comput. Syst. **24**, 342–353 (2008)

16. Duggan, J., Cetintemel, U., Papaemmanouil, O., Upfal, E.: Performance prediction for concurrent database workloads. In: SIGMOD, pp. 337–348, Athens (2011)

17. Albuitiu, M.C., Kemper, A.: Synergy based workload management. In: Proceedings of the VLDB Ph.D. Workshop, Lyon (2009)

18. Paton, N.W., Buenabad, J.C., Chen, M., Raman, V., Swart, G., Narang, I., Yellin, D.M., Fernandes, A.A.A.: Autonomic query parallelization using non-dedicated computers: an evaluation of adaptivity options. VLDB **18**, 119–140 (2009)

19. Shah, M.A., Hellerstein, J.M., Chandrasekaran, S., Franklin, M.J.: Flux: an adaptive partitioning operator for continuous query systems. In: 19th International Conference on Data Engineering, pp. 25–36. IEEE Press (2003)

20. Raman, V., Han, W., Narang, I.: Parallel querying with non-dedicated computers. In: 31st international conference on Very large databases, pp. 61–72. Trondheim, Norway (2005)

21. Ganapathi, A., Kuno, H., Daval, U., Wiener, J., Fox, A., Jordan, M., and Patterson, D.: Predicting multiple performance metrics for queries: better decisions enabled by machine learning. In: Proceedings of International Conference on Data Engineering, Shanghai, pp. 592–603, Mar 2009

22. Luo, G., Naughton, J. F., Yu, P. S.: Multi-query SQL progress indicators. In: Proceedings of the 10th International Conference on Extending Database Technology, pp. 921–941, Munich, March 2006

23. Duggan, J., Papaemmanouil, O., etintemel, U. C., Upfal, E.: Contender: a resource modeling approach for concurrent query performance prediction. In: Proceedings of International Conference on Extending Database Technology (EDBT) (2014)

24. Li, J., König, A. C., Narasayya, V., Chaudhuri, S.: Robust estimation of resource consumption for Sql queries using statistical techniques. In: Proceedings of the VLDB Endowment, Istanbul, vol. 5, pp. 1555–1566, July 2012

25. Avnur, R., Hellerstein, J.M.. Eddies.: continuously adaptive query processing. In: Proceedings of the 2000 ACM SIGMOD international conference on Management of Data, vol. 29, pp. 261–272 (2000)

26. Tian, F., DeWitt, D.J.: Tuple routing strategies for distributed eddies. In: Aberer, K., Koubarakis, M., Kalogeraki, V. (eds.) Proceedings of the 29th International Conference on Very Large Data Bases, LNCS, vol. 2944, pp. 333–344. Springer, Heidelberg (2004)

27. Jorg, S., Jens, D., Jorge-Arnulfo, Q.: Runtime measurements in the cloud: observing, analyzing, and reducing variance. In: 36th International Conference on Very Large Data Bases, vol. 3, Singapore (2010)

28. Transaction Processing and Database Benchmark.: http://www.tpc.org/tpch/

Compressive Data Recovery in Wireless Sensor Networks—A Matrix Completion Approach

Maha A. Maged, Haitham M. Akah, Bassant Abdelhamid and Salwa H. El-Ramly

Abstract Energy-consumption of the wireless sensor networks(WSNs) is proportional to the *sampling rate*. The need for more energy-efficient methods for WSNs data gathering is greater demand since most of the energy is consumed in sampling and transmission. Recently, *compressive sensing* (CS) and *matrix completion* (MC) have been earning increasing interests in the area of WSNs. Both of CS and MC exploit the sparsity presents in sensing environment to reduce sampling rate needed for data acquisition and processing. In this paper, a new scheme based on MC is proposed to reduce such sampling rate. The new scheme is evaluated in a real-data domain and its performance is compared with CS techniques in unified framework. The results show the superiority of MC over the CS techniques in the accuracy of data recovery with different sampling rate.

Keywords Compressive sensing · Matrix completion · Wireless sensor networks · Energy-saving

1 Introduction

In recent years, the development of WSNs has received much attention both in the academia and in the industry. Due to its ability to operate at inaccessible areas. While WSNs are equipped to handle some of complex functions and in-network processing,

M.A. Maged (✉) · H.M. Akah
Space Communications Department, NARSS, Cairo, Egypt
e-mail: maha_maged@yahoo.com; maha_maged@narss.sci.eg

H.M. Akah
e-mail: haitham_akah@narss.sci.eg

B. Abdelhamid · S.H. El-Ramly
Faculty of Engineering, Ain Shams University, Cairo, Egypt
e-mail: bassant.abdelhamid@eng.asu.edu.eg

S.H. El-Ramly
e-mail: salwa.elramly@eng.asu.edu.eg

© Springer International Publishing Switzerland 2016
T. Gaber et al. (eds.), *The 1st International Conference on Advanced Intelligent System and Informatics (AISI2015), November 28–30, 2015, Beni Suef, Egypt,*
Advances in Intelligent Systems and Computing 407, DOI 10.1007/978-3-319-26690-9_32

357

these sensor nodes are required to use their energies efficiently in order to extend their effective network lifetime [1–3]. It has recognized the sparseness in signal structures in a surplus of applications where, various signals of interest may be sparsely modelled and that sparse modelling is often beneficial, or even essential to signal recovery. The rich theory of sparse and compressible signal recovery has recently been developed under the name compressed sensing (CS). This revolutionary research has demonstrated that many signals can be recovered from severely under-sampled measurements by taking advantage of their inherent *low-dimensional structure*. More recently, an offshoot of compressive sensing has been a focus of research on other low-dimensional signal structures such as *matrices of low rank*. A subclass of low rank matrices problems called *matrix completion* (MC). To further extend the sparse problem from the vector to the matrix, the matrix completion can complete a matrix from a small set of corrupted entries based on the assumption that the matrix is essentially of low rank. The compressive sensing involves in the vectors l_1-norm minimization, which is the number of non-zero items of a vector. In matrix recovery problem, the signal to be recovered is a low-rank matrix $M \in R^{m \times n}$, about which we have information supplied by means of a linear operator $P_\Omega : R^{m \times n} \to R^{m \times n}$. In this paper, a new method based on the matrix completion for WSNs is proposed to reduce such sampling rate based on non pre-defined probability (i.e., *random-distribution*). The purposed method is evaluated in a real-data WSNs domain and its performance is compared to two compressive sensing methods in unified framework. The key contributions of this paper are concluded as follows: (1) We propose MC technique for data recovery to improve the lifetime of wireless sensor networks by reduce such sampling rate, (2) We evaluate the proposed techniques accuracy through experimental results, and (3) We demonstrate the superiority of our MC data recovery scheme through experiments on extensive datasets compared with other CS-algorithms, our MC-scheme achieves significant energy-saving while delivering data to the end with high fidelity.

The rest of the paper is organized in the following manner. In Sect. 2, the related works are surveyed. In Sect. 3, we describe a brief mathematical introduction on CS and MC theory that will be applied in this paper. In Sect. 4, a new algorithm is proposed for data aggregation and recovery in WSNs. Detailed experiments results are provided in Sect. 5. Finally, Sect. 6 concludes the paper and discuss the future work.

2 Related Works

The traditional approach to data recovery in WSNs is to collect data from each sensor to one central server, and then process and compress the data centrally. However, WSNs sensing data often have high spatial or temporal correlation [4]. These sparsity in structure of the signals, channels and sensors can be explored and utilized by many techniques such as distributed source coding techniques [5], in-network collaborative wavelet transform methods [6] and clustered data aggregation methods [7]. Recently, both compressive sensing and matrix completion help to significantly reduce the number of transmissions required to have an efficient and reliable data communications [8]. This is because, the compressive sensing (CS) [9]; is compressing the sensed data in each sensor prior to transmission to the fusion center [10, 11]. Recently, Candes and Recht [12] describes the matrix completion (MC) which has emerged as an extend to compressive sensing and promising technology to enable efficient data gathering in WSNs [10, 11, 13, 14]. The matrix representations have many advantages over vector representations. The matrix completion proposed approaches are taking the advantage of the *intra-temporal* correlation of the sensor measurements and the data matrix filled by sensory readings exhibits low rank feature [15]. Therefore, the fusion center can only collect part of the sensory readings and use matrix completion algorithm to recover the missing data, due to data communication losses results from bandwidth limitations and the resource-constrained nature of the sensors, by using the matrix completion principles [16]. Both compressive sensing and matrix completion based data compression are moves most computation from sensor nodes to the fusion center, which makes it a good fit for in-network data suppression and compression. Cheng et al. [17] are proposed an Efficient Data Gathering Approach (EDCA) which takes advantage of the low-rank feature to achieve both less traffic and high-level accuracy. However, suffer from a low sampling rate tends to result in insufficient measurements, leading to large recovery errors. Recently, a Spatio-Temporal Compressive Data Collection (STCDG) is an extend to (EDCA) data gathering scheme which takes advantage of both the low-rank and short-term stability features to reduce the amount of traffic in WSNs and achieves a reasonable recovery accuracy [18]. One of the major disadvantages of these previously successful attempts to implement the matrix completion for WSNs, is to forward its readings to the fusion center according to a *pre-set probability*. Another issue is does not evaluate its performance against compressive sensing-based approaches in low level sampling rate.

3 Preliminaries

3.1 Compressive Sensing Technique

Compressive sensing has emerged as a new framework for signal acquisition and sensor design that enables a potentially large reduction in the sampling and computation costs for sensing signals that have a sparse or compressible representation. The concept of compressive sensing is based on the fact that there is a difference between the rate of change of a signal and the rate of information in the signal [19]. A conventional compression algorithm would then be applied to all of these samples taken to remove any redundancy present, giving a reduced number of bits that represent the signal. In contrast, compressive sensing exploits the information rate within a particular signal. In the context of a WSN, suppose that $\mathbf{x} \in R^N$ is a signal referred to the sensed data of an individual sensor. compressive sensing theory proves that if $\mathbf{x}$ is sparse in some domain, it can be *reconstructed exactly* with high probability from $\mathbf{M}$ randomized linear projections of signal $\mathbf{x}$ into a measurement matrix $\Phi \in R^{M \times N}$, where $M \ll N$. When a signal is sparse in the domain it is samples in, e.g. a high sample-rate time-series sampling infrequently occurring events, one can make a *condensed representation* of this signal by simple dimensionality reduction.

$$y = \Phi x \tag{1}$$

where in (1) $\mathbf{x}$ is an N-dimensional K-sparse vector of the original signal, Φ is an $M \times N$ dimensionality reduction matrix and $\mathbf{y}$ is an M-dimensional dense vector containing a condensed representation of $\mathbf{x}$, where:

$$K \leq M \ll N \tag{2}$$

It can then be proven that using *random projection* i.e. using a random dimensionality reduction matrix Φ, will allow for recovery of $\mathbf{x}$ with little to no information loss, with overwhelming probability. When a signal is sparse in a domain different from the sampled domain, one can make a transformation, by way of a transformation matrix.

$$y = \Phi \Psi b \tag{3}$$

where in (3) Ψ is an $N \times N$ transformation matrix e.g. Wavelets, Discrete Fourier Transform (DFT) or Discrete Cosine Transform (DCT). The dimensionality reduction matrix and the transformation matrix can be combined to an augmented dimensionality reduction matrix.

$$y = \Phi \Psi x = \Theta b \tag{4}$$

where $\Theta = \Phi \Psi$. This, combined with the fact that random projection is an option negates the need for a transformation matrix, as a random projection of any matrix is just another random matrix. It is then only necessary to be aware of the domain in which the signal is sparse when one is reconstructing the signal, not when sampling or transmitting. One needs only to be able to know or construct the random projection matrix used to compress the signal. This can be done by using pseudo-random generator with a known or predictable seed. Essentially, each sensor instead of transmitting a signal $\mathbf{x} \in R^N$, it performs compressive sensing, and it finally transmits a smaller signal $\mathbf{y} \in R^M$. The original vector b and consequently the sparse signal $\mathbf{x}$ is estimated by solving the following $\ell_1 - norm$ constrained optimization problem:

$$\min_{\mathbf{b}} \| \mathbf{b} \|_1 \quad \text{subject to} \quad y = \Theta b \tag{5}$$

3.2 Matrix Completion Technique

The job of matrix completion is to recover a matrix $\mathbf{M}$ from only a small sample of its entries. Let $\mathbf{M} \in R^{m \times n}$ be the matrix we would like to know as precisely as possible while only observing a subset of its entries $(i, j) \in \Omega$. It is assumed that observed entries in Ω are uniform randomly sampled. Low-rank matrix completion is a variant that assumes that $\mathbf{M}$ is low-rank. The tolerance for "low" is dependant upon the size of $\mathbf{M}$ and the number of sampled entries. For further explanation let us define the sampling operator $P_\Omega : R^{m \times n} \rightarrow R^{m \times n}$ as

$$P_\Omega(\mathbf{A}) = \begin{cases} A_{ij} & : (i,j) \in \Omega \\ 0 & : \text{otherwise} \end{cases} \tag{6}$$

Let an opposite operator $P_{\bar{\Omega}}$ is defined which keeps those outside Ω unchanged and sets values inside Ω (i.e. $\bar{\Omega}$) to 0. Since we assume that the matrix to recover is low-rank, one could find the unknown matrix by solving

$$\begin{aligned} & \text{minimize } \text{rank}(\mathbf{M}) \\ & \text{subject to } M_{ij} = A_{ij} \qquad (i,j) \in \Omega \end{aligned} \tag{7}$$

where matrix $\mathbf{M}$ is the partially observed matrix. Unfortunately, this optimization problem has been shown to be NP-hard and all known algorithms which provide exact solutions require exponential time complexity in the dimension of the matrix both in theory and in practice [12]. Also, the $rank(\mathbf{M})$ makes the problem non-convex and several approximations to the problem exists. One, the tightest convex relaxation of $rank(\mathbf{M})$ [20] is the following problem

$$\begin{aligned} & \text{minimize } \tau \|\mathbf{M}\|_* := \sum_{i=1}^{r} \sigma_i(\mathbf{M}) \\ & \text{subject to } P_\Omega(\mathbf{M}) = P_\Omega(\mathbf{A}) \end{aligned} \tag{8}$$

where $\sigma_i(\mathbf{M})$ denotes the ith largest singular value of M and $\|M\|_*$ is called the nuclear norm. The main point of this relaxation is that the nuclear norm is a convex function and thus can be optimized efficiently via semi-definite programming or by iterative soft thresholding algorithms [21, 22]. The case where observed entries may contain a limited amount of noise $\mathbf{E}$ is considered. The corresponding objective is considered as following:

$$\min_{\mathbf{M}} \tau \|\mathbf{M}\|_* + \frac{\lambda}{2} \|P_\Omega(\mathbf{E})\|_F^2 \qquad \text{subject to } P_\Omega(\mathbf{A}) = P_\Omega(\mathbf{M}) + P_\Omega(\mathbf{E}) \tag{9}$$

where τ and λ are regularisation parameters. The matrix completion theory proves that if this matrix (denoted by $\mathbf{M} \in R^{m \times n}$) has a low rank, it can be recovered with high probability. More interestingly, one can recover ($\mathbf{M}$) from $s \geqslant Cd^{6/5}r\log(d)$ random measurements, where C is a positive constant, $d = max(m, n)$, and r is the rank of the matrix [23].

4 Proposed Approach

In this section, a new data aggregation method for WSNs which is based on matrix completion is proposed. Consider a sensor network consisting of N nodes. Each node samples at a fixed rate and forwards the data to the fusion center through a single-hop way. Where at the fusion center an $N \times T$ matrix can be constructed, where T means the maximum number of samples at each node, i.e., $M \in R^{N \times T}$. However, due to the sampling scheduling or lossy transmission and the failure of sensor nodes, some data are missing and the matrix is incomplete and only S samples are received by fusion center. Thus adopting a linear operator $P_\Omega(\cdot)$ to represent such lossy or selection effect(s) (see Sect. 3). The implemented matrix completion algorithm using the Interest-Zone Matrix-Approximation [24]. The representation of the operations of proposed system is shown below in Fig. 1 and the proposed procedure is listed in Algorithm 1. As demonstrated in Fig. 1, the data sampled or collected and transmitted to the fusion center which recovers the missing or lost samples based on the linear operator $P_\Omega(\cdot)$.

Algorithm 1 WSN data recovery using matrix completion (MC) algorithm

---

for *sensor* = 1 to N
    for *time* = 1 to T
        1. Apply uniform data sampling or selection for S samples
        2. Send the S samples to fusion center
        3. The fusion center form a $\mathbf{M} = N \times T$ matrix
        4. Recover the missing samples values at fusion center
        $\min_{\mathbf{M}} \ \| P_\Omega(\mathbf{M}) - P_\Omega(\mathbf{A}) \|_F$ subject to $f(\mathbf{M}) \leq 0$
        5. If $\| \mathbf{M} - \hat{\mathbf{M}} \| < \epsilon$, where ϵ is a specified stopping criteria, stop.
    end
end

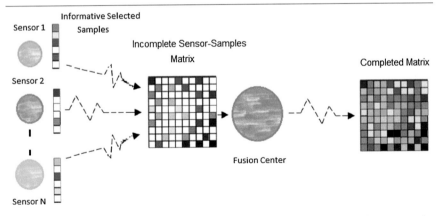

Fig. 1 Representation of the operations of matrix completion system in a WSN. The sensor nodes send the measured data to the fusion center based on the sampling or lossy communication, then the fusion center complete these entries using matrix completion algorithm

By applying the same concept using the compressive sensing theory for both the *orthogonal matching pursuit* (OMP) algorithm and *adaptive principle component analysis* (PCA) [14], where the signal model is learned from the measurements.

5 Implementation and Evaluation

In this section, the performance of discussed methods are evaluated and compared empirically; the applied algorithms is verified in the multi-sensor case where nearby sensor nodes measure spatially correlated signals and a real-world meteorological datasets are used as the ground truth, namely Valais [14]. The Valais is provided by a micro-climate monitoring service provider [25]. A total of six stations are deployed in a mountain valley, covering an area of around $18 \, \text{km}^2$. The deployments were started in March 2012 and collected 125 days of continuous temperature measurements. The data set is 144 uniformly sampled data points for each day. The real world trace forms a 144×6 matrix which represents the temperature values sampled. A key issue regarding the evaluation of performance of different techniques, is related to the specific metric that we use in order to quantify the results. The commonly used root mean squared error (RMSE) is employed which given by,

$$\text{RMSE} = \frac{1}{\sqrt{m \times n}} \parallel \mathbf{M} - \hat{\mathbf{M}} \parallel_{l_2} \tag{10}$$

defined as the root mean squared error between the fully populated and the reconstructed measurements matrix, normalized with respect to the l_2 norm. where, $\mathbf{M}$ is the actual data matrix, $\hat{\mathbf{M}}$ is the recovered data matrix and $m \times n$ are the dimensions of the matrices. In our paper, we implemented the proposed data recovery methods using MATLAB. Table 1 summarize the reconstruction error under a fixed data selection or sampling rate (η) with respect to the estimation error of SNR, whereas the true SNR is 30 dB. According to results given in Table 1 and Fig. 2, MC is both more accurate and robust when compared to the state-of-the-art CS methods.

Then, we analyze the influence of the sampling rate. Figure 3, comparing the recovery error (RMSE) against different sampling rate (η) shows that the MC-scheme has the minimum error value compared to other techniques. The experiments have been carried out with 5–50 % of the samples. It has been empirically observed that

Table 1 Evaluation of different approaches at sampling rate $\eta = 20\%$

	CS-OMP	CS-PCA	MC
RMSE	2.237493	1.522985	0.957022
Average error	1.441720	1.173088	−0.068085
Maximum error	56.238366	7.542841	6.387068

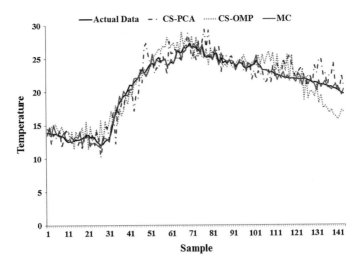

Fig. 2 The recovery of the last day for first sensor using orthogonal matching pursuit (OMP), adaptive principle component analysis (PCA) and matrix completion (MC) algorithms

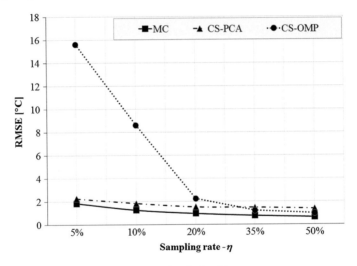

Fig. 3 The sampling rate (η) versus RMSE of each algorithm

if the number of samples is higher, it slightly reduces the error of the recovered value. The results shows that both of MC and PCA recovery rate is more accurate in low sampling rate than OMP algorithm. However, at higher sampling rate the OMP algorithm converges nearly to the MC and PCA results. An estimations of the SNR for the noisy measurement is necessary to avoid degradations of the recovery

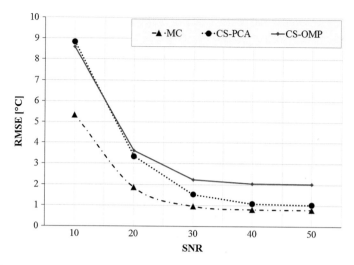

Fig. 4 Recovery error (RMSE) versus SNR (dB) of the measurements under sampling rate ($\eta = 20\%$)

quality. Figure 4 shows the comparison results of the three algorithms. In the high SNR regime ($\geqslant 30$ dB), the three algorithms tend to perform the similar trend and are stable w.r.t. errors in the SNR estimation. However, it can be noted that MC generally performs better than the two compressive sensing-based schemes, in high and low SNR regime. also, it can be see that matrix completion performs the best and is more accurate and robust when compared to the other sparse sensing methods.

6 Conclusions and Future Work

In this paper, a new data aggregation technique for WSNs based on MC is proposed, which obtains successful results with a uniform random selection of samples. The detailed algorithm is explained, and experiments are carried out to show its effectiveness. The potential of both CS and MC techniques are investigated by using real sensor data. The results of our simulations on real data is promising and confirming that utilizing spatio-temporal correlation in MC increases the performance, which is introduced in terms of reconstruction error and the number of measurements taken. The fact that a minimum *mutual coherence* between the measurement matrix Φ and representing matrix Ψ in CS-schemes is not guaranteed which is a requirement for achieving a successful CS recovery even with the using of adaptive PCA scheme. The main contribution is to show the feasibility of employing the MC for WSN data aggregation purposes in the under-sampled case which achieves significant energy-saving while delivering data to the end with high fidelity. In the future, the studies will be aimed to improve the performance of proposed method and study the effect of different parameters such as *communication loss* and increasing *number of sensors* for very large-scale WSN problems.

References

1. Krishnamachari, B.: Networking Wireless Sensors. Cambridge University Press, New York (2005)
2. Anastasi, G., Conti, M., Francesco, D., Passarella, A.: Energy conservation in wireless sensor networks a survey. Ad Hoc Netw. **7**(3), 537–568 (2009)
3. Duarte, M., Shen, G., Ortega, A., Baraniuk, R.: Signal compression in wireless sensor networks. Philos. Trans. R. Soc. A **370**, 118–135 (2012)
4. Vuran, M., Akan, O., Akyildiz, I.: Spatio-temporal correlation: theory and applications for wireless sensor networks. Comput. Netw. J. **45**, 245–259 (2004)
5. Yuen, K., Liang, B., Li, B.: A distributed framework for correlated data gathering in sensor networks. IEEE Trans. Veh. Technol. **45**, 578–593 (2008)
6. Ciancio, A., Pattem, S., Ortega, A., Krishnamachari, B.: Energy efficient data representation and routing for wireless sensor networks based on a distributed wavelet compression algorithm. In: Proceedings of the Fifth International Conference on Information Processing in Sensor Networks, USA, pp. 309–316 (2006)
7. Xu, X., Li, X., Wan, P., Tang, S.: A distributed framework for correlated data gathering in sensor networks. IEEE/ACM Trans. Netw. **20**, 690–698 (2012)
8. Fragkiadakis, A., Askoxylakis, I., Tragos, E.: Joint compressed-sensing and matrix-completion for efficient data collection in WSNs. In: 18th IEEE International Workshop on CAMAD, pp. 79–83 (2013)
9. Candes, E., Wakin, M.: An introduction to compressive sampling. IEEE Signal Process. Mag. **25**(2), 21–30 (2008)
10. Zhu, H., Li, H., Yin, W.: Compressive Sensing for Wireless Networks. Cambridge University Press, New York (2013)
11. Abdur-Razzaque, M., Dobson, S.: Energy-efficient sensing in wireless sensor networks using compressed sensing. Sensors **14**, 2822–2859 (2014)
12. Candes, E., Recht, B.: Exact matrix completion via convex optimization. Found. Comput. Math. **9**(6), 717–772 (2009)
13. Huang, H., Misra. S., Tang, W., Baran, H., Al-Azzawi, H.: Applications of compressed sensing in communications networks. IEEE (2014)
14. Chen, Z., Ranieri, J., Zhang, R., Vetterli, M.: DASS: Distributed Adaptive Sparse Sensing. IEEE proceedings (2010)
15. Piao, X., Hu, Y., Sun, Y., Yin, B., Gao, J.: Correlated spatio-temporal data collection in wireless sensor networks based on low rank matrix approximation and optimized node sampling. Sensors **14**, 23137–23158 (2014)
16. Candes, E., Plan, Y.: Matrix completion with noise. Proc. IEEE **98**(6), 925–936 (2010)
17. Cheng, J., Jiang, H., Ma, X., Liu, L., Qian, L., Tian, C., Liu, W.: Efficient data collection with sampling in WSNs: making use of matrix completion techniques. In: IEEE Proceedings Globecom 2010 (2010)
18. Cheng, J., Ye, Q., Jiang, H., Wang, C., Wang, D.: STCDG: an efficient data gathering algorithm based on matrix completion for wireless sensor networks. IEEE Trans. Wirel. Commun. **12**(2), 850–861 (2013)
19. Donoho, D.: Compressive sampling. IEEE Trans. Inf. Theory **52**(4), 1289–1306 (2006)
20. Fazel, M.: Matrix rank minimization with applications. Ph.D. thesis, Stanford University (2002)

21. Cai, J.F., Candes, E., Shen, Z.: A singular value thresholding algorithm for matrix completion. SIAM J. Optim. **20**(4), 1956–1982 (2010)
22. Goldfarb, D., Ma, S.: Convergence of fixed-point continuation algorithms for matrix rank minimization. Found. Comput. Math. **11**(2), 183–210 (2011). Springer
23. Candes, E., Tao, T.: he power of convex relaxation: near-optimal matrix completion. IEEE Transactions on Information Theory **56**(5), 2053–2080 (2010)
24. Shabat, G., Averbuch, A.: Interest zone matrix approximation. Electron. J. Linear Algebra **23**(1), 50 (2012)
25. Ingelrest, F., Barrenetxea, G., Schaefer, G., Vetterli, M.: Sensorscope: application specific sensor network for environmental monitoring. ToSN **6**(2), 17 (2010)

Big Data Classification Using Belief Decision Trees: Application to Intrusion Detection

Mariem Ajabi, Imen Boukhris and Zied Elouedi

Abstract Over the past few years, the data volume explosion fueled by exciting progression in computer technologies, made the Big Data the focus of widespread attention. Big Data is nebulous since it is an interaction result of several dimensions of scale, among them the veracity which refers to biases and noise in data. Therefore, Big Data veracity is a challenge because it requires a different approach in order to cope with this imperfection. We propose to involve the belief function theory and the belief decison tree as a classification technique to accommodate large applications where the uncertainty reigns. In this paper, we will be firstly concerned with the construction of the belief decision tree, using MapReduce programming model and the averaging approach as a classification method under uncertainty. Then, we will conduct experiments on intrusion detection massive data set, to distinguish between attacks and normal connections in such uncertain context.

Keywords Big Data · MapReduce · Hadoop · Veracity · Belief function theory · Classification · Belief decision tree · Intrusion detection

1 Introduction

Through the recent decades, many supervised learning techniques are developed to ensure classification [1]. Among them decision trees which are one of the most popular methods for classification in several data mining applications. Decision trees are simple to use and easy to understand [2]. However, this approach exhibits some drawbacks related basically to its inability to ensure classification in an uncertain envi-

M. Ajabi (✉) · I. Boukhris (✉) · Z. Elouedi
LARODEC, Institut Supérieur de Gestion de Tunis, Université de Tunis, Tunis, Tunisia
e-mail: ajabi.mariem@gmail.com

I. Boukhris
e-mail: imen.boukhris@hotmail.com

Z. Elouedi
e-mail: zied.elouedi@gmx.fr

© Springer International Publishing Switzerland 2016
T. Gaber et al. (eds.), *The 1st International Conference on Advanced Intelligent System and Informatics (AISI2015), November 28–30, 2015, Beni Suef, Egypt,*
Advances in Intelligent Systems and Computing 407, DOI 10.1007/978-3-319-26690-9_33

ronment. Moreover, ignoring such uncertainty may lead to bad results. To overcome these issues, we shed more light on belief decision trees (BDT) [3] as a well-known approach adapted to handle such uncertainty about objects' classes in the training set. In fact, BDTs are based on both decision trees as a classification technique and belief function theory as an uncertainty theory. In addition, building a decision tree can be very time consuming especially when the data set is extremely huge which involves the need to parallel computing. Indeed, with the continual evolution in technology, the amount of available data has rapidly sprung up. This makes the Big Data a fact of today's world as well as a challenge to deal with. In this paper, we propose firstly a distributed adaptation of belief decision trees using MapReduce as a distributed framework in order to achieve the Big Data classification in an uncertain context. The second aim of this paper is to provide experimental results using KDD'99 intrusion detection data set. We will be concerned by several experimentations to distinguish between attacks and normal connections in an uncertain context of Big Data security. Furthermore, some related works such as [4] and [5] have used a sample of this data set to ensure the classifcation in a certain environment using the standard decision trees. The rest of this paper is structured as follows: Sect. 2 provides a review of the Big Data. Section 3 presents the belief decision trees. Our proposed solution related to Big Data classification using BDTs is introduced in Sect. 4. Section 5 exposes its experimental results and analysis conducted on KDD'99 massive data set. Finally, the contribution is summarized and the paper is concluded in Sect. 6.

2 Big Data

The world contains an unimaginably large amount of data which is getting ever vaster rapidly. Therefore, any application must be able to scale up to voluminous real-world data sets. Big Data is defined as an evolving term for huge data sets containing varied and complex structure and having the potential to be mined.

2.1 Big Data Dimensions

Nowadays, the world is awash with data that can be analyzed. However, scale alone does not tell the whole story of what makes Big Data a moving target. Big Data is characterized by several dimensions (the V's) among them we define [6]:

- **Volume**: 90 % of all available data have been collected only between 2011 and 2013. That is why, Petabyte data sets are rapidly becoming the norm.
- **Velocity**: it refers to the increasing speed of data generation, process and storage.
- **Variety**: data today come from a great variety of sources and in all shapes: structured, semi structured and unstructured format.

- **Veracity**: this dimension refers to the credibility, noise or doubt in data. The data deluge from multiple sources increases data imperfection. Thus, the trustworthiness of data becomes questionable [7].

2.2 Hadoop

Big Data requires exceptional and scalable tools to deal efficiently with it, among them Hadoop [8]. Hadoop is an open source programming framework that supports the processing of large data sets in a distributed computing environment. Thus, Hadoop enables a computing solution that is scalable, cost effective and flexible. This powerful plateform provides two major components:

- **MapReduce**: it is both a programming model for depicting distributed computations on huge data, and a powerful execution framework for processing these data on clusters. It is based on the divide and conquer method and it is implemented by two primitives: the first one is the Map function that takes a set of data and converts it to a set of intermediate (key, value) pairs associated to the same key. The second one is the Reduce function that retrieves the output from the previous task and merges these values together to form a possibly smaller set of values.
- **Hadoop Distributed File System (HDFS)**: it is a highly fault tolerant storage system that stores voluminous data while overcoming failures during the storage without losing data [9].

3 Belief Decision Trees

In this section, we will recall the belief decision trees as well as the belief function theory. More details can be found in [3].

3.1 Belief Function Theory

The belief function theory or Dempster-Shafer theory [10] as interpreted in the Transferable Belief Model (TBM) [11] is a useful theory to represent uncertain knowledge [12]. Suppose that Θ is a finite set of elementary events applied to a given problem, we call it the frame of discernment. The impact of a piece of evidence on the subsets of Θ is represented by the basic belief assignement (*bba*) [3], defined as follows:

$$m : 2^{\Theta} \rightarrow [0, 1]$$

$$\sum_{A \subseteq \Theta} m(A) = 1 \tag{1}$$

In this case, $m(A)$ represents the basic belief mass (*bbm*) which defines the part of belief assigned exactly to the event A. Shafer has proposed:

- Normalized *bba* expressed by: $m(\emptyset) = 0$
- Total ignorance described by: $m(\Theta) = 1$
- Certain *bba* defined by: $m(A) = 1$ *and* $m(B) = 0$ *for all* $B \neq A, B \subseteq \Theta$ and $A \in \Theta$

In the Transferable Belief Model, we distinguish between holding beliefs and making decisions. Hence, TBM is based on two levels mental model:

- **The credal level**: this level tends to represent knowledge.
- **The pignistic level**: beliefs are used to make decisions and represented by probability functions called pigninstic probabilities which can be defined as:

$$BetP(A) = \sum_{B \subseteq \Theta} \frac{|A \cap B|}{|B|} \frac{m(B)}{(1 - m(\emptyset))} \; where \; A \in \Theta \tag{2}$$

3.2 Belief Decision Trees

A belief decision tree is a decision tree in an uncertain environment [3]. There are two approaches of building BDT: the averaging approach and the conjunctive approach. In this paper, we deal only with the avaraging one. It is an extension of the classical approach based on the gain ratio criterion [2]. There are two fundamental aims of this classification technique under uncertainty:

- **Building belief decision trees**: the purpose of this step is to find at each decision node of the tree, the best appropriate attribute that can reduce the heterogeneity between classes. We assume that each class of a training object may be uncertain and represented through a *bba*, whereas the features' values are known with certainty.
- **Ensuring the classification**: using a top down approach, we repeat testing the attribute in each node until reaching a leaf. As a leaf is characterized by a *bba* on classes, the pignistic transformation is applied. Indeed, we choose the class holding the highest value of pignistic probability. Due to the incompleteness related to the instances' classes, the structure of leaves changes. Instead of assigning a unique class to each leaf, it will be labeled by a *bba* expressing the part of belief on the actual class of objects belonging to this leaf.

Before building a belief decision tree, we must first of all define some parameters:

- **The attribute selection measure**: this step tends to determine the best attribute for each node in order to have the best feasible partition of the training set T. It is based on the entropy [13] computed from the average of the pignistic probabilities relative to each instance in the node. The following steps are proposed to choose the appropriate attribute:

1. Compute the pignistic probability of each object I_j by applying the pignistic transformation to $m^{\Theta}\{I_j\}$
2. Compute the average pignistic probability function $BetP^{\Theta}\{S\}$ taken over the set of objects S. For each $C_i \in \Theta$:

$$BetP^{\Theta}\{S\}(C_i) = \frac{1}{|S|} \sum_{I_j \in S} BetP^{\Theta}\{I_j\}(C_i) \qquad (3)$$

3. Compute the entropy Info(S) of the average pignistic probabilities in the set S. This Info(S) corresponds to:

$$Info(S) = - \sum_{i=1}^{n} BetP^{\Theta}\{S\}(C_i) \log_2 BetP^{\Theta}\{S\}(C_i) \qquad (4)$$

4. Select an attribute A. Then, collect the subset S_v^A made with cases of S having v as a value for the attribute A. The next step need the computation of the average pignistic probability for objects in subset S_v^A. The result is denoted $BetP^{\Theta}\{S_v^A\}$.
5. Compute $Info_A(S)$ [2]:

$$Info_A(S) = - \sum_{v \in D(A)} \frac{|S_v^A|}{|S|} Info(S_v^A) \qquad (5)$$

where D(A) is the domain of the possible values of the attribute A and $Info(S_v^A)$ is computed using $BetP^{\Theta}\{S_v^A\}$.
6. Compute the information gain provided by the attribute A in the set of objects S as follows:
$$Gain(S, A) = Info(S) - Info_A(S) \qquad (6)$$

7. Using the *SplitInfo*, compute the gain ratio relative to attribute A:

$$Gainratio(S, A) = \frac{Gain(S, A)}{SplitInfo(S, A)} \qquad (7)$$

where

$$SplitInfo(S, A) = - \sum_{v \in D(A)} \frac{|S_v^A|}{|S|} \log_2 \frac{|S_v^A|}{|S|} \qquad (8)$$

8. Repeat the same process for every attribute A belonging to the set of attributes that can be selected. Next, choose the attribute that maximizes the gain ratio.

- **The partitioning strategy**: the current training set will be divided by taking into account the selected test attribute.

- **The stopping criteria**: we stop the partitioning process if we are faced with one of these four situations: if the treated node includes only one instance or if it includes only instances that have the same $m^\Theta\{I_j\}$ or if all the attributes are split or finally if the value of the applied attribute selection measure has no more informational contribution.

The leaf in a BDT will be labeled by a *bba*. If only one object belongs to the leaf, the leaf's *bba* is equal to the instance's *bba*. Otherwise, the leaf's *bba* is equal to the average of the different instance's *bba* belonging to this leaf.

4 Big Data Classification Using Belief Decision Trees

In this paper, we propose a MapReduce adaptation of belief decision trees for several reasons. Firstly, this solution takes into account the uncertainty of the environmental context during the classification of voluminous data which was not possible in the standard decision trees. Second, it is based on parallel computing in order to deal with time consuming problem. Indeed, the BDT is unable to perform the classification of huge data set such as the KDD'99. Third, the proposed algorithm is general since it deals with categorical as well as numerical attributes. To this end, we need to firstly define several data structures and then expose the proposed solution.

4.1 Handling Data Structure Within the Proposed Method

The data set is generally composed by attributes in columns, instances in rows denoted by a unique identifier and a value of each feature belonging to an object in cells. The last column corresponds to the class attribute. However, due to the uncertainty caused by the environment context, the structure of the data set is different from the traditional one. Note that, the proposed distributed parallel of BDTs is typically dynamic. Indeed, there is an exchange between MapReduce nodes in order to retrieve information. This information exchange is beneficial since it can be used to avoid common troubles. For instance, if one node detects that another fails then it must spread the detection to other nodes to cope this problem. This fact causes communication costs which will be resolved by defining three data structures. The first one is the attribute table, which stores the basic attribute information A, including the row identifier of instance called *row_id*, feature values and instances' classes C labeled by a *bba* expressing a belief. The second one is the count table, which computes the number of instances given specific class label described by a *bba* if split by attribute A. This table contains two fields: class C labeled with *bba* and a counting number CT. The last one is the hash table, which stores the link information between tree nodes *node_id* and *row_id*, as well as the link between parent node *node_id* and its branches *subnode_id* [14].

4.2 MapReduce Adaptation

The whole process is composed of four steps with a sequential execution of Map and Reduce procedures namely data preparation, attribute selection, tables updating and tree growing. Note that the last three steps are repetitive.

1. **Data Preparation**: the BDTs are essentially built to handle uncertain classes where their uncertainty is represented by a *bba*. In the training set, attributes values are certain and class labels are uncertain. In other words, we have doubts only on class labels. These *bba*'s characterizing the class labels are created artificially. They consider the real class C of an instance and the degree of uncertainty denoted by D divided into 4 groups as follows:

   - High degree of uncertainty: $0 < D \leq 0.3$
   - Middle degree of uncertainty: $0.3 < D \leq 0.6$
   - Low degree of uncertainty: $0.6 < D \leq 0.9$
   - No uncertainty: $D = 1$

   Each *bba* has 2 focal elements: the first one is the actual class of the object characterized by a *bbm* and the second one is Θ representing ignorance. Likewise, we have to convert this table into the three data structure mentioned above for further MapReduce processing. At this level, a Map procedure computes the pignistic probability of each instance and transforms the instances into attribute table containing attribute A_j as key, and *row_id* and a set of classes C labeled each one by a *bba*, both as values. Then, a Reduce procedure computes the number of objects with each class C_i characterized by a *bba* if split by attribute A_j, which forms the count table.

2. **Attribute Selection**: a Reduce procedure retrieves the number of instances for each pairs of (attribute, value) to aggregate the total size of records for a given attribute A_j. Next, the a Map procedure computes the information and split information of A_j. After that, another Reduce procedure computes the average pignistic probability function and then the information gain ratio. Hence, the feature having the highest GainRatio value will be selected as the most relevant attribute A_{best}. In other words, it depicts a decision node.

3. **Update Tables**: this step turn around updating count table and hash table. It is composed of two procedures: the Map procedure reads a record from attribute table with (key, value) equals to A_{best}, and emits the count of class labels. Then, the second Map procedure assigns *node_id* based on a hash value of A_{best} to make sure that records with same values are split into the same partition.

4. **Tree Growing**: the last step consists on growing the induced BDT by building linkage between nodes. Hence, we compute *node_id* and we test: if the value remains to the same, it means that *node_id* is a leaf node, otherwise, a new sub node is linked. The final condition imposes that all *node_id* become leaf nodes, after that the BDT is built.

We recall that Hadoop implements MapReduce paradigm. Indeed, MapReduce splits the whole input data set into sub data sets executed as small blocks of work in parallel in order to reduce time consumption. Thereafter, to ensure reliability, HDFS establishes multiple replicas of data blocks and assigns them to the concerned nodes. Hence, this framework can handle these data on their associated nodes. Thus, the MapReduce solution of belief decision trees can ensure the classification of Big Data while reducing the execution time and the costs of Input/Output operations.

5 Experimental Setup

In order to assess the performance and scalability characteristics of our MapReduce adaptation of BDTs, we conduct some experiments on KDD'99,[1] a massive data set of interest regarding the intrusion detection. Indeed, this dataset has been the point of attraction for many security researchers from the last decade.

5.1 Intrusion Detection

The intrusion detector learning task is to build a predictive model able to distinguish between malicious connections called intrusions or attacks, and good or normal connections [15]. The intrusion detection is a part of networks security which is strongly related to the concept of Big Data for several reasons. On one side, in a controlled network, there are a wide variety of devices and technologies. It may be a sensor that identifies the alerts tripped due to an attack, or that reports in intervals summaries, a network traffic that have been issued by different equipments. The recorded information reported from the various sensors, can be uncertain. Consequently, the need to relate the network security with the Big Data arises. On the other side, Big Data is considered actually as a currency since it is changing our life, models and applications. Moreover, the security of these data has always been an important concern for the data owners. This increases the need to machine learning techniques which provide a comprehensive security solution in order to detect and prevent hazardous attacks. In this way, the decision trees methods under uncertainty can be useful to achieve that goal. All these reasons mentioned above motivated us to apply our proposed solution within the averaging approach to intrusion detection data set in experiments. A connection is labeled as either normal, or as an attack. Attacks fall into four main categories DoS (Denial of Service), R2L (Remote to Local), U2R (User to Root) and Probe (surveillance and other probing). The proposed solution can be applied to a data set including a blending of categorical as well as numerical attributes. This latest stems from the 1999 KDD-Cup [16] and contains real network data, named KDD'99 data set, described in Table 1.

In this paper, we are will prove experimentally the BDT efficiency with adaptation to Big Data environment and MapReduce distributed framework.

---

[1] https://archive.ics.uci.edu/ml/datasets/KDD+Cup+1999+Data.

Table 1 KDD'99 description

Training set size	Test set size	Numerical attributes	Categorical attributes	Classes
4,898,431	2,984,154	34	7	5

5.2 Performance for Different Uncertainty Degrees

Among the advantages of accomplishing the classification under the belief function theory and using the belief decision trees as a classifier, we recall that one extreme case expresses a certain *bba* according to the training instance classes. In other words, the classes of all training instances I_j are known with certainty. Therefore, our proposed solution within the averaging approach is appropriate to describe the total knowledge case of standard decision trees handled by Quinlan [2] adapted to Big Data. In the following, we will consider the Percent of Correct Classification (PCC) as an accuracy measurement defined as:

$$PCC = \frac{Number\ of\ correctly\ classified\ instances}{Total\ number\ of\ classified\ instances} \times 100 \qquad (9)$$

Using this evaluation metric, experiments show that 96.94 % of instances are correctly classified. Also, in these experiments, we evaluate the classification results accuracy by varying the uncertainty degree. Hence, different results were carried out from these experiments to deal with the uncertain cases.

Table 2 demonstrates that the PCC increases from the higher to the lower uncertainty degree. For instance, it rises from 89.63 % with high uncertainty to 90.74 % with middle uncertainty and reaches 92.07 % with low uncertainty.

5.3 Scalability

Now, we perform the scalability of our MapReduce adaptation of BDTs. We mark the execution time of this algorithm by varying the uncertainty degree (*D*) with different nodes number (*N*) (see Table 3).

By increasing the number of nodes, we notice that the execution time decreases. For instance, the mean execution time decreases from 47 with one node to 21 min

Table 2 Experimental results for the PCC (%) (uncertain case)

	High uncertainty (%)	Middle uncertainty (%)	Low uncertainty (%)	Mean PCC (%)
PCC	89.63	90.74	92.07	90.81

Table 3 Experimental results for the time execution (minutes)

D N	High	Middle	Low	No	Mean per N
1	51	53	52	32	47
2	41	39	37	35	38
3	29	25	26	24	26
4	24	22	22	16	21
Mean per D	36	35	34	27	–

with four nodes. Besides, by decreasing the uncertainty degree, the execution time decreases too. For example, the mean execution time decreases from 36 with high uncertainty to 34 min with middle uncertainty and reaches 27 min in a certain case.

6 Conclusion and Future Works

In this paper, we propose a MapReduce adaptation of belief decision trees. Our algorithm take into account the variety of attributes' types and can also scale up to massive data sets of interest. The experiments realized on KDD'99 to evaluate the efficiency of our method, prove that the process of building BDTs applied to Big Data can be time-economical and efficient. Besides, our solution provides a parallel and distributed version based on MapReduce framework in order to alleviate the high cost of Input/Output operations. The main aim of this work is to perform the classification task in an uncertain environment within the averaging approach. The intrusion detection classification allow us to enhance the Big Data security. To summarize the contributions involved by this paper, our MapReduce adaptation of belief decision trees exhibit time efficiency, scalability and deal with massive data veracity. In future works, we will propose a MapReduce adaptation of belief decision trees using the conjunctive approach applied to Big Data in order to be close in spirit to the transferable belief model.

References

1. Wu, X., Kumar, V., Quinlan, J. R., Ghosh, J., Yang, Q., Motoda, H.: Top 10 algorithms in data mining. Int. J. Knowl. Info. Syst. **14**(1), 1–13 (2008)
2. Quinlan, J.R.: C4. 5: Programs for Machine Learning, vol. 1. Morgan Kaufmann, San Mateo (1993)
3. Elouedi, Z., Mellouli, K., Smets, P.: Belief decision trees: theoretical foundations. Int. J. Approximate Reasoning **28**(2), 91–124 (2001)

4. Stein, G., Chen, B., Wu, A.S., Hua, K.A.: Decision tree classifier for network intrusion detection with GA-based feature selection. In: Proceedings of the 43rd Annual Southeast Regional Conference, pp. 136–141 (2005)
5. Ben Amor, N., Benferhat, S., Elouedi, Z.: Naive bayes versus decision trees in intrusion detection systems. In: Proceedings of ACM SIG Symposium on Applied Computing, pp. 420–424 (2004)
6. Sagiroglu, S., Sinanc, D.: Big Data: a review. In: Proceedings of IEEE International Conference on Collaboration Technologies and Systems, pp. 42–47 (2013)
7. Lukoianova, T., Rubin, V.L.: Veracity roadmap: is big data objective, truthful and credible? Adv. Classif. Res. Online **24**(1), 4–15 (2014)
8. Narasimhan, R., Bhuvaneshwari, T.: Big data—a brief study. Int. J. Sci. Eng. Res. **5**, 1–4 (2014)
9. Shvachko, K., Kuang, H., Radia, S., Chansler, R.: The hadoop distributed file system. In: Proceedings of IEEE Symposium on Mass Storage Systems and Technologies, pp. 1–10 (2010)
10. Dempster, A.P.: A generalisation of Bayesian inference. J. R. Stat. Soc. **30**, 205–247 (1968)
11. Smets, P.: Non-Standard Logics for Automated Reasoning. Academic Press, London, pp. 253–286 (1988)
12. Shafer, G.: A mathematical Theory of Evidence, vol. 1. Princeton University Press, Princeton (1976)
13. Shannon, C.E.: A mathematical theory of communication. Bell Syst. Tech. J. **27**, 379–423 (1948)
14. Dai, W., Ji, W.: A mapreduce implementation of C4.5 decision tree algorithm. Int. J. Database Theory Appl. 49–60 (2014)
15. Lee, W., Stolfo, S.J., Mok, K.W.: A data mining framework for building intrusion detection models. In: Proceedings of IEEE Symposium on Security and Privacy, pp. 120–132 (1999)
16. Suthaharan, S.: Big Data classification: problems and challenges in network intrusion prediction with machine learning. Proc. ACM SIGMETRICS Perform. Eval. **41**(4), 70–73 (2014)

Hierarchical Attribute-Role Based Access Control for Cloud Computing

Alshaimaa Abo-alian, Nagwa L. Badr and M.F. Tolba

Abstract With the rapid and wide adoption of cloud computing, data outsourcing in cloud storage is gaining attention due to its cost effectiveness, reliability and availability. However, data outsourcing introduces new data security and privacy issues, therefore access control and cryptography are essential ingredients in a cloud computing environment to assure the confidentiality of the outsourced data. Existing access control systems suffer from manual user role and role permission assignments that impose online and computational burdens on the data owner in large scale cloud systems. In this paper, a hierarchical attribute driven role based access control system is proposed, such that the user role assignments can be automatically constructed using policies applied on the attributes of users and roles. The proposed access control system consequently solves the scalability and key management problems in cloud storage systems.

Keywords Role based access control · Attribute based encryption · Cloud storage · Confidentiality

1 Introduction

A cloud storage service (CSS) can be defined as a network of distributed data centers which typically uses cloud computing technologies like virtualization and offers some kind of interface for storing data [2]. A CSS offers unlimited storage on a subscription basis which reduces the management costs of underlying

A. Abo-alian (✉) · N.L. Badr · M.F. Tolba
Faculty of Computer and Information Sciences, Ain Shams University, Cairo, Egypt
e-mail: a_alian@cis.asu.edu.eg; a_alian2010@yahoo.com

N.L. Badr
e-mail: nagwabadr@cis.asu.edu.eg

M.F. Tolba
e-mail: fahmytolba@gmail.com

© Springer International Publishing Switzerland 2016   381
T. Gaber et al. (eds.), *The 1st International Conference on Advanced Intelligent System and Informatics (AISI2015), November 28–30, 2015, Beni Suef, Egypt,*
Advances in Intelligent Systems and Computing 407, DOI 10.1007/978-3-319-26690-9_34

infrastructure [14]. Moreover, a CSS offers more reliability and availability, as data owners can access their data from anywhere and at any time.

Despite the appealing advantages of cloud storage services, they also bring challenging security threats towards user's outsourced data. Since cloud service providers (CSPs) are separate administrative entities and are not in the same security domain with the cloud users, the correctness and the confidentiality of the data in the cloud is at risk due to the following reasons [1]: First, since the cloud infrastructure is shared between organizations, it is facing the broad range of both internal and external threats for data integrity. Second, cloud users no longer physically possess the storage of their data, i.e., data is stored and processed remotely, So cloud users may worry that their data could be misused or accessed by unauthorized users. As a result, the CSP should adopt data security practices to guarantee the privacy and confidentiality of their users' data.

Traditional access control methods such as Access Control List (ACL) [4] cannot be directly applied to cloud storage systems because data owners do not fully trust the cloud service providers. Additionally, traditional cryptographic solutions such as encryption schemes [8] cannot be applied directly while sharing data on cloud servers because these solutions require complicated key management and high storage overheads on the server. Thus, considerable research studies have recently been directed towards deploying access control policies to the cloud.

Role Based Access Control (RBAC) becomes a popular access control model that reflects organizational structure and easy to review and administer [7, 13], however, RBAC has frequently been criticized for the difficulty of setting up an initial role structure and for inflexibility in rapidly changing domains [9]. A pure RBAC may provide inadequate support for dynamic attributes and policies that should be considered when determining user permissions. Therefore, it is impractical to manually assign users to roles for large scale cloud systems with a huge number of users. Recently, attribute based access control (ABAC) [10, 15, 19], with its two variants; Key-Policy based ABE (KP-ABE) [5, 12] and Ciphertext-Policy based ABE (CP-ABE) [16–18] are considered as complimentary to RBAC because they provide a flexible and fine grained access policy. However, ABAC is more complex than RBAC in constructing and reviewing the access policy.

In this paper, a new access control system that integrates attribute based access control with hierarchical role based access control is proposed for cloud environments. The proposed access control system automatically assigns users to roles based on policies applied on the attributes of users and roles. So it relieves the data owner from the online and computational burdens of user-role assignment processes, especially, for large scale systems with a huge number of users and continuously changing user role policies. In addition, hierarchical roles are applied in order to solve key management problems in a decentralized environment. Attribute based policies support a fine grained and flexible access control.

The rest of the paper is organized as follows. Section 2 overviews related work. Section 3 provides the detailed description of the proposed access control scheme. Then, Sect. 4 illustrates the performance analysis of the proposed scheme in comparison with state-of-the-art and finally, we conclude in Sect. 5.

2 Related Work

Zhou et al. [20] proposed a role-based encryption scheme for cloud storage. In their scheme, the decryption key size still remains constant regardless of the number of roles that the user has been assigned to. However, the decryption cost is proportional to the number of authorized users in the same role and it also suffers from manual user role assignments.

Recently, Zhou et al. [21, 22] proposed an administrative model to manage and enforce role based access policies in cloud systems. Unfortunately, they used identity-based signature schemes to certify the authority of administrative roles that are inexpressive and coarse grained access control model, moreover, the proposed model suffers from high cost user revocation.

Liu et al. [11] proposed a hierarchical CP-ABE scheme that supports a time based proxy re-encryption to allow a user's access right to expire automatically after a predetermined period of time. However, their scheme has two drawbacks: First, it assumes that there is a global time among all entities. Second, the user secret key size is O (mn), where m is the number of nodes in the time tree corresponding to the user's effective time period and n is the number of user attributes, so the user secret key size can be large if the user has many attributes which affects the communication cost.

Hohenberger and Waters [6] proposed a KP-ABE scheme in which ciphertexts can be decrypted with a constant number of pairings without any restriction on the number of attributes. However, the size of the user's private key is increased by a factor of number of distinct attributes in the access policy. Furthermore, there is a trusted single authority that generates private keys for users which violates the user privacy and causes a key-escrow problem.

Deng et al. [3] proposed a ciphertext policy hierarchical attribute based encryption (CP-HABE) scheme. Their scheme supports key delegation that enables efficient data sharing among hierarchically organized large enterprises. In addition, the ciphertext size is independent of the number of users or the user hierarchy. However, the CP-HABE has three limitations: First, it does not support user revocation. Second, it suffers from high decryption costs. Third, it assumes a single attribute authority that is responsible for adding, assigning and revoking the attributes.

3 The Proposed Approach

The proposed access control system consists of four main entities as shown in Fig. 1; CSP, data owner, domain manager and user. The cloud service provider (CSP) stores the owners' data and provides data access service to users. In this paper, a tree is used to represent the access policy structure. Let τ be an access tree in which each interior node is a threshold gate, i.e. AND/OR gate while the leaves

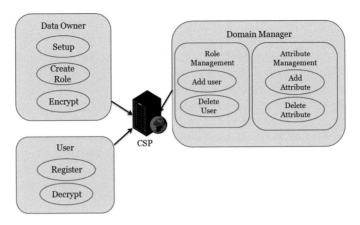

Fig. 1 The proposed access control system

are associated with attributes. A user can be assigned to a role and consequently decrypt ciphertexts corresponding to that role if and only if there is an assignment of the attributes from the user's private key to the leaf nodes of the tree such that the tree is satisfied.

The data owner outsources some data to the cloud and the data is encrypted before being outsourced. So the data owner is responsible for determining the access policy for each data, determining the roles, encrypting the data under the access policy and generating the master public and private keys. More formally, the data owner is responsible for three processes:

1. Setup: This process takes as input the security parameter λ and outputs a master key mk and a group public key pk.
2. Create Role: This process takes as input of a role with its identity ID_R and the role policy tree τ_R, generates the role secret key sk_R and returns a set of public parameters pub_R of the role and an empty user list RUL that will list all the users whose attributes satisfy the role policy tree. Figure 2 illustrates a detailed description of the "Create Role" process.
3. Encrypt: This process is executed by the data owner to encrypt the private data M by using the public information of the role R and outputs the ciphertext C to be outsourced to the cloud.

The Domain Manager is a trusted entity per role and it manages the set of users and their corresponding attributes in a given role. It can assign a role to a user if the user's attributes satisfy the role policy, or exclude a user if the user's attributes change. It can also issue, delete and update user's attributes according to their role or identity in its domain. The Domain Manager comprises two modules: The role management and the attribute management. The role management module contains two processes:

Input: role identity ID_R, the role policy tree τ_R
Output: role secret sk_R, role public parameters pub_R

To specify a role policy tree τ_R over a set of attributes Y:
For each node x in the tree τ_R
  Let k_x is the threshold value of the node .

  IF the non-leaf node has "AND" logical gate
  THEN k_x = *No. of its children*.
  IF the non-leaf node has "OR" logical gate
  THEN $k_x = 1$.

  Choose a polynomial q_x starting from the root node X and its degree
    $d_x = k_x - 1$.

  IF the node x is the root
     THEN
       Choose a random value $t \in \mathbb{z}_P^*$.
     Set $q_X(0) = t$.
  ELSE
     Set $q_x(0) = q_{p(x)}(index(x))$ where $p(x)$ is the parent of   node x and $index(x)$ is a unique index number assigned   to each node in τ_R .
Let Y be the set of leaf nodes in τ_R .
Generate the role secret $sk_R = (K_R, \{\forall\, y \in Y : C_y = g^{q_y(0)}, C'_y = H_2((y)^{q_y(0)t})\})$ where $K_R = g^{\frac{1}{s+H_1(ID_R)}}$
Compute the role public parameters as follows:
$pub_R = (A, B)$ where $A = h^{(s+H_1(ID_R))\prod_{i=1}^{m}\left(s+H_1\left(ID_{R_i}\right)\right)}$, $B = A^k$, where $\{ID_{R_1}, ..., ID_{R_m}\}$ are the identities of all the predecessor roles of role R.

Fig. 2 The "create role" process

1. Add User: This process takes as input the role secret key sk_R (with the access policy embedded in it) and the user identity ID_U, checks whether the user attributes qualify the access policy of the role and returns the secret decryption key dk_U for that user if the user's attributes satisfy the role access policy. Additionally, the process updates the role's public parameters pub_R and adds the user ID_U to the user list *RUL*.
2. Delete User: This process takes as input the identity ID_U of the user U and the role secret key sk_R, removes the user U from the user list *RUL* of the role R and updates the role's public parameter pub_R.

The attribute management module monitors a disjoint subset of attribute types. For example, in the university domain, the staff affair may issue "TA," "lecturer," "Assistant professor," etc., the specialties manager may issue "medicine," "commerce," "computer sciences," etc. The attribute management contains two processes:

1. Add attribute: After user's registration, the domain manager can assign some attributes to the user. This process takes as input the user identity ID_U and updates the user's secret key SK_U.
2. Delete Attribute: This process can be used to revoke some user's attributes. This process takes as input the user identity ID_U and updates the following:

   - User's secret key SK_U.
   - All roles' public parameters $\{pub_R\}$ in which the user is involved.

The user is a data consumer that downloads encrypted data files from the CSP and then decrypt them if they have the required permissions. Each user has an ID and a set of descriptive attributes and can execute two processes:

1. Register: In this process, the user registers to the system and provides the basic information. It outputs the user identity ID_U, then the domain manager is responsible for assigning attributes such as years of experience, education, etc., generating a public attribute key for each user's attribute it manages and a secret key for the user associates with their attributes.
2. Decrypt: This process is executed by the user to decrypt the ciphertext C outsourced in the CSP. It takes as input the user decryption key dk_U and outputs the plain text M if the user has the permission to access the data and fails if otherwise.

4   Performance Analysis

Table 1 evaluates the performance of the proposed system in comparison with the state-of-the-art. As shown in Table 1, the proposed outperforms the state-of-the art as it achieves constant ciphertext size and constant user secret key size, irrespective of the number of users in the system and the number of attributes per user.

Table 1 Performance analysis of the proposed system and the state-of-the-art

Access control method	Approach	Ciphertext size	User key size	Decryption cost	User revocation cost	Automatic user-role assignment
[3]	Hierarchical CP-ABE	Constant	Linear with the depth of the user in the hierarchy and the no. of attributes	Linear with no. of attributes satisfying the access structure	N/A	No
[11]	CP-ABE + time-based proxy re-encryption	Linear with the number of clauses in the access structure	Linear with the no. of nodes in the time tree and the no. of user attributes	Constant	Constant	No
[20]	Hierarchical RBAC	Constant	Constant	Linear with no. of users in the same role.	Linear with no. of roles	No
The proposed access control system	Integrating CP-ABE with hierarchical RBAC	Constant	Constant	Linear with no. of users in the same role.	Constant	Yes

On the other hand, the user secret key size in [3] grows linearly with the depth of the user in the hierarchy and the number of user attributes which causes high communication overhead in the case of large user hierarchy size and enormous number of user attributes. The ciphertext size in [11] is linear with the number of clauses in the access policy structure which imposes an increasing storage overheard at the CSP.

In the proposed system, the constant ciphertext size and user secret key size reduce the storage overhead at the CSP side and the communication costs between the user and the CSP. Furthermore, the proposed system supports the automatic user role assignments based on the user's attributes and role's access policy which relieve the data owner from the online and computational burdens of adding and removing users in large scale organizations with a huge and continuously changing number of users. In the case of user revocation process, instead of re-encrypting the related files and distributes the new keys to the non-revoked users via the CSP, the domain manager only updates the public parameters of the corresponding roles. Thus, the complexity of user revocation process is constant, i.e., independent of the number of users and the role hierarchy size, which makes the system scalable.

5   Conclusion

Despite of the cost effectiveness and reliability of cloud storage services, data owners may consider them as an uncertain storage pool outside of their enterprise. Therefore, access control is an effective way to ensure data security by restricting the access to data to privileged/authorized entities. In this paper, we proposed a scalable access control system for cloud storage systems. The proposed system integrates the attribute based encryption into the role based access control to support automatic user role assignments and that feature makes the proposed access control system more scalable, especially, for large scale organizations with a huge number of users. The performance analysis shows that the proposed system achieves constant ciphertext size, user secret key size and user revocation costs.

References

1. Abo-alian, A., Badr, N.L., Tolba, M.F.: Auditing-as-a-service for cloud storage. In: Intelligent Systems' 2014, pp. 559–568. Springer International Publishing (2015)
2. Borgmann, M., Hahn, T., Herfert, M., Kunz, T., Richter, M., Viebeg, U., et al.: On the Security of Cloud Storage Services. Fraunhofer-Verlag, Stuttgart (2012)
3. Deng, H., Wu, Q., Qin, B., Domingo-Ferrer, J., Zhang, L.L., Shi, W.: Ciphertext-policy hierarchical attribute-based encryption with short ciphertexts. Inf. Sci. **275**, 370–384 (2014)
4. Glasser, D.S., Zaner-Godsey, M., Gates, W.H., Cheng, L., Meijer, H.J., Snyder, I.L.: Cloud-based Access Control List. U.S. Patent Application 11/536, 457 (2006)

5. Goyal, V., Pandey, O., Sahai, A., Waters, B.: Attribute-based Encryption for fine-grained access control of encrypted data. In: The 13th ACM Conference on Computer and Communications Security, pp. 89–98. ACM (2006)
6. Hohenberger, S., Waters, B.: Attribute-based encryption with fast decryption. In: Public-Key CryptographyPKC 2013, pp. 162–179. Springer, Berlin (2013)
7. Huang, J., Nicol, D.M., Bobba, R., Huh, J.H.: A framework integrating attribute-ased policies into role-based access control. In: The 17th ACM Symposium on Access Control Models and Technologies, pp. 187–196. ACM (2012)
8. Kamara, S., Lauter, K.: Cryptographic Cloud Storage. Financial Cryptography and Data Security, pp. 136–149. Springer, Berlin (2010)
9. Kuhn, D.R., Coyne, E.J., Weil, T.R.: Adding attributes to role-based access control. Computer **6**, 79–81 (2010)
10. Li, J., Chen, X., Li, J., Jia, C., Ma, J., Lou, W.: Fine-Grained access control system based on outsourced attribute-based encryption. In: Computer Security–ESORICS 2013, pp. 592–609. Springer, Berlin (2013)
11. Liu, Q., Wang, G., Wu, J.: Time-Based proxy re-encryption scheme for secure data sharing in a cloud environment. Inf. Sci. **258**, 355–370 (2014)
12. Li, Q., Xiong, H., Zhang, F., Zeng, S.: An expressive decentralizing KP-ABE scheme with constant-size ciphertext. Int. J. Netw. Secur. **15**(3), 161–170 (2013)
13. Ni, Q., Lin, D., Bertino, E., Lobo, J.: Conditional privacy-aware role based access control. In: Computer Security–ESORICS 2007, pp. 72–89. Springer, Berlin (2007)
14. Pervez, Z., Khattak, A.M., Lee, S., Lee, Y.K., Huh, E.N.: Oblivious access control policies for cloud based data sharing systems. Computing **94**(12), 915–938 (2012)
15. Sahai, A., Waters, B.: Fuzzy identity-based encryption. In: Advances in Cryptology–EUROCRYPT 2005, pp. 457–473. Springer, Berlin (2005)
16. Wan, Z., Liu, J., Deng, R.H.: HASBE: a hierarchical attribute-based solution for flexible and scalable access control in cloud computing. IEEE Trans. Inf. Forensics Secur. **7**(2), 743–754 (2012)
17. Waters, B.: Ciphertext-Policy attribute-based encryption: an expressive, efficient, and provably secure realization. In: Public Key Cryptography–PKC 2011, pp. 53–70. Springer, Berlin (2011)
18. Xie, X., Ma, H., Li, J., Chen, X.: New ciphertext-policy attribute-based access control with efficient revocation. In: Information and Communication Technology, pp. 373–382. Springer, Berlin (2013)
19. Yang, K., Jia, X.: DAC-MACS: Effective data access control for multi-authority cloud storage systems. In: Security for Cloud Storage Systems, pp. 59–83. Springer, New York (2014)
20. Zhou, L., Varadharajan, V., Hitchens, M.: Enforcing role-based access control for secure data storage in the cloud. Comput. J. **54**(10), 1675–1687 (2011)
21. Zhou, L., Varadharajan, V., Hitchens, M.: Secure administration of cryptographic role-based access control for large-scale cloud storage systems. J. Comput. Syst. Sci. **80**(8), 1518–1533 (2014)
22. Zhou, L., Varadharajan, V., Hitchens, M.: Cryptographic role-based access control for secure cloud data storage systems. In: Security, privacy and trust in cloud systems, pp. 313–344. Springer, Berlin (2014)

A Peer-to-Peer Architecture for Cloud Based Data Cubes Allocation

**Mohammed Ezzat, Rasha Ismail, Nagwa Badr
and Mohamed Fahmy Tolba**

Abstract Large amounts of data are generated daily, according to the wide usage of social media websites and scientific data. These data need to be stored and analyzed to help decision makers but the traditional database concepts are insufficient. Data warehouse and OLAP are useful technologies in the storage and analysis of big data. Using MapReduce will help to save processing time, using cloud computing will help in saving resources and storage. In this paper, we propose a system that integrates the OLAP and MapReduce over cloud (considering workload balance) in order to enhance the performance of query processing over big data. The proposed system is applied to large amounts of data stored in cubes located in a Peer-to-peer cloud; this process is done using an allocation approach to save resources and query processing times. The proposed system achieves enhancements as time saving in query processing and in resources usage.

Keywords Big data · Cloud computing · Data warehouse · Peer-to-peer cloud

M. Ezzat (✉) · R. Ismail · N. Badr · M.F. Tolba
Faculty of Computer and Information Systems, Ain Shams University, Cairo, Egypt
e-mail: mohammed_ezzat@live.com

R. Ismail
e-mail: rashaismail@yahoo.com

N. Badr
e-mail: dr.nagwabadr@gmail.com

M.F. Tolba
e-mail: fahmytolba@gmail.com

© Springer International Publishing Switzerland 2016   391
T. Gaber et al. (eds.), *The 1st International Conference on Advanced Intelligent
System and Informatics (AISI2015), November 28–30, 2015, Beni Suef, Egypt,*
Advances in Intelligent Systems and Computing 407, DOI 10.1007/978-3-319-26690-9_35

1   Introduction

There are different definitions of Cloud Computing [1] according to the objectives or the usage. The definition of (NIST) is, "a model for enabling ubiquitous, convenient, on-demand network access to a shared pool of configurable computing resources that can be rapidly provisioned and released with minimal management effort or service provider interactions" [2, 3]. According to this definition, the main benefit of cloud computing is to save resources by using just what need to use [4]. Cloud computing merged with data warehouse and other technologies utilizes the computing resources [5–7].

These terabytes of data are generated daily by scientific researches, and social networks [4]. Big data is new expression of huge amounts of data to be analyzed [4, 8, 9]. Big Data most probably is unstructured data [10, 11] so using the relational databases will not be the efficient [12]. The best is using non-relational database like MongoDB [13, 14], especially when using in the cloud [15].

MapReduce is a parallel processing method produces by google to access big data by splitting the data into small pieces, processing then collecting the values [16].

Data warehouse is the base data storage for any DDS or KBS [17]. It may be centralized or distributed. Instead of distributing all the data, the data marts may be distributed and the data warehouse be centralized, which make the process faster [5–7].

Big Data and Cloud Computing are the fastest developing technologies and the integration between them is very useful. Some of the solutions to deal with big data are; distributed processing, data partitioning, and as a software; MapReduce [4–6].

Analysis process used in decision making and predictions. OLAP tools are used in a data warehouse to retrieve data and make the analysis process [18].

Data cube is a data modeling technique, each data cube constructed based on measures. Using cube dimensions users can access the data faster; and accurate [19].

Data access in the cloud means using resources and network traffic, these resources should be minimized, so access time should be minimized.

In this paper we introduce an approach to access big data "social network" stored in data cubes allocated in a peer-to-peer cloud based on the enhanced cube allocation approach [7] and the queries are posed on these cubes to be answered, and the work load balance are managed.

2   Related Work

GridMinor [20] is a data warehouse and OLAP project it was integrated with Grid technology, which saves storage, and workloads. They presented an indexing approach to accelerate the cube access and they presented architecture about the integration between OLAP and GridMiner, but this architecture was built on a specific toolkit. In [16] the authors discussed the management of data warehouse over cloud computing and how they get benefits using OLAP. They mentioned that the benefits of this merge were cost reduction, scalability, and easy to collaborate and access anywhere.

The authors of [21] proposed a cloud based mobile system to access cubes. Some benefits of their proposed system were: saving infrastructure and maintenance, and easy access. They saved time on; database storage, and query response. However using the system on android means a compatibility problem with other platforms.

In [22] the authors proposed TrackerCache; a system to help in OLAP query answering. The proposed system uses a subsystem called TrackerServer; this system split the files into chunks. In query answering the proposed system allows data exchange and query aggregation. The main idea of this work is to cache metadata, which saves time. They did not consider the data allocation, this process done randomly.

In [23] authors proposed a social media analysis system based on data warehouse and cloud computing. Also, they applied a BSP with MapReduce techniques on cloud computing. The queries were searching for URLS as a result.

In [7] authors presented a materialized views allocation approach over peer-to-peer cloud, the allocation done after the most beneficial views selected to be materialized. The main enhancement was saving time, and utilizing the storage space on the cloud.

In this paper we introduce a system that integrates MapReduce which will be used in the query answering, OLAP is used to construct the cubes and search for answers and cube allocation approach applied over cloud peers, which processes the load balance. The results show a saving in query processing time, query answering time, workload balance between the cloud peers, and cloud resources utilization.

3   The Proposed System Architecture

We propose a peer-to-peer cloud architecture [24]. The architecture consists of 4 modules. Module 1: cube construction, Module 2: cube allocation based on [7]. Module 3: query processing, the queries forwarded from peers to be answered and the last one is query answering considering load balance. Figure 1 shows the architecture.

- **Module 1: Cube Construction**.
  We applied the proposed system on social media data [25]. Generally, the data is questions and answers stored as tags. By extracting the data to its usual form (text files), the first step is cube construction and the data is represented in three dimensions: Tag of the post, answered or not and the month of the post.

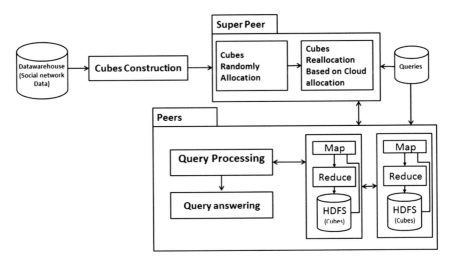

Fig. 1 The proposed architecture

- **Module 2: Cube allocation**.
  All the cubes are allocated randomly on peers, then the queries posed and
  re-directed to the peers which contain the required cube. After that the system
  will count the frequencies of answered queries for each peer. The re-allocation
  process will be applied for each cube to the peer with the highest frequency of
  queries to these cubes. The cube re-allocation process is based on cloud allocation [7].
- **Module 3: Query Processing**.
  The queries are posed to any peer and the peer will check if the cube contains
  the answer for the query, if not found it will redirect the query to the next peer. If
  not found, the peer will forward the query to the next peer and so on.
- **Module 4: Query Answering**.
  After the query processing is done the required cube is found, the results of the
  query will be forward to the master peer to dispatch the results to the corresponding peer. Searching for the answers done based on the MapReduce
  technique over the Hadoop which is installed at peers. In case the selected peer
  is busy (as it is a peer side process) the master peer will assign another peer to
  get part of the answer.

4 The Proposed Cloud Based Big Data Cubes Allocation and Processing Algorithm

The proposed algorithm will be executed at both master peer and other peers, at first
data cubes will be constructed and then they are allocated randomly to the peers.
The cubes will be re-allocated to the most beneficial peers and finally the queries

will be posed and answered using the MapReduce algorithm. The algorithm consists of three main functions that are:

The first function (Cube Construction). It takes a string from the data warehouse and returns a cube in the text file, which contains the data as separated text. The string will contain the cube dimensions. As the following pseudo code:

```
Cube CubeConstruction(string line)
Begin
Define values as an array of string lines separated by '@'
Construct Cube
```

The second part (Re-allocation): Before the allocation there is a function called CreateQueries, it generates list of queries and send them randomly to the peers.

- The AllocatedQueries function will run the queries over the peers. And count the frequency for each answered query.
- The Re-allocation function will run and each cube will be re-allocated at the peer that has the highest frequency of answered queries.
- The Re-allocatedQueries function will run the queries after the re-allocation of the cubes. As this pseudo code:

```
Return ListQueries
Run AllocatedQueries()
Begin
Set CubTable randomly
        For each QueryPeer in ListQueries
        Send Query to peer
        Define QueryPeer QueryAnswer
Receive QueryAnswer from Peer
 Calculate Time
End
Re-allocation ()
Define Count = No. of answered queries
        If Count >Cube Count
            Cube Count= Count
            Cube Table (peer) = PeerCubeName
 End
Run Re-Allocated Queries()
Set CubeTable Re-allocated
For each Query Peer in List Queries
    Send Query Peer to peer
    Define Query Peer Query Answer
    Receive QueryAnswer from Peer
    Calculate Time
    QueryAnswer.Time Of Query = Time
    Re- allocated Query Results and (QueryAnswer)
```

The third part (query processing and answering). Any peer will receive a new query and a cube name. If this peer is free and contains the cube, it will search for the answer then send it to the master peer, if the peer is busy and the current query requires another peer to get answers or if the selected peer doesn't contain the answer, it will forward the query to the next peer. As shown in pseudo code:

```
Query Processing
Receive AllPeers ()
While (true)
   If (GotNewQuery)
       Receive Current Query from peer
       If (peer is Master Peer)
       Call MapReduceAgent ()
       Else
          Call Forward MapReduceAgent()
Query Answering:
MapReduceAgent ()
   If the selected peer is busy or required cube not found
       Send Current Query to another Peer
       If (Result Status is Answered)
          Receive Query Result from this Peer
          Send Query Result to Master Peer, Exit
       Else if (Result Status is Not Answered)
          Continue
   Else
       MapReduce(Current Query)
       Send CurrentQuery to Master Peer
   End
Forward MapReduceAgent()
   If the selected peer is busy or required cube not found
       Result Status = Not Answered
       Send Result Status to Previous Peer, Exit
   Else
       MapReduce(Current Query)
       Send Current Query to PreviousPeer
Map Reduce
Define Current Query
Define Class Cube (String Tag, Int Month, Boolean Is Answered, Int Count)
Map (String InputLine, MapperContext context)
   Define array SearchTags = CurrentQuery.Tags
   Define array SearchMonths = CurrentQuery.Months
   Define Cube Value = Cube Parse (InputLine)
   If (SearchTags.Contains (Value.Tag) & SearchMonth.Contains (Value.Month)
   Define String Key = Value.Tag
   Context.EmitKeyValue (Key, value.Count)
Reduce (string key List<string>, valuesMapperContext)
   Define Int Count=0
   Foreach value in values
       Define IntIntValue = Value converted to integer
       Count += IntValue
   Context.EmitKeyValue (key, Count)
```

5   The Experimental Environment

The implementation experiment has 3 cloud peers. Each peer has Virtual Machine [28], Windows 7, Microsoft VS 2012 and Hadoop [26].

The used technologies are: Microsoft VS 2012 and C#, and Hadoop SDK. MongoDB [12, 13] was used to store data as text. We also used MongoVue [27], as it is a MongoDB desktop application. Also MongoSharp used to access MongoDB using C#. We applied the cloud peers as virtual machines using VMware ver. 8 [28].

6 Experimental Results

To evaluate the proposed system we applied these experiments: The first was posting 10, 20 and 30 queries randomly over 3 peers. Each peer contains cubes. The first measure is the resources usage, before the allocation approach, the 10 queries required 16 visits to the peers to be answered, the 20 queries required 32 visits and the 30 queries required 50 visits. After the re-allocation, the results were: The 10 queries required 14 visits (12.5 % enhancement), the 20 queries required 27 visits (15.6 % enhancement) and the 30 queries required 37 visits (26 % enhancement). As in Fig. 2.

The second measure was the workload, based on the previous experiment, when the 10 queries posed, the workload enhanced but with the same percentage of workload balance, before there were 16 visits to the peers, the workload was 8 for each peer, and after there were 14 visits, the workload was 7 for each peer. When the 20 queries, before the allocation there were 32 visits and the workload was 15

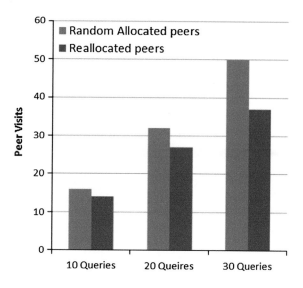

Fig. 2 Resource saving experiment

Fig. 3 Workload balance

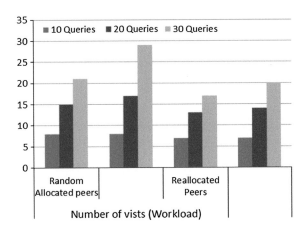

Number of vists (Workload)

Fig. 4 Processing time

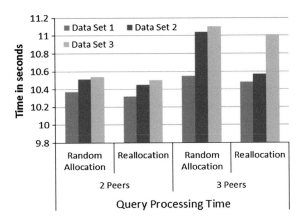

Query Processing Time

and 17 (46.8 and 53.2 %), after there were 27 visits, and the workload was 13 and 14 (48.1 and 51.8 %). For the 30 queries before the allocation there were 50 visits, and the workload was 21 and 29 (42 and 58 %), after there were 37 visits, and the workload was 17 and 20 (46 and 54 %). As in Fig. 3. The average enhancements is 18 %.

The second experiment was evaluating the query processing times by using 3 different data sets of social media [25], and running the same queries, this experiment was done over 2 peers and 3 peers. As Fig. 4 show that at the 2 peers the query processing time was 10.37, 10.51 and 10.54 min for the three data sets, with an average time of 10.47 min, but after reallocation the query processing time became 10.32, 10.45 and 10.5 min for the three data sets, with an average time of 10.42 min. At the 3 peers architecture, the query processing time was 10.55, 11.04

Fig. 5 Average time (Query processing)

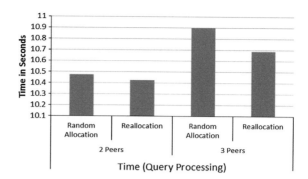

Time (Query Processing)

and 11.1 min for the three data sets, with an average time of 10.89 min after reallocation, the query processing time became 10.48, 10.57 and 11.01 min for the three data sets, with an average time of 10.68 min.

As shown in Fig. 5 the experiments proved saving time by 2 %.

7 Conclusion

Big Data such as social network data need to be analyzed to help in decision making, and other business processes and also making predictions. Big data analysis requires more resources and time to be completed accurately so, it needs a huge storage and processing resources to make the analysis. Cloud computing is a solution for solving both the storage and processing problems, it provides the required storage and computational resources with many features like; scalability and elasticity and which are essential for big data storage and analysis. So, in this paper we proposed an approach to deal with big data in the cloud, query answering and in this approach we applied MapReduce to answer the queries, OLAP to construct the cubes and cloud cube allocation technique to allocate the cubes in the cloud peers and we applied the system to the [25] data sets. Our system improves the time of query processing, saves cloud resources and improves the workload balance in the cloud architecture.

Cloud computing has many architectures but peer-to-peer is more suitable for social media big data as it should be distributed. Big data should get advantages from cloud computing features to allow easy and fast accessing and loading data from the databases/data warehouses as done in our approach based on a social media big data. When using cloud computing to deal with big data it should be utilized and the resources should be saved as possible, as it costs a lot. The results show that time (processing and transfer time) and resources (storage and processing resources) are saved.

References

1. Verma, H.: Data-warehousing on cloud computing. Int. J. Adv. Res. Comput. Eng. Technol. (IJARCET) **2**(2), 411 (2013)
2. Branch, R., Tjeerdsma, H., Wilson, C., Hurley, R., McConnell, S.: Cloud computing and big data: a review of current service models and hardware perspectives. J. Softw. Eng. Appl. (2014)
3. Mell, P., Grance, T.: The NIST definition of cloud computing, pp. 800–145, NIST Special Publication(2011)
4. Ji, C., Li, Y., Qiu, W., Awada, U., Li, K.: Big data processing in cloud computing environments. In: 12th International Symposium onPervasive Systems, Algorithms and Networks (ISPAN), pp. 17–23. IEEE, (2012)
5. Aloisioa, G., Fiorea, S., Foster, I., Williams, D.: Scientific big data analytics challenges at large scale. In: Proceedings of Big Data and Extreme-scale Computing (BDEC) (2013)
6. Fiore, S., Palazzo, C., D'Anca, A., Foster, I., Williams, D.N., Aloisio, G.: A big data analytics framework for scientific data management. In: IEEE International Conference on Big Data (2013)
7. Megahed, M.E., Ismail, R.M., Badr, N.L., Tolba, M.F.: An enhanced cloud-based view materialization approach for peer-to-peer architecture. In: Advanced Machine Learning Technologies and Applications, pp. 401–412. Springer Berlin Heidelberg (2012)
8. Brezany, P., Zhang, Y., Janciak, I., Chen, P., Ye, S.: An elastic OLAP cloud platform. In: IEEE Ninth International Conference on Dependable, Autonomic and Secure Computing (DASC) (2011)
9. Al-Atroshi, A.M., Abdullah, F.M.: Design a Distributed Data Warehousing based ROLAP with Materialized Views, (2013)
10. Guo, M.: Financial system analysis and research of OLAP and data warehouse technology. Inf. Technol. J. **13**(3), 522–528 (2014)
11. Najafabadi, M.M., Villanustre, F., Khoshgoftaar, T.M., Seliya, N., Wald, R., Muharemagic, E.: Deep learning applications and challenges in big data analytics. J. Big Data (2015)
12. Cuzzocrea, A., Song, I.Y., Davis, K.C.: Analytics over large-scale multidimensional data: the big data revolution!. In: ACM 14th International Workshop on Data Warehousing and OLAP. ACM, (2011)
13. Bhatewara, A., Waghmare, K.: Improving network scalability using NoSQL database. Int. J. Adv. Comput. Res. (IJACR) **2**(6), 4 (2012)
14. Patel, M.P., Hasan, M.I., Vasava, H.D.: Performance improvement of sharding in MongoDB using k-mean clustering algorithm. In: International Journal of Advance Engineer ing and Research Development (IJAERD), vol. 1 (2014)
15. Liu, Y., Wang, Y., Jin, Y.: Research on the improvement of MongoDB auto-sharding in cloud environment. In: 7th International Conference on Computer Science and Education (ICCSE). IEEE, (2012)
16. Ene, S., Nicolae, B., Costan, A., Antoniu, G.: To overlap or not to overlap: optimizing incremental MapReduce computations for on-demand data upload. In: Proceedings of the 5th International Workshop on Data-Intensive Computing in the Clouds, pp. 9–16. IEEE Press (2014)
17. Gandhi, V.C., Prajapati, J.A. Darji, P.A.: Cloud computing with data warehousing In: International Journal of Emerging Trends and Technology in Computer Science (IJETTCS) (2012)
18. Fišer, B., Onan, U., Elsayed, I., Brezany, P.: On-line analytical processing on large databases managed by computational grids. In: 15th International Workshop on Database and Expert Systems Applications, Proceedings, pp. 556–560. IEEE (2004)
19. Alrayes, N., Luk, W.S.: Automatic transformation of multi-dimensional web tables into data cubes, pp. 81–92. Springer Berlin (2012)

20. Brezany, P., Janciak, I., Min Tjoa, A.: GridMiner: a fundamental infrastructure for building intelligent grid systems. In: Web Intelligence, Proceedings. IEEE/WIC/ACM International Conference (2005)
21. Rajankar, M.R., Jasutkar, R.W.: Cubic Approach to Mobile Cloud Computing. In: International Journal of Advanced Research in Computer Engineering and Technology (IJARCET), (2012)
22. Ke-hua, Y., Manirakiza, A.: Efficient and semantic OLAP aggregation queries in a peer to peer network. Int. J. Inf. Electron. Eng. **2**(5), 697–701 (2012)
23. Kossmann, D., Kraska, T., Loesing, S.: An evaluation of alternative architectures for transaction processing in the cloud. In: SIGMOD International Conference on Management of Data. ACM, (2010)
24. Kalnis, P., Ng, W.S., Ooi, B.C., Papadias, D., Tan, K.L.: An adaptive peer-to-peer network for distributed caching of olap results. In: Proceedings of the ACM SIGMOD International Conference on Management of Data, pp. 25–36. ACM, (2002)
25. Stackoverflow.com: Stack Overflow, (2015) http://www.stackoverflow.com. Accessed 01 April 2015
26. (2015) http://www.hadoop.apache.org. Accessed 01 March 2015
27. Mongovue.com: (2015) http://www.mongovue.com. Accessed 01 March 2015
28. Vmware.com: VMware virtualization for desktop & server, application, Public & hybrid clouds I United States, (2015), http://www.vmware.com. Accessed 03 Feb 2015

Enabling Cloud Business by QoS Roadmap

Sameh Hussein

Abstract Day after day, global economy becomes tougher. It is a fact in which Cloud Computing business gets impacted the most. When it comes to cost, Cloud Computing is the most attractive paradigm for IT solutions. It is because Cloud Computing relaxes cost constraints. This relaxing enables Cloud Customers to operate their business through Cloud Computing under such economic pressure. On the other side, Cloud Computing solutions should deliver IT services at the agreed and acceptable Quality of Service levels. This moves the economic challenges from Cloud Customer premises to Cloud Provider premises. Accordingly, Cloud Providers are in dire need to deliver high quality levels using less resources and IT equipment. Therefore, Cloud Provider management and decision makers need to optimize their Cloud resources. This can be achieved by paying their great attention to critical resources. Listing and extracting these critical resources is the new challenge which takes place. Thus, our contribution allows Cloud Provider management and decision makers to focus on certain resources to guarantee the needed quality levels of delivery. Our proposed roadmap lists Cloud elements in certain order, each according to its importance and priority.

Keywords Cloud Computing (CC) · Cloud Quality of Service · Quality of Service (QoS) · Cloud Computing business · Cloud economic impacts · Cloud economic challenges and Quality of Service roadmap

1 Introduction

Being the most desirable IT solution, Cloud Computing (CC) suffers new challenges with every day. CCs new challenges get the flavor of the new solutions it offers, new scale it becomes fitting and new regulations it complies with [1]. Based on our previously proposed IT/Legal Framework, QoS always appears with Legal Aspects (LA)

S. Hussein (✉)
School of Communications and Information Technology, Nile University, Cairo, Egypt
e-mail: sameh.hussein@nileu.edu.eg

© Springer International Publishing Switzerland 2016
T. Gaber et al. (eds.), *The 1st International Conference on Advanced Intelligent System and Informatics (AISI2015), November 28–30, 2015, Beni Suef, Egypt,*
Advances in Intelligent Systems and Computing 407, DOI 10.1007/978-3-319-26690-9_36

Fig. 1 Cloud Computing
IT/Legal architecture

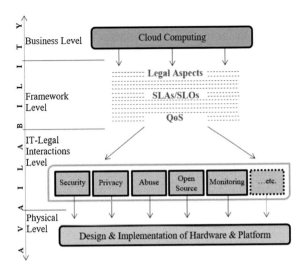

and Service Level Agreement (SLA) [2]. However, the economic pressure on Cloud Providers is one of the new challenges because it destabilizes QoS levels. Then, to exactly figure out what impacts Cloud QoS and to what extent, it is necessary to find its exact position within Cloud Computing. Our proposed IT/Legal Architecture Fig. 1, shows the exact location of Cloud QoS. The IT/Legal Architecture is created based on the understanding of Cloud Computing and its IT/Legal related definitions [2]. In progress, we have another paper to be published which deeply describes this IT/Legal architecture.

Cloud QoS is the lower layer in framework level of the CC IT/Legal architecture. Cloud QoS was just above IT-Legal interaction challenges level. This is because technical considerations are representing the biggest portion of Cloud QoS. The other portions are related to LA and SLA in the same level. Thus, Cloud QoS has a link with Cloud LA as well as a link with Cloud SLA as described within IT/Legal Framework [2]. However, addressing QoS within Cloud Computing is a bit tough. It mandates to look at Clouds available designs, mechanisms and approaches for monitoring and measuring Cloud QoS. Cloud QoS has a location in which it implements the junction between Cloud business and Cloud technicalities. On the other side, challenge levels below QoS are also directly linked to it. If both Cloud parties have reached a robust Cloud QoS monitoring, then SLAs/SLOs will be amended to be stronger. Consequently, LA and Cloud contracts can be adapted. After-word, adaptable Cloud contract can drive the business to success. It is like the falling dominoes game, Fig. 2, as monitoring is the trigger where dominoes pieces start to fall sequentially. This game reflects the importance of monitoring Cloud QoS which can threaten Cloud business.

The same as many Cloud related terms, Cloud QoS has many definitions according to how Cloud parties understand the nature of the delivered service [3]. Generally, most of the definitions are either stated by Cloud Providers or Cloud Customers

Fig. 2 Cloud QoS
dominoes game

[4]. Cloud Providers uses technical terms such as throughput, delay, jitters, and loss rate [5]. On the other side, Cloud Customers discussing SLOs and its related metrics such as image resolution, sound quality, and appropriate language [6]. There is often a missing part which gathers both sides. Thus, the definition that this paper builds on is a modified version of the one stated by PIREGRID corporation [7]. Our modification was made to reflect the location of Cloud QoS within the IT/Legal architecture. Then, it is the Clouds ability to allocate resources and prioritize users, applications and data flows to guarantee Cloud service delivery at the performance agreed through quality levels.

Throughout the rest of this paper, we address the main concern for Cloud QoS as well as todays related work in Sect. 2. Then, we explain and demonstrate the suffered deficiencies for Cloud QoS to in Sect. 3. This is in addition to describing our proposed Cloud QoS roadmap in Sect. 4. After this, we formulate the roadmap into tables in Sect. 5. Finally we conclude and expect the future in Sect. 6.

2 Cloud QoS Main Concern

Cloud QoS suffers many technical and non-technical challenges [8]. Those challenges are used to create argues between Cloud Providers and Cloud Customers [9]. They are stemming from one main concern which is guaranteeing and maintaining the QoS at agreed levels regardless of obstacles and difficulties [10]. However, these levels should be continuously conducted even in tough situations such as data overloading and economic pressure [11]. Therefore, this main concern is mostly driven by Cloud readiness capacity and surrounding economic factors [12]. Both can represent two bullets from a dual shotgun, Fig. 3. Both bullets are racing to hit a flying bird which tries to deliver the service safely. Who-ever does, the bird gets hurt. This scenario Fig. 3, shows how the main concern impacts Cloud QoS.

'Cloud Capacity Overload' is one of the technical challenges where a Cloud needs to be ready for any dramatic demand. A Cloud optimization is to deliver Cloud services to all Cloud Customers and Cloud Clients using the minimum possible resources. In other words, it is a process to offer maximum possible production with minimum possible utilization. However, this process should be a real-time process to have the Cloud ready for increased demands. Cloud Providers usually feel happy when they can seize more customers and clients with the same infrastructure. In fact, this allows Cloud Providers to gain more revenue at the same investment levels. On the other side, the more Cloud capacity gets utilized, the less Cloud QoS gets delivered. It is the nature of any shared resources environment. Then, it affects Cloud capability to guarantee and maintain current services to be delivered at the agreed levels. This challenge needs strategic decisions to be faced, such as pouring more money to invest in Cloud infrastructure so the Cloud get more capacity. However, this decision may cost a lot of money where other strategic decisions can handle it smartly. For example, a Cloud Provider may release small business customers like those who outsource minimum functionalities. When doing this, a Cloud can retrieve small but many resources to be assembled and allocated for medium or large business channels.

'Economic Factors' are representing non-technical challenges. As its bullet got shot from the same used gun, economic factors are more or less linked to Cloud capacity. The more global economic circumstances get tough, the fewer budgets get offered. This philosophy takes place in different business sectors, including Cloud IT environment. Thus, recent years already have shown the massive migrations of applications and IT services to CC. These migrations are used to target any reduced cost solution such as shared resources. Although economic factors lead to cost cutting, but it also mandates more productivity to provide the business a chance to survive and compete. It means that by time passing, Cloud Providers are required to offer more IP Band Width (IPBW), less packets loss and less packets delay as well as less prices. At this point, Cloud optimization and capacity play their role.

Based on the previous scenario, Fig. 3, Cloud Providers have to do their best to deliver the agreed service targets. This can be achieved just when the two bullets miss the bird or at least not killing QoS bird. This sequence of thinking leads to create a spectrum of Cloud QoS. It is need-ed to measure how successfully the Cloud is able to deliver service targets. Figure 4 shows how a Cloud Provider can do its best to fit the agreed QoS. The best effort indicator shows how good the Cloud Provider can perform its committed QoS. The sliding of the best effort indicator depends on

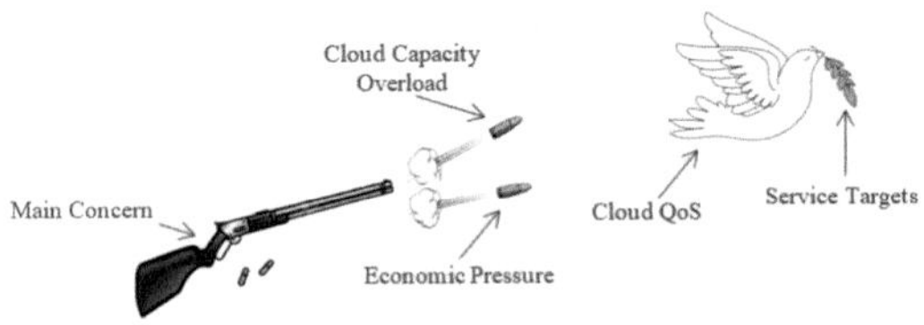

Fig. 3 Cloud QoS main concern

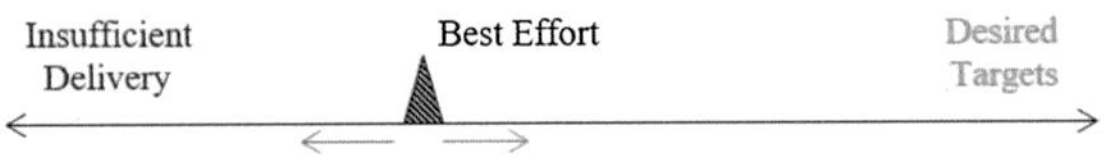

Fig. 4 Cloud QoS spectrum

many external and internal factors such as installing new equipment, and customers demand.

Before sailing into our proposed Cloud QoS roadmap, it is necessary to visit todays approach and related work. Navigating through todays situation will show what our proposed roadmap can add. Monitoring Cloud QoS is an area where a Conflict of Interest (CoI) takes place between Cloud Providers and Cloud Customers. This conflict usually appears with every report which measures Cloud QoS [13]. Cloud Providers are supposed to provide such report of monitoring service metrics [13]. In some cases, reporting Cloud QoS figures is considered as a separate service which Cloud Customers need to pay for [14]. In the light of the economic challenges, the delivered Cloud QoS is subjected to best effort changes, Fig. 4. Consequently, reporting Cloud QoS findings are subjected to misleading figures according to inaccurate readings. Reporting service figures that meet the agreed service quality levels is the main driver of the CoI [14]. Where Cloud Providers prefer to escape from penalties and compensations, Cloud Customers firmly apply such terms and conditions [15]. Cloud Customers cannot prove the mismatch between the agreed levels and the delivered metrics. A possible approach for Cloud Customers is to monitor themselves then compare them with what got reported by Cloud Providers [15]. This approach adds more headache to management. However, this approach is not always feasible as it grants Cloud Customers the access to monitor Cloud elements. The access can be granted in case of private Clouds, but never possible for public Clouds [16]. This is because public Clouds carry confidential information for other Cloud Customers. However, monitoring from different places, does not guarantee similar readings. It is because technical difficulties are location based [16]. This leads to have CoI very often. According to real-life best practice, another possible approach for both Cloud parties is to hire a third-party just for monitoring and reporting service metrics [17]. This approach has been commonly used. In some cases a joint interest between the Cloud Provider and the third-party can takes place. It might lead to manipulated reports and biased figures [17]. In some other cases, third-party may use low quality monitoring mechanisms in which the issued reports got unrealistic readings. In both approaches there is always an unexpected possibility to have inaccurate reports. Our proposed roadmap encourages and facilitates accurate monitoring operation for Cloud Providers. It lists the most important Cloud elements to be monitored closely even in tough economic situations.

3 Cloud QoS Stabs

As mentioned before, Cloud QoS bird either gets killed or gets hurt by bullets. In case of Cloud QoS gets killed, it means that the Cloud is no more able to deliver the desired service levels. Then, Cloud business continuity is not achievable. However, if it gets hurt, this means that the bird still has a chance to survive and resist delivering the desired service. In the second case, the carried out damage extends in depth to cause five main stabs. They represent the technical impacts which are cre-

Fig. 5 Stabs treatment

Treatment

Table 1 Impacting stab metrics

QoS stab	What capacity and economic factors can impact?
Performance	Processing latency and system delay
Reliability	Network connectivity efficiency
Compliance	Collect information and prevent incidents
Economic goals	Inter-operability
Information security	Information availability, confidentiality and integrity

ated by Cloud capacity and economic factors. These stabs need to be treated firmly, so the Cloud keeps delivering the agreed QoS. The treatment should be continuously performed as best effort to slide the indicator to desired targets. In Fig. 5, the green arrow shows the direction where the treatment pulls the best effort towards the desired targets.

These stabs are Cloud service performance, Cloud service compliance, Cloud service reliability, Cloud service information security and Cloud service economic goals. Table 1, shows what metrics can both capacity and economic factors impact.

4 Our Proposed Cloud QoS Roadmap

The roadmap represents a course of advices for what and where to measure metrics of the stabs. Each of these stabs represents a road hole where the proposed advices represent the road caution signs, Fig. 6. The proposed work scope includes Cloud service performance and Cloud service information security. Other stabs are left for future research and farther analysis. Service performance is used to cover throughput, voice delay and video streaming. The scope of work addresses throughput and voice delay where video streaming is left for future research and farther analysis. Before revealing the proposed course of advices, it is necessary to mention that this roadmap was built on an example of public Cloud shown in Fig. 7. It shows some common elements such as servers and network components which are used as monitoring points. They represent measuring nodes where information get collected then get sent to analysis and reporting department.

Regardless the delivery model (IaaS, PaaS, SaaS, etc.), in public Clouds there is always a need to specify the locations where service metrics are monitored. In case of private Clouds, Cloud elements are different. Accordingly, our proposed roadmap needs to be recomputed to cover the desired metrics within the available elements.

Fig. 6 Stabs treatment

Fig. 7 Public cloud
elements

To support Cloud QoS main concern and it two drivers, Cloud QoS roadmap is trying to help treating the five QoS stabs (Parameters). This helps is provided by the roadmap via allocating the impacted metrics so a Cloud Provider can keep an eye to monitor and measure.

'Information Security' has its Authorization and authentication are impacted metrics which can be monitored at servers like Apps, Web, Service Node, Portal and Single Sign On (SSO). This is because these servers are responsible for granting access to users. Non-repudiation as a metric can be measured at Message Queue, Data Base and Storage servers. It is because of these servers are accepting data transactions and queries by users. Infrastructure Architects have to be certified and accredited to perform their role. Integrity and information availability have to be measured at servers like DNS, DHCP and Internal LAN records. It is vital to guarantee integrity and information availability at these elements because lack of information availability and integrity will influence service delivery.

'Performance—Throughput' it has metrics like data refresh frequency, data quantity and data transfer time have to be measured at Cache server. This is because Cache server saves data of recently used service and application to avoid fetching or creating it again. Offered IPBW can be measured at Provider Edge router. The delivered IPBW can be measured at Customer Edge router. The difference between both IPBW represents connectivity quality of the Public Switching Telephone Network (PSTN) channel. Internal routing speed is also a throughput metric that can be measured by measuring Internal LAN and Core Network. Performance of both networks and Cache server affect the data readiness to be transferred to the customer. Operation Engineers have to be ready to support networks latency.

'Performance—Voice Delay' it is mostly measured by user-side delay metric that can be monitored at User VOIP device. An early detection can be obtained by monitoring cache hit ratio at Cache server as well as dial wait time at Data Base and Discovery servers. Service response time is also a metric which can be monitored at Service Node server. Disk busy time gets watched at DNS, Data Base and SSO serves. Internal LAN and Core Network are used to monitor Portal response time, DNS lockup time, round trip delay, one-way-delay and transaction duration.

5 Roadmap Formulated Tables

Tables 2, 3 and 4 help Cloud parties to review the impacted metrics and easily get the appropriate location for each of them. These tables are built on the assumed Cloud and customer elements as show in Fig. 7. Simply, these formulated tables map monitoring node locations versus the desired metrics. These locations can be changed according to the service nature, Cloud infrastructure, Cloud type and monitoring

Table 2 Information security monitoring node locations

Public Cloud Elements	Information Security Metrics						
	Authorization	Authentication	Nonrepudiation	Certification	Accredited	Integrity	Information Availability
Data Base Server	-	-	●	-	-	-	-
Web Server	●	●	-	-	-	-	-
Discovery Server	-	-	-	-	-	-	-
Apps Server	●	●	-	-	-	-	-
DNS Server	-	-	-	-	-	●	●
DHCP Server	-	-	-	-	-	●	●
Service Node Server	●	●	-	-	-	-	-
SSO Server	●	●	-	-	-	-	-
Portal Server	●	●	-	-	-	-	-
Message Query Server	-	-	●	-	-	-	-
Internal LAN	-	-	-	-	-	●	●
Storage Server	-	-	●	-	-	-	-
Infrastructure Architects	-	-	-	●	●	-	-

Table 3 Performance—throughput monitoring node locations

Public Cloud Elements	Performance - Throughput Metrics							
	Data Refresh Frequency	Data Transfer Time	Data Quantity	Offered IPBW	Delivered IPBW	Internal Routing Speed	Data Readiness	Network Support Latency
Internal LAN	-	-	-	-	-	●	●	-
Core Network	-	-	-	-	-	●	●	-
Cache Server	●	●	●	-	-	-	●	-
Operation Engineers	-	-	-	-	-	-	-	●
Provider Edge	-	-	-	●	-	-	-	-
Customer Edge	-	-	-	-	●	-	-	-

Table 4 Performance—voice delay monitoring node locations

Public Cloud Elements	Performance - Voice Delay Metrics									
	User Side Delay	Cache Hit Ratio	Dial Wait Time	Service Response Time	Disk Busy Time	Portal response time	Round trip delay	DNS lookup time	One way delay	Transaction duration
Data Base Server	-	-	•	-	•	-	-	-	-	-
Discovery Server	-	-	•	-	-	-	-	-	-	-
DNS Server	-	-	-	-	•	-	-	-	-	-
Service Node Server	-	-	-	•	-	-	-	-	-	-
SSO Server	-	-	-	•	-	-	-	-	-	-
Internal LAN	-	-	-	-	-	•	•	•	•	•
Core Network	-	-	-	-	-	•	•	•	•	•
Cache Server	-	•	-	-	-	-	-	-	-	-
User VOIP	•	-	-	-	-	-	-	-	-	-
User	•	-	-	-	-	-	-	-	-	-

purposes. Tables 2, 3 and 4 are mentioning Cloud elements which can carry monitoring nodes for each metrics.

6 Conclusions and Future Works

These elements can be removed or replaced by other elements. It depends on factors like nature of Cloud type, Cloud architecture, monitoring mechanisms and service nature. Like such roadmap can be developed more to cover all Cloud QoS metrics as well as service parameters. According to the resulted tables, Cloud elements can be categorized upon the number of monitoring nodes each one has. It leads to determine how important an element is. Then, Cloud business could be enabled when attention gets granted to the vital elements.

Further ahead more metrics to be added, more elements to be added and more QoS stabs to be considered. Our proposed roadmap can be customized to fit future tougher conditions as well as new economic considerations. The same concept can be applied for private Clouds and other types. Our proposed roadmap can be more complex in case of outsourcing to multiple Clouds or in case of a Cloud interoperates with another Clouds. This level of complexity is kept for future research.

References

1. Rahman, Z., Hussain, O., Hussain, F.: Time series QoS forecasting for management of cloud services. In: IEEE 9th International Conference on Broadband and Wireless Computing, Communication and Applications (BWCCA), pp. 183–190 (2014)
2. Sameh, H., Nashwa, A.: Towards IT-Legal framework for cloud computing. In: Springer Conference, SecNet'13, Cairo, pp. 122–130 (2013)

3. Frey, S., Luthje, C., Reich, C., Clarke, N.: Cloud QoS scaling by fuzzy logic. In: IEEE International Conference on Cloud Engineering (IC2E), pp. 343–348 (2014)
4. Christoph, K., Alexander, R., Thorsten, S.: QoS-aware storage virtualization for cloud file systems. In: ACM Proceedings of the 1st International Workshop on Programmable File Systems, pp. 19–26 (2014)
5. Misra, S., Krishna, P., Kalaiselvan, K., Saritha, V.: Learning automata-based QoS framework for cloud IaaS. IEEE Trans. Netw. Serv. Manag. **11**(1), 15–24 (2014)
6. Hisham, A., Hala, H., Amany, M., Abdelkarim, E., Sherif, A.: QoS Optimization for Cloud Service Composition Based on Economic Model. Springer Lecture Notes of the Institute for Computer Sciences, Social Informatics and Telecommunications Engineering, pp. 355–366. Springer, Berlin (2014)
7. Rabi, P., Manas, P., Suresh, S.: SLAs in cloud systems: the business perspective. IEEE Int. J. Comput. Sci. Technol. **3**(1), 481–488 (2014)
8. TechTarget. http://searchunifiedcommunications.techtarget.com/definition/QoS-Quality-of-Service (2015)
9. Calheiros, R., Ranjan, R., Buyya R.: Virtual machine provisioning based on analytical performance and QoS in cloud computing environments. In: IEEE International Conference on Parallel Processing (ICPP), pp. 295–304 (2011)
10. Buyya, R., Chee, S., Venugopal, S.: Market-oriented cloud computing: vision, hype, and reality for delivering IT services as computing utilities. In: IEEE 10th International Conference on High Performance Computing and Communications, pp. 5–13 (2014)
11. Rajinder, S., Sandeep, K.: Scheduling of big data applications on distributed cloud based on QoS parameters. Springer J. Clust. Comput. **18**(2), 817–828 (2015). Springer
12. Shangguang, W., Zhipiao, L., Qibo, S., Hua, Z., Fangchun, Y.: Towards an accurate evaluation of quality of cloud service in service-oriented cloud computing. J. Intell. Manuf. **25**(2), 283–291 (2014). Springer
13. Liang, Z., Sherif, S., Anna, L., Athman, B.: QoS-aware service compositions in cloud computing. Cloud Data Management, pp. 119–133. Springer, Switzerland (2014)
14. Zhi-Hui, Z., Xiao-Fang, L., Yue-Jiao, G., Jun, Z., Henry, S., Yun, L.: Cloud computing resource scheduling and a survey of its evolutionary approaches. ACM J. Comput. Surv. (CSUR) **47**(4), 25–39 (2015)
15. Valeria, C., Vincenzo, G., Francesco, P., Matteo, N.: Distributed QoS-aware scheduling in storm. In: ACM Proceedings of the 9th International Conference on Distributed Event-Based Systems, pp. 344–347 (2015)
16. Yong, Z., Jia, Z., Yan, G.: Study of local monitoring system based on SMS under the cloud environment. Ubiquitous Computing Application and Wireless Sensor. Lecture Notes in Electrical Engineering. Springer, The Netherlands (2015)
17. Kui, R., Cong, W., Qian, W.: Security challenges for the public cloud. IEEE Mag. Comput. Soc. **16**(1), 69–73 (2012)

Assistive Technology Solution for Blind Users Based on Friendsourcing

Dina A. Abdrabo, Tarek Gaber and M. Wahied

Abstract In this paper, IT-based solution was proposed an assistance tool for blind users to enable them to deal with their daily activities while supporting their privacy. This solution is based on the friend-sourcing concept through the integration between the capabilities of both the smartphone and the social networking website (Twitter). The proposed solution not only supports the English but also the both Arabic language which is highly needed for the Arabic world. The first stage of this solution (capturing and saving the video by the blind user) was implemented and evaluated through a field study consisting of a number of blind users who were happy with the idea and found the current stage of the prototype easy of use.

Keywords Crowdsourcing · FriendSourcing · Social networking · Twitter · Privacy · Talkback · Twitter API

1 Introduction

The World Health Organization (WHO) reported that 285 million people around the world are visually impaired (39 million are blind and 246 have low vision) [1]. Blinds cannot be self-independent on their life activities. Nowadays, responsibilities of life become complex especially for disabled people. The advances in information technology has inspired researchers and developers to direct the technology to serve disabled people by suggesting solutions to assist them in their life. Examples of these solution include OCR [2] and screen readers such as (Window-Eyes or JAWS) [3] which reads the text from the cell phone screen for the blind users. However, those solutions were not enough for blinds to recognize their surroundings and to meet all

D.A. Abdrabo · T. Gaber (✉) · M. Wahied
Faculty of Computers and Informatics, Suez Canal University, Ismailia, Egypt
e-mail: tmgaber@gmail.com; tarek.gaber@ci.suez.edu.eg
URL: http://www.egyptscience.net

T. Gaber
Scientific Research Group in Egypt (SRGE), Cairo, Egypt

© Springer International Publishing Switzerland 2016
T. Gaber et al. (eds.), *The 1st International Conference on Advanced Intelligent System and Informatics (AISI2015), November 28–30, 2015, Beni Suef, Egypt,*
Advances in Intelligent Systems and Computing 407, DOI 10.1007/978-3-319-26690-9_37

their needs. They are still expensive, (typically hundreds of dollars [2]) and are often not working effectively in real-world situations.

On the other hand, the existing applications of the social network (e.g., Facebook, Twitter, and Yahoo) are initially designed for sighted users and the blinds' needs are overlooked. In addition, despite the effort of hardware manufacturers to include accessibility in their touch based cell phone devices, these efforts are not enough to assist the low vision users because of their low accessibility on sighted application's controls and buttons [4]. Thus, the accessibility features for unsighted users are not always adequate to obtain a reliable result.

Helal et al. in [4], proposed approach called LOW VISION MOBILE APP PORTAL that collects all applications designed specifically for low vision users. This portal enables them to easily access a wide variety of specified applications designed for them [4]. Nevertheless this great effort, this portal is still inefficient for blind users to recognize their surroundings, since the existence of these applications on this portal is not guaranteeing that they are useful for blind users in their everyday activities. Other efforts employed a new concept (friend sourcing [5] and crowdsourcing [5] ideas) which merge the human factor assistance with the technology in identifying the surrounding environment to get accurate responses for the blind person [3, 6, 7].

Nowadays, responsibilities of life become complicated for the normal people and indeed it is more complicated for disabled people. As one type of these disabled, the blind people face numerous difficulties in their live, especially recognizing their surroundings. The advance of the information technology (IT) could provide a solution to those people to enable them to engage and contribute in their societies. In this paper, a blind society application is proposed. It is android application which allows blind users to recognize their unknown surroundings by utilizing two concepts: online social networks and friendsourcing. These concepts were employed by allowing the family members and the trusted friends who are registered on Twitter to answer the blind's question on a realtime. The solution also supports two languages (Arabic/English).

The rest of the paper is organized as follows. Section 2 gives a brief background about the friendsourcing and Twitter while Sect. 3 highlight the related work. In Sect. 4 the proposed solution is presented and the discussion is given in Sect. 5.1. Finally, the paper is concluded in Sect. 6.

2 Background

Blind people often ask for assistance from their family and friends or from anonymous people in order to help them for their accessibility problems in the physical world. Similar to this way, the friendsourcing [5] which is way of gathering information from people one knows and trusts rather than from strangers, can be used for blind to get good recommendations and save their privacy. In other cases, blind people may rely on or recruit strangers to assist them in their everyday activities, e.g., information access or object recognition (e.g., recognizing currency while paying for their buyings). This way is known as crowdsourcing [8].

2.1 Friendsourcing and Crowdsourcing

When information is available to someone, people seek to ask others to get help. In other words, they depend on each other by asking them for help in everyday situation, e.g., asking for an address by a passenger in the street. Concept of get help from unknown anonymous people is called Crowd sourcing [8]. In the field of information technology, the crowdsourcing refers to the online recruitment of people to achieve certain tasks which is too large or too difficult for a single person. Utilizing this concept, a number of application/solution have been proposed: VizWiz [7] and Chorus:view [3] which depend on anonymous people to provide assistance to blind users.

Similar to the crowdsourcing, the friendsourcing is introduced but it only make use of the friends or the family members to gather or to get accurate information for someone. The friendsourcing is based on the potential of relatively small networks of friends where there is no wisdom of the crowd, but the wisdom of a carefully collected network of friends, which becomes a high quality source of information [5].

When one uses the social network to ask his/her friends a question, either the crowdsourcing or the friendsourcing can be used. However, there are private questions that make users uncomfortable to ask strangers (crowdsourcing). Thus, they seek help from known friends to feel free in asking and to some extant, protecting their privacy [5].

2.2 Twitter

It is a free online social networking service which is based on short text messages called *tweets*. Every tweet can contain only 140 character. There are two types of tweets: (1) *Public Tweet* which is visible to anyone, whether or not they have accounts at Twitter, (2) *Protected Tweet* which may only be visible to your approved Twitter followers. Also, Twitter gives its users the ability to send private text messages, create group and lists of selected users, participate in conversations, follow and search for hashtags according to personal interests.

Twitter rapidly became one of the most popular social networking services in the world, after publishing it in 2006 by Jack Dorsey. This service has more than 100 million users who posted 340 million tweets per day in 2012. Later in May 2015, the Twitter web-service reached more than 500 million users [9]. Its API is considered very powerful and well structures tool for building different Apps or web-based services.

3 Related Work

A number of solutions aiming to help blind people have been proposed and they include Legion [6], VIZWIZ [7] and Chorus:View [3], and others [10, 11]. These solutions make use of the integration between human resource and the information technology. An overview of these solution is given below.

VizWiz is an iPhone application that allows blind users to send pictures along with visual questions about their surroundings to some workers to receive quick answers which is read to the blind users sing the text to speech technology. VizWiz works as described in the following steps [7].

Firstly, a picture is captured as follows. When VizWiz is opened, a camera button will be appeared on the screen for the user to presses double-taps on the button of the camera till he/she hears a sound of the camera taking the picture. Secondly, once the image is taken, a different screen will be loaded and by a double-tap from the user, the phone will vibrate indicating the user to speak his/her question clearly. To end-up his recording, the user needs to double tap the screen. Thirdly, another screen will be loaded enabling the user to choose an answering party (anonymous workers, Facebook, Twitter, or an email contact) about his question. Double tapping for one option will activate it and the send button will be activated for the user to send his question to the distention. Fourthly, short time after sending the image and the question, VizWiz will load the "View Answers" screen and the received answers will be spoken as they appear. Also, the source of the answer and its rate will be read loudly [7].

As reported in [3], VizWiz is not a time efficient solution as (1) sometimes the picture is not clear, thus a helper asks the blind to recapture the image, and repeating the processing again and (2) a helper sometimes explains the image content without listening to the question carefully, thus the blind would need repeat the questions again. In addition, VizWiz do not protect the users' privacy.

Chorus:View [3] is another assistive a system that enables blind users to recognize the surrounding objects through recording a video for these objects. This system makes use of the crowdsourcing to help the blind to recognize their surroundings. This allows answering sequential questions quickly and effectively for blind users involved in a continuous conversation. This is considered the main strength of Chorus:View as the continuous interaction makes the request and the conversation more reliable. In addition, the continuous interaction allows workers to maintain and recognize the context in a real time. Chorus:View works as described in the following steps. Firstly, a video is captured by a users phone and then the user records an audio message stating the question about this video. The user then sends the audio message to anonymous workers who watch the video stream and then provide the feedback with the answer to the users question.

Also the Chorus:View is better than VizWiz in terms of giving accurate answer about the captured object as the latter depends on video which gives more information about the object. However, Like VisWiz, the Chorus:View does not considered the privacy of the user as when the user sends a video to the workers or the anonymous helpers, these helpers can see and know whatever is in the received video. For example, imagine the blind user wants to recognize the information in his bank statement, once he sends a video about it to the workers, all the confidential information in this statement will be disclosed.

4 The Proposed Solution: Trust Blind Society Application

To overcome the limitation of VizWiz (i.e., not providing enough information about the unknown object as only support images) and Chorus:View (i.e., not support privacy issue). We proposed assistive Mobile App called: Trust Blind Society (TBS). The idea of this App is based on the friend-sourcing concept while providing help through the use of the video recording received over the social website Twitter. The use of friendsourcing (family members and close friends) would help to protect the blind users' privacy such as credit card numbers, emails passwords, accounts, private situations that may embarrassing him and any private data in his surrounding environment.

To achieve this aim, TBS integrates the features of (a) Android OS (i.e., Talkback), (b) Twitter (i.e., posting video link), and (c) friend sourcing (one's friends using Twitter). Using TBS application, a blind user can record a video for a given object (something in his/her house or at the street) and send it to his/her friends on Twitter who will reply by the details of this object. Figure 1 summarizes the idea of the TBS application.

4.1 System Architecture

The TSB application consists of three main components: User Interface representing the blind user, Video server hosting the videos sent by the blind users, and friends' interface representing the blind user's friends who will help to answer the questions.

User Interface: This interface designed such that when a user opens the TBS Application, the user will be asked to select either the Arabic or the English language. Such selection will be saved as the default language. Then the user will be then asked to setup a list of his friends who will help him. This will be achieved by allowing the App to fetch his/her friends' names from his Twitter account. If he has a single friend list, the application will ask the user to choose whether he wants to select all the list or certain friends from the list. If he has multiple friend lists, a pager view (slider) of the name's list is opened and when he hovers over any name, the application will narrate the name and if he wants to select a certain list, he will click double click. If the blind user wants to add a new friend to the list, he can go to Edit from the settings and then add his details. In case no one of the selected list replied to his question posted on Twitter, the blind user will choose only one friend for *urgent phone calls*.

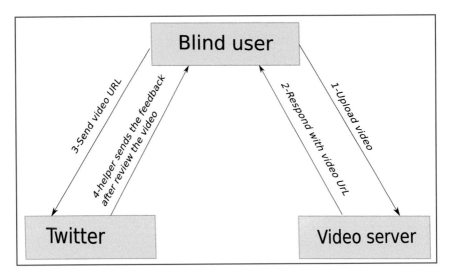

Fig. 1 block diagram of the TBS application solution

After setting up his target friends, the blind user can record a video of any object around him. (An autofocus camera [12] should be used to help the blind to capture the entire object.) The user can then start recording the video for only 30 s by a double-tap on the screen and stop recording by a three-tap. Otherwise after 30 s, the application will vibrate and the video will be uploaded to video server to be stored for later use.

The Video Server: When the user finished his video recording, the application will upload that video to a remote and secured server at which the blind user should have already registered and received a user name and password to be used for any future access. Upon receiving the video, the sever will respond with the video link to the application which will save this link and then send it to the list of his friend on Twitter.

Friends Interface: Once the video is sent to the predefined list on Twitter and they are online, they can watch the video and respond with the answer. If all of them are not online, the application will send a text SMS to the friend who is selected for urgent calls/SMS. Then application will work in the background and check if there were new messages or not, if there is, the application will loudly read the messages to the blind user.

5 Case Study and Discussion

This case study demonstrate the developed beta version of the suggested system that just captures a video stream by the blind user and save it to his phone. This development has done using **Twitter API**: which is the tool that will enable us to send the

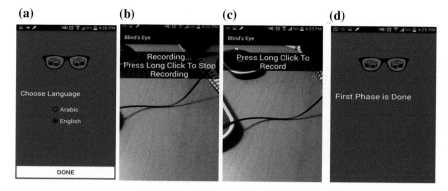

Fig. 2 Screen shots of the user interface

video link through twitter as a private message to the blind user's friends, **Android Talkback** which is a feature (screen reader) that will read all data and instructions for the user from the screen. Talkback [13] is an accessibility service that helps blind and vision-impaired users interact with their devices. Talkback adds spoken, audible, and vibration feedback to your device. This application is only activated if you explicitly turn the accessibility on.

Figure 2 shows the general user interface of the implemented system and the following steps describing how the blind user use it.

- The blind user can open the TBS application by clicking on it.
- Welcome message enabling the user to select his language is prompted as shown in Fig. 2(a)
- The camera is then opened and the application asks the user to press the screen long press to start a video recording, also the application will vibrate once its start the record as can be seen in Fig. 2(b)
- To help the blind to record the full object, an autofocus camera is used and the flash will be enabled while recording
- If the user wants to stop recording before the allowed 30 seconds, he can do so by pressing on the screen another long press, the application then will vibrate indicating stop recording (see Fig. 2(c))
- Once the recording is stopped, the video will be saved on the blind's mobile for future usage (e.g. send it to friends)

5.1 Discussion

In order to examine the first stage of our solution (capturing the video and storing it in the mobile), we presented the application to a group of 20 blind volunteers, describing the idea of the TBS application (illustrated in Fig. 3) and its usability, in

420D.A. Abdrabo et al.

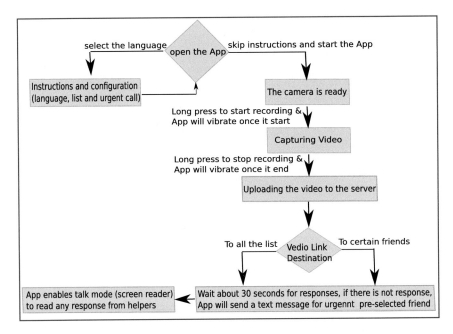

Fig. 3 Data flow diagram of the TBS Application

order to check the usability of the application. Generally, the volunteers were asked to use the application by recording a video for random object. The flash of the mobile is automatically turned on if the surrounding area is dark.

While calculating the number of the successful trials, we found that 17 of them were successfully recorded a video stream about the desired object from the first trial, and 2 of them were successfully completed this cycle from the second trial and only one blind user who successfully recored a video after three trials. In another hand, the blind volunteers said that the TBS application is easy to use and has friendly interface.

In addition, the volunteers appreciated the fact that the TBS is an android-based smartphone as it is now available with low prices which support the economic case of most blind peoples in Egypt and most Arabic countries. So, the TBS application will be much affordable than VizWiz application which is only working at iOS-based smart phones. Moreover, the proposed TBS application is supported with Arabic language which is provided neither in VizWiz nor Chours-View. This means that the TBS application has the potential to be used for all Arabic countries.

6 Conclusion

In this paper, we introduced a proposed solution aiming to help the blind users to recognize their surroundings. This solutions integrates the social networking websites (Twitter) with the widely used smartphones. It made use of the advance in both of the speed of the Internet and the capabilities of the Android-based smartphones to allow the blind users to capture a short video of their surroundings and send it to their friends on Twitter (i.e. utilizing the friendsourcing concept) who responds with a text describing this object. In this way, only the close friends can know what is this object, thus protect the blind's privacy. In addition, the proposed solution supports both Arabic an English language which is very needed for the Arabic world. The first stage of this solution (capturing and saving the video) was implemented and tested by a number of blind users who were happy with the idea and asking for the complete application and found it a friendly access application. The future include assessing the performance of this solution in the real time in terms of the transmission rate, and the response time and the quality of the video.

References

1. Martínez-Pérez, B., de la Torre-Díez, I., López-Coronado, M.: Mobile health applications for the most prevalent conditions by the world health organization: review and analysis. J. Med. Internet Res. **15**(6), e120 (2013)
2. Kane, S.K., Jayant, C., Wobbrock, J.O., Ladner, R.E.: Freedom to roam: a study of mobile device adoption and accessibility for people with visual and motor disabilities. In: Proceedings of the 11th International ACM SIGACCESS Conference on Computers and Accessibility, ACM, pp. 115–122 (2009)
3. Lasecki, W.S., Wesley, R., Nichols, J., Kulkarni, A., Allen, J.F., Bigham, J.P.: Chorus: a crowd-powered conversational assistant. In: Proceedings of the 26th Annual ACM Symposium on User Interface Software and Technology, ACM, pp. 151–162 (2013)
4. Helal, A.S., Moore, S.E., Ramachandran, B.: Drishti: an integrated navigation system for visually impaired and disabled. In: Proceedings of the Fifth International Symposium on Wearable Computers, IEEE, pp. 149–156 (2001)
5. Bernstein, M.S., Tan, D., Smith, G., Czerwinski, M., Horvitz, E.: Personalization via friendsourcing. ACM Trans. Comput. Hum. Interact. **17**(2), 6:1–6:28 (2008)
6. Lasecki, W.S., Murray, K.I., White, S., Miller, R.C., Bigham, J.P.: Real-time crowd control of existing interfaces. In: Proceedings of the 24th Annual ACM Symposium on User Interface Software and Technology, ACM, pp. 23–32 (2011)
7. Bigham, J.P., Jayant, C., Ji, H., Little, G., Miller, A., Miller, R.C., Miller, R., Tatarowicz, A., White, B., White, S., et al.: Vizwiz: nearly real-time answers to visual questions. In: Proceedings of the 23rd Annual ACM Symposium on User Interface Software and Technology, ACM, pp. 333–342 (2010)
8. Brady, E.L., Zhong, Y., Morris, M.R., Bigham, J.P.: Investigating the appropriateness of social network question asking as a resource for blind users. In: Proceedings of the 2013 Conference on Computer Supported Cooperative Work, ACM, pp. 1225–1236 (2013)
9. Makice, K.: Twitter API: Up and Running: Learn How to Build Applications with the Twitter API, 1st edn. O'Reilly Media, Inc., Sebastopol (2009)
10. Ghali, N.I., Soliuman, O., El-Bendary, N., Nassef, T.M., Ahmed, S.A., Elbarawy, Y.M., Hassanien, A.E.: Virtual reality technology for blind and visual impaired people: reviews

and recent advances. In: Gulrez, T., Hassanien, A.E. (eds.) Advances in Robotics and Virtual Reality. ISRL, vol. 26, pp. 363–385. Springer, Heidelberg (2012)

11. El-Gayyar, M., ElYamany, H.F., Gaber, T., Hassanien, A.E.: Social network framework for deaf and blind people based on cloud computing. In: The proceedings of the Federated Conference on Computer Science and Information Systems (FedCSIS), pp. 1313–1319 (2013)

12. White, P., Podaima, B., Friesen, M.: Algorithms for smartphone and tablet image analysis for healthcare applications. IEEE Access 2, 831–840 (2014)

13. Zhong, Y., Raman, T.V., Burkhardt, C., Biadsy, F., Bigham, J.P.: Justspeak: enabling universal voice control on android. In: Proceedings of the 11th Web for All Conference, W4A'14, ACM, New York, pp. 36:1–36:4 (2014)

Part VI
Intelligent Systems and Informatics (II)

Belief Function Combination: Comparative Study Within the Classifier Fusion Framework

Asma Trabelsi, Zied Elouedi and Eric Lefèvre

Abstract Data fusion under the belief function framework has attracted the interest of many researchers over the past few years. Until now, many combination rules have been proposed in order to aggregate beliefs induced form dependent or independent information sources. Although the choice of the most appropriate rule among several alternatives is crucial, it still requires non-trivial effort. In this investigation, we suggest to evaluate and compare some combination rules when dealing with independent information sources in the context of the classifier fusion framework.

Keywords Data fusion · Belief function theory · Combination rules · Independent information sources · Classifier fusion

1 Introduction

Pattern recognition has been widely studied to solve classification problems owing to its capacity to achieve the greatest possible classification accuracy [15]. One of the proposed solutions is based on an advanced method named ensemble classifiers. Hence, various combination approaches have been proposed to combine multiple classifiers such as voting-based systems, plurality, Bayesian theory, belief function theory [8]. This latter, also known as Dempster-Shafer theory, is regarded as a convenient method for representing and managing different kinds of imperfect data [14] and has proved to be an efficient approach for combining a set of classifiers. Thus, it

A. Trabelsi (✉) · Z. Elouedi
Institut Supérieur de Gestion de Tunis, LARODEC, Université de Tunis, Tunis, Tunisia
e-mail: trabelsyasma@gmail.com

Z. Elouedi
e-mail: zied.elouedi@gmx.fr

E. Lefèvre
Laboratoire de Génie Informatique et d'Automatique de l'Artois (LGI2A),
University Artois, EA 3926, F-62400 Béthune, France
e-mail: eric.lefevre@univ-artois.fr

© Springer International Publishing Switzerland 2016
T. Gaber et al. (eds.), *The 1st International Conference on Advanced Intelligent System and Informatics (AISI2015), November 28–30, 2015, Beni Suef, Egypt,*
Advances in Intelligent Systems and Computing 407, DOI 10.1007/978-3-319-26690-9_38

425

provides several combination rules which mainly differ according to the way of managing the mass assigned to the empty set also called conflict [1, 2, 5, 7, 12]. Basically, in this paper, we are interested in the Dempster rule [1], the conjunctive rule [12], the combination with adapted conflict rule (CWAC rule) [5] and the improved CWAC rule [2]. It is noteworthy that the combination rule of Dempster does not support the value of the conflict generated when combining pieces of evidence and consequently this latter should be proportionally distributed over all focal elements. However, in the conjunctive combination rule the mass allocated to the empty set should be kept in the purpose of reflecting the degree of conflict between the combined sources. Nevertheless, this conflict has an absorption effect: when we apply a large number of conjunctive combinations, the mass assigned to the conflict tends towards 1 and hence the conflict loses its initial role. The CWAC and the improved CWAC rules are defined by an adaptive weighting between the Dempster and the conjunctive rules in order to give the conflict its paramount role as an alarm signal. With this diversity of combination rules, the choice of the most efficient one becomes really a challenging task. So, in this work, we propose to compare the CWAC and the improved CWAC rules with the conjunctive and the Dempster rules within the classifier fusion framework in order to pick out the most appropriate combination rule. The rest of this paper is structured as follows. Section 2 provides a brief overview of the basic concepts of the belief function theory. We outline some combination rules dealing with distinct pieces of evidence in Sect. 3. Section 4 is devoted to discussing our comparative approach. The experiments and the results are presented in Sect. 5. The conclusion is reported in Sect. 6.

2 Belief Function Theory: Basic Concepts

Let Θ be a finite non-empty set of N elementary events related to a given problem, these events are assumed to be exhaustive and mutually exclusive. Such Θ is called the frame of discernment. The power set of Θ, denoted by 2^Θ, is composed of all the subsets of Θ.

The impact of evidence assigned to each subsets of the frame of discernment Θ is named basic belief assignment (bba). It is defined as:

$$m : 2^\Theta \rightarrow [0, 1]$$

$$\sum_{A \subseteq \Theta} m(A) = 1 \tag{1}$$

The amount $m(A)$, known as basic belief mass (bbm), expresses the degree of belief committed exactly to the event A.

To make decision within the belief function framework, we must transform the bba into a probability measure called pignistic probability denoted $BetP$ and defined as follows [13]:

$$BetP(A) = \sum_{B \subseteq \Theta} \frac{|A \cap B|}{|B|} \frac{m(B)}{1 - m(\emptyset)} \quad \forall A \in \Theta \tag{2}$$

where $|B|$ denotes the cardinality of B.

The reliability of each information source S can be quantified. In fact, if S is not fully reliable then the bba provided by S should be discounted using a reliability factor denoted $1—\alpha$ [11]. The discounted bba is obtained as follows:

$$m^\alpha(A) = (1 - \alpha)m(A) \quad \forall A \subset \Theta \tag{3}$$
$$m^\alpha(\Theta) = \alpha + (1 - \alpha)m(\Theta)$$

Given two bbas m_1 and m_2, according to [4], the distance measure between them is computed as follows:

$$d(m_1, m_2) = \sqrt{\frac{1}{2}(m_1 - m_2)^T D(m_1 - m_2)} \tag{4}$$

with D is the Jaccard index matrix, the elements of which are calculated as follows:

$$D(A, B) = \begin{cases} 1 & \text{if } A = B = \emptyset \\ \dfrac{|A \cap B|}{|A \cup B|} & \forall A, B \in 2^\Theta \end{cases} \tag{5}$$

3 Combination Rules Dealing with Independent Information Sources

As mentioned earlier, there exist several combination rules assuming items of evidence combined to be independent. In this section, we present only the conjunctive rule [12], the Dempster rule [1], the CWAC rule [5] and the improved CWAC rule [2].

1. The conjunctive rule, proposed by Smets, is used to combine two bbas provided by reliable and distinct information sources [12]. The resulting bba, denoted $m_1 \textcircled{O} m_2$, is defined by:

$$(m_1 \textcircled{O} m_2)(A) = \sum_{B,C \subseteq \Theta : B \cap C = A} m_1(B).m_2(C) \tag{6}$$

The mass assigned to the empty set $(m_1 \textcircled{O} m_2(\emptyset))$ quantifies the degree of disagreement between the two combined sources.
2. The Dempster rule, based on the orthogonal sum, is a normalized version of the conjunctive rule where the mass of the empty set must be reallocated over all

focal elements in the case where $m_1 \bigcirc m_2(\emptyset) \neq 0$ thanks to a normalization factor, denoted K [11]. This rule, assuming pieces of evidence combined to be reliable and distinct, is defined as follows:

$$(m_1 \oplus m_2)(A) = K(m_1 \bigcirc m_2)(A) \tag{7}$$

and

$$(m_1 \oplus m_2)(\emptyset) = 0 \tag{8}$$

where

$$K^{-1} = 1 - (m_1 \bigcirc m_2)(\emptyset) \tag{9}$$

3. In [5] the authors have proposed the CWAC combination rule which is defined by an adaptive weighting D between the conjunctive and the Dempster rules. This adaptive weighting offers an effective way to obtain the same behavior as the conjunctive rule when the bbas are contradictory and the same behavior as the Dempster rule when the bbas are similar. The CWAC rule uses the Jousselme distance to measure the dissimilarity between sources. Assume we have M bbas, denoted as $m_1,....,m_M$, the result of their combination using the CWAC operator is noted as $m_{\ominus}$ and is defined as follows:

$$m_{\ominus}(A) = D m_{\bigcirc}(A) + (1 - D) m_{\oplus}(A) \tag{10}$$

and

$$m_{\ominus}(\emptyset) = 1 \text{ when } m_{\bigcirc}(\emptyset) = 1 \tag{11}$$

with

$$D = \max_{i,j}[d(m_i, m_j)] \quad \forall i, j \in [1, M] \tag{12}$$

$$m_{\bigcirc}(A) = (\bigcirc_i m_i)(A) \text{ and } m_{\oplus}(A) = (\oplus_i m_i)(A) \quad \forall i \in [1, M] \tag{13}$$

4. The improved CWAC rule [2], inspired from the spirit of the CWAC rule, is employed to combine reliable and distinct pieces of evidence. Authors in [2] have proved that the improved CWAC rule enhances the ability of the CWAC rule to preserve the conflict as an alarm signal and also it truly reflects the opposition between bbas in the combination. Assume we have M bbas, denoted as $m_1,.....,m_M$, the result of their combination using the improved CWAC operator, denoted $\ominus^I$, is defined as follows:

$$m_{\ominus^I}(A) = \bar{D} m_{\bigcirc}(A) + (1 - \bar{D}) m_{\oplus}(A) \tag{14}$$

and

$$m_{\ominus'}(\emptyset) = 1 \text{ when } m_{\bigcirc}(\emptyset) = 1 \tag{15}$$

with

$$\bar{D} = \frac{\sum_{i=1,j>i}^{M} d(m_i, m_j)}{\frac{M(M-1)}{2}} \quad \forall i,j \in [1,M] \tag{16}$$

$$m_{\bigcirc}(A) = (\bigcirc_i m_i)(A) \text{ and } m_{\oplus}(A) = (\bigoplus_i m_i)(A) \quad \forall i \in [1,M] \tag{17}$$

As the CWAC and the improved CWAC rules are defined by an adaptive weighting between the conjunctive and the Dempster rules, we suggest to make a comparative approach that allows to select the most efficient rule within the classifier fusion framework.

4 Comparative Approach

Ensemble classifier systems, also known as multiple classifiers, is considered as an efficient way to solve pattern recognition issues. As well, the fusion of a set of classifiers in the context of the belief function framework has been extensively explored in several studies [6, 10]. In this paper, ensemble classifier systems will be used as a way for evaluating and comparing some fusion rules dealing with independent information sources. The process of combining classifiers within the belief function framework composed of two distinct parts. The first one consists of the construction of mass functions from classifiers' outputs and the second one focuses on the combination of these mass functions across some combination rules.

4.1 Mass Functions Construction from Classifiers' Outputs

Consider a pattern recognition problem where $B = \{x_1, ..., x_n\}$ is a database with n instances, $C = \{C_1, ..., C_M\}$ is a set of M classifiers and $\Theta = \{w_1, ..., w_N\}$ is a set of N class labels. B should be randomly split into learn and test sets. We first construct classifiers from the learning set and then we apply them to predict the label class of all test patterns. In order to combine classifiers within the belief function framework, classifiers outputs must be transformed into bbas. Since classifiers outputs may differ from one classifier to another, each pattern test should have M bbas obtained as follows:

$$m_i(\{w_j\}) = 1 \tag{18}$$
$$m_i(A) = 0 \quad \forall A \subseteq \Theta \text{ and } A \neq \{w_j\}$$

where $m_i(\{w_j\})$ expresses the part of belief assigned exactly to the predicted class w_j through the classifier C_i.

Results supplied by classifiers can be unreliable, therefore the bbas generated must be discounted by taking into consideration the reliability rate of each classifier. The reliability r_i of a classifier C_i is computed as follows:

$$r_i = \frac{\text{Number of well classified instances}}{\text{Total number of classified instances}} \tag{19}$$

If r_i equals 1 then the classifier C_i is absolutely reliable, by against the classifier C_i is totally unreliable in the case where r_i is equal to 0. The discounted mass functions, using Eq. 3, are obtained as follows:

$$m_i^{\alpha_i}(\{w_j\}) = r_i \tag{20}$$
$$m_i^{\alpha_i}(\Theta) = 1 - r_i$$

with $\alpha_i = 1 - r_i$.

5 Classifier Fusion

Let us remind that the process of combining classifiers within the belief function framework consists on two main steps: classifiers' outputs modeling and classifiers' combination. So, if the outputs of all classifiers are converted into bbas, we move on to classifier fusion using combination rules mentioned in Sect. 3. Combination results allows us to assess and compare these alternative rules in order to select the most efficient one. Thus, we rely on two assessment criteria: the PCC and the distance.

- The PCC criterion that represents the percent of correctly classified instances will be used to compare the CWAC and the improved CWAC rules of combination with the Dempster one. Such case demands the use of three variables n_1, n_2, n_3 which respectively correspond to the number of well classified, misclassified and rejected instances. For each combination rule, we proceed as follows:
  1. Firstly, we set a tolerance thresholds $S = [0.1, 1]$. For any threshold $s \in S$, we examine the mass of the empty set ($m(\emptyset)$) induced by each model test as follows:
     – if $m(\emptyset) > s$, the classifier chooses to reject instance instead of misclassifying it. As a result, we increment the number of rejected instances n_3.
     – if $m(\emptyset) \leq s$, we calculate the pignistic probability (*BetP*) in order to choose the most probable class. Accordingly, the chosen class will be compared to

the real one: in the case where the chosen class is similar to the real one, we increment the number of well classified instance n_1, inversely we increment the number of the misclassified instances n_2.

2. Secondly, having the well classified, misclassified and rejected instances, we calculate the PCC for each threshold $s \in S$ using the following formula:

$$PCC = \frac{n_1}{n_1 + n_2} * 100 \qquad (21)$$

The most appropriate combination rule is the one that has the highest value of PCC $\forall \, s \in S$.

- The distance criterion, which corresponds to the Jousselme distance between two bbas, will be employed to compare the CWAC and the improved CWAC rules with the conjunctive one. Thus, for each combination rule, we track the following steps to compute the distance criterion:

1. The real class w_j of each pattern test must be transformed into a mass function. It is obtained as follows:

$$m_r(\{w_j\}) = 1 \qquad (22)$$
$$m_r(A) = 0 \quad \forall A \subseteq \Theta \text{ and } A \neq \{w_j\}$$

2. Then, we compute for each test pattern the Jousselme distance (See Eq. 4) between the mass function relative to its real class (m_r) and the mass function generated when combining bbas induced from M classifiers.

3. At last, we sum the Jousselme distances obtained by all test patterns in order to get the total distance.

The best combination rule is the one that has the minimum total distance.

6 Simulation and Experimentations

6.1 Experimental Setup

To evaluate our alternative combination rules, our experiments are performed using several real world databases obtained from the U.C.I repository [9] described in Table 1. These databases have different number of instances and different number of attributes. However, their classe numbers are equal to 2 or 3. It is noteworthy that our alternative combination rules can support databases with a number of classes greater than 3.

We have carried out experiments using four machine learning algorithms implemented in Weka [3]: the Naive Bayes (NB), the Decision tree (DT), the k-Nearest Neighbor where k equals 1 (1-NN), and the Neural Network (NN) algorithms. These

Table 1 Description of databases

Databases	#Instances	#Attributes	#Classes
Diabetes	768	2	2
Fertility	100	10	2
Heart	270	13	2
Hepatitis	155	19	2
Iris	150	4	3
Parkinsons	195	23	2

Table 2 Single classifier accuracies (%)

Databases	NB	DT	1-NN	NN
Diabetes	75.65	70.57	73.82	74.21
Fertility	87.00	83.00	85.00	87.00
Heart	82.96	75.55	75.18	78.88
Hepatitis	83.22	81.29	80.00	79.35
Iris	95.33	95.33	95.33	96.66
Parkinsons	69.74	96.41	87.17	91.79

latter were run based on the leave one out cross validation approach. The accuracy values of the single classifiers are given in Table 2.

6.2 Results and Discussion

Let us start by comparing the CWAC and the improved CWAC rules with the Dempster one in term of the PCC criterion. Figure 1 illustrates the PCCs for the Dempster, the CWAC and the improved CWAC rules relative to all the mentioned databases. The results, as seen in Fig. 1, indicate that for any threshold s the value of PCC relative to the Dempster rule is constant due to the fact that n_3 is always equal to 0 (no rejected instances). It should be noted that the PCC values of the CWAC and the improved CWAC rules are greater or equal to those corresponding to the Dempster rule for the different databases. For instance, the values of PCC relative to the Diabetes database for both CWAC and improved CWAC rules varies from 84 to 76 % with the variation of s, whereas they are equal to 76 % for the Dempster rule $\forall s \in S$. This result might be due to the fact that the average number of rejected instance correspond to the CWAC and the improved CWAC rules are greater than 0, whereas that correspond to the Dempster rule is equal to 0. It should be emphasized that this interpretation is available for the remaining databases. Then, we can conclude that the CWAC and the improved CWAC rules are more efficient than the Dempster rule according to the PCC criterion. To this end, we move to the comparison of the

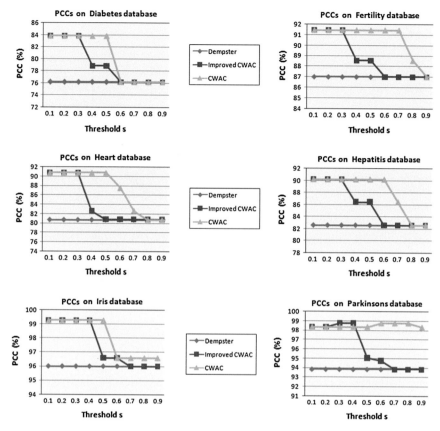

Fig. 1 PCC values for all databases

CWAC and the improved CWAC rules with the conjunctive one. Figure 2 illustrates the conjunctive, the CWAC and the improved CWAC distances correspond to the different databases.

From Fig. 2, we can notice that the CWAC and the improved CWAC rules achieve best results compared with the conjunctive rule. In fact, the distances correspond to the CWAC and the improved CWAC rules are lower than those correspond to the conjunctive rule. For example, for Pima Indian Database, the CWAC and the improved CWAC distances are respectively 270.39 and 226.60, whereas the conjunctive distance is equal to 270.42. For Hepatitis database, we have 48.62 as conjunctive distance, 43.63 as CWAC distance and 34.75 as improved CWAC distance. As far, we can assume that this result can be applied to all the other databases. Thus, we can conclude that the CWAC and the improved CWAC rules are more adequate than the conjunctive one according to the distance criterion. Also, from Fig. 1, we can remark that the PCCs obtained by the CWAC rule are higher than those obtained by the improved CWAC rule (for all $s \in S$). Moreover, the total distances of the

Fig. 2 Distance results of
the conjunctive, CWAC and
improved CWAC rules

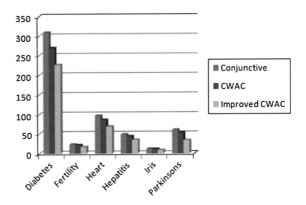

improved CWAC rule are less than those relative to the CWAC rule. Therefore, we
can note that the improved CWAC rule is better than the CWAC rule according to
the distance criterion but worse than the CWAC rule according to the PCC criterion.

7 Conclusion

In this paper, we have outlined some combination rules assuming the pieces of evidence combined to be distinct. Then, we relied on the ensemble classifier system
to carry out some experimental tests in the purpose of comparing these alternative
rules. The obtained results show the efficiency of the CWAC rule and the improved
CWAC rule compared with the conjunctive and Dempster ones. Results of experimentations show also that the improved CWAC rule is the best combination rule
according to the distance criterion, whereas the CWAC rule is considered as the best
rule of combination in term of the PCC criterion.

References

1. Dempster, A.P.: Upper and lower probabilities induced by a multivalued mapping. Ann. Math.
   Stat. **38**, 325–339 (1967)
2. Deng, X., Deng, Y., Chan, F.T.S.: An improved operator of combination with adapted conflict.
   Ann. OR **223**(1), 451–459 (2014)
3. Hall, M., Frank, E., Holmes, G., Pfahringer, B., Reutemann, P., Witten, I.H.: The WEKA data
   mining software: an update. ACM SIGKDD Explor. Newsl. **11**(1), 10–18 (2009)
4. Jousselme, A., Grenier, D., Bossé, E.: A new distance between two bodies of evidence. Inf.
   Fusion **2**(2), 91–101 (2001)
5. Lefèvre, E., Elouedi, Z.: How to preserve the conflict as an alarm in the combination of belief
   functions? Decis. Support Syst. **56**, 326–333 (2013)
6. Mandler, E.: Combining the classification results of independent classifiers based on the
   Dempster-Shafer theory of evidence. Pattern Recogn. Artif. Intell. 381–393 (1988)

7. Martin, A.: About conflict in the theory of belief functions. In: Belief Functions: Theory and Applications, pp. 161–168. Springer (2012)
8. Mercier, D., Cron, G., Denœux, T., Masson, M.: Fusion of multi-level decision systems using the transferable belief model. In: 7th International Conference on Information Fusion, FUSION'2005, vol. 2, pp. 655–658. IEEE (2005)
9. Murphy, P., Aha, D.: UCI repository databases. http://www.ics.uci.edu/mlearn (1996)
10. Ponti, M.: Combining classifiers: from the creation of ensembles to the decision fusion. In: 24th Conference on Graphics, Patterns and Images Tutorials (SIBGRAPI-T), pp. 1–10. IEEE (2011)
11. Shafer, G.: A Mathematical Theory of Evidence. Princeton University Press, New Jersey (1976)
12. Smets, P.: The application of the transferable belief model to diagnostic problems. Int. J. Intell. Syst. **13**, 127–157 (1998)
13. Smets, P.: The transferable belief model for quantified belief representation. Handb. Defeasible Reason. Uncertain. Manage. Syst. **1**, 267–301 (1998)
14. Vatsa, M., Singh, R., Noore, A., Singh, S.K.: Belief function theory based biometric match score fusion: case studies in multi-instance and multi-unit iris verification. In: 7th International Conference on Advances in Pattern Recognition (ICAPR), pp. 433–436 (2009)
15. Weiss, S.M., Kapouleas, I.: An empirical comparison of pattern recognition, neural nets, and machine learning classification methods. In: 11th International Joint Conference on Artificial Intelligence (IJCAI), pp. 781–787. Morgan Kaufmann (1989)

Genetic Algorithms for the *Tree T-Spanner* Problem

Riham Moharam, Ehab Morsy and Ismail A. Ismail

Abstract The *tree t-spanner* problem is one of the most important spanning tree optimization problems and has different applications in communication networks and distributed systems. Let $G = (V, E)$ be an undirected edge-weighted $G = (V, E)$ with vertex set V and edge set E. We consider the problem of constructing a *tree t-spanner* T in G in the sense that the distance between every pair of vertices in T is at most t times the shortest distance between the two vertices in G. The value of t, called the stretch factor, quantifies the quality of the distance approximation of the corresponding *tree t-spanner*. The problem of finding a *tree t-spanner* with the smallest possible value of t is known as the Minimum Maximum Stretch Spanning Tree (MMST) problem. It is well known that, for any $t \geq 1$, the problem of deciding whether G contains a *tree t-spanner* is NP-complete, thus, the MMST problem is NP-complete. In this paper, we present a genetic algorithm that returns a high quality solution for the MMST problem.

Keywords Tree Spanner · Stretch Factor · Minimum Maximum Stretch Spanning Tree · Genetic Algorithms · Graph Algorithms

This work is partially supported by Alexander von Humboldt foundation.

R. Moharam (✉) · E. Morsy
Department of Mathematics, Suez Canal University, Ismailia 41522, Egypt
e-mail: Riham.Sci@gmail.com

E. Morsy
e-mail: ehabmorsy@gmail.com

I.A. Ismail
Faculty of Computer and Information, Department of Computer Sciences,
6 October University, Giza, Egypt
e-mail: amr442-2@hotmail.com

© Springer International Publishing Switzerland 2016
T. Gaber et al. (eds.), *The 1st International Conference on Advanced Intelligent System and Informatics (AISI2015), November 28–30, 2015, Beni Suef, Egypt,*
Advances in Intelligent Systems and Computing 407, DOI 10.1007/978-3-319-26690-9_39

437

1 Introduction

Let $G = (V, E)$ be an undirected edge-weighted graph with vertex set V and edge set E such that $|V| = n$ and $|E| = m$. A spanning tree T in G is said to be a *tree t-spanner* if the distance between every pair of vertices in T is at most t times the shortest distance between the two vertices in G. For a given spanning tree T in G, the goodness of the distance approximation of T is estimated by the value of the stretch factor t of T [1]. The problem of finding a *tree t-spanner* with the smallest possible value of t is known as the Minimum Maximum Stretch Spanning Tree (MMST) problem [2].

The *tree t-spanner* problem is widely applied in communication networks and distributed systems. For example, it is applied to the arrow distributed directory protocol that supports the mobile object routing [3]. In particular, the MMST is used to minimize the delay of mobile object routing from the source node to every client node in case of concurrent requests through a routing tree. The worst case overhead the ratio of the protocol is proportional to the maximum stretch factor of T (see [4]). Kuhn and Wattenhofer [5] showed that the arrow protocol is a distributed ordering algorithm with low maximum stretch factor. Another application of the MMST is in the analysis of competitive concurrent distributed queuing protocols that intend to minimize the message transit in a routing tree [6].

For any $t \geq 1$, the problem of deciding whether G contains a *tree t-spanner* is NP-complete [7]. Consequently, the MMST problem, the problem of finding *a tree t-spanner* that minimizes t, is NP-complete. In this paper we present an efficient genetic algorithm to the MMST problem. Our experimental results show that the proposed algorithm returns high quality *tree t-spanner*.

The rest of this paper is organized as follows. Section 2 reviews some results on the *tree t-spanner* and related problems. Section 3 presents the proposed genetic algorithm. Section 4 evaluates our algorithm by applying it to randomly generated instances of a *tree t-spanner* problem. Section 5 makes some concluding remarks.

2 Related Work

In this section, we present results on related problems.

For an unweighted graph G, Cai and Corneil [7] produced a linear time algorithm to find a *tree t-spanner* in G for any given $t \geq 2$. Moreover, they showed that, for any $t \geq 4$, the problem of finding a *tree t-spanner* in G is NP-complete. Brandstädt et al. [8, 9] improved the hardness result in [7] by showing that a *tree t-spanner* is NP-complete even over chordal graphs which each created a cycle with length 3 whenever $t \geq 4$ and chordal bipartite graphs which each created a cycle with length 4 whenever $t \geq 5$.

Peleg and Tendler [10] proposed a polynomial time algorithm to determine a minimum value for t for the *tree t-spanner* over outerplanar graphs. In [11], Fekete and

Kremer showed that it is NP-hard to determine a minimum value for t for which a *tree t-spanner* exists even for planar unweighted graphs. They designed a polynomial time algorithm that decides if the planar unweighted graphs with bounded face length contains a *tree t-spanner* for any fixed t. Moreover, they proved that for $t = 3$, it can be decided whether the unweighted planar graph has a *tree t-spanner* in polynomial time. The problem was left open whether a *tree t-spanner* is polynomial time solvable in case of $t \geq 4$. Afterwards, this open problem is solved by Dragan et al. [12]. They proved that, for any fixed t, the *tree t-spanner* problem is linear time solvable not only for a planar graphs, but also for the class of sparse graphs which include graphs of bounded genus. Emek and Peleg [2] presented an $O(\log n)$-approximation algorithm for finding the *tree t-spanner* problem in a graph of size n. Moreover, they established that unless P = NP, the problem cannot be approximated additively by any o(n) term.

Recently, Dragan and Köhler [13] examined the *tree t-spanner* on chordal graphs, generalized chordal graphs and general graphs. For every n-vertex m-edge unweighted graph G, they proposed a new algorithm constructs a tree $(2\lfloor \log_2 n \rfloor)$-spanner in $O(m \log n)$ time for chordal graphs, a tree $(2\rho \lfloor \log_2 n \rfloor)$-spanner $O(m \log^2 n)$ time or a tree $(12\rho \lfloor \log_2 n \rfloor)$-spanner in $O(m \log n)$ time for graphs that confess a Robertson-Seymour's tree-decomposition with bags of radius at most ρ in G and a tree $(2\lceil t/2 \rceil \lfloor \log_2 n \rfloor)$-spanner in $O(mn \log^2 n)$ time or a tree $(6t \lfloor \log_2 n \rfloor)$-spanner in $O(m \log n)$ time for graphs that confess *a tree t-spanner*. They produced the same approximation ratio as in [2] but in a better running time.

3 Genetic Algorithm

In this section, we propose a genetic algorithm for the MMST problem, the problem of finding a *tree t-spanner* in a given edge-weighted graphs that minimizes the stretch factor t (see Sect. 1).

We first introduce some terminologies that will be used throughout this section. Let G' be a subgraph of G. The sets $V(G')$ and $E(G')$ denote the set of vertices and edges of G', respectively. The shortest distance between two vertices u and v in G' is denoted by $d_{G'}(u, v)$. For two subgraphs G_1 and G_2 of G, let $G_1 \cup G_2$, $G_1 \cap G_2$, and $G_1 - G_2$ denote the subgraph induced by $E(G_1) \cup E(G_2)$, $E(G_1) \cap E(G_2)$, and $E(G_1) - E(G_2)$, respectively.

3.1 Algorithm Overview

The Genetic Algorithm (GA) is an iterative optimization approach based on the principles of genetics and natural selection [14]. We first have to define a suitable data structure to represent individual solution (chromosomes), and then construct a set of candidate solutions as an initial population (first generation) of an appropriate cardinality *pop − size*. The following typical procedure is repeated as long as a predefined stopping criteria are met. Starting with the current generation, we use a predefined selection technique to repeatedly choose a pair of individuals (parents) in the current generation to reproduce, with probability p_c, a new set of individuals (offsprings) by exchanging some parts of between the two parents (crossover operation). To avoid local minimum, we try to keep an appropriate diversity among different generations by applying mutation operation, with specific probability p_m, to genes of individuals of the current generation. Finally, based on the values of an appropriate fitness function, we select a new generation from both the offspring and the current generation (the more suitable solutions have more chances to reproduce).

Note that, determining representation method, population size, selection technique, crossover and mutation probabilities, and stopping criteria in genetic algorithms are crucial since they mainly affect the convergence of the algorithm (see [15–19]).

The rest of this section is devoted to describe steps of the above algorithm in details.

3.2 Representation

Let $G = (V, E)$ be a given undirected graph such that each vertex in V is assigned a distinct label from the space $1, 2, \ldots, n$, i.e., $V = \{1, 2, \ldots, n\}$. Clearly, each edge $e \in E$ with end points i and j is uniquely defined by the unordered pair i, j. Moreover, every subgraph of G is uniquely defined by the set of unordered pairs of all its edges. In particular, every spanning tree T in G is induced by a set of exactly $n − 1$ unordered pairs corresponding to its edges since T is a subgraph of G that spans all vertices in V and has no cycles. Therefore, each chromosome (tree t-spanner) can be represented as a set of unordered pairs of integers each of which represent a gene (edge) in the chromosome.

3.3 Initial Population

Constructing an initial generation is the first step in typical genetic algorithms. We first have to decide the population size *pop − size*, one of the decisions that affect the convergence of the genetic algorithm. It is expected that small population size may lead to weak solutions, while, large population size increases the space and time complexity of the algorithm. Many literatures studied the influence of the population size to the performance of genetic algorithms (see [18] and the references therein). In this paper, we discuss the effect of the population size on the convergence time of the algorithm (cf. Sect. 4).

On of the most common methods is to apply random initialization to get an initial population. Namely, we compute each chromosome in the initial population by repeatedly applying the following simple procedure as long as the cardinality of the set of visited vertices is less than n (or as long as the cardinality of the set of traversed edges is less than $n - 1$). Let T denote the tree constructed so far by the procedure (initially, T consists of a random vertex from $V(G)$). We first select a random vertex $v \notin V(T)$ from the set of the neighbors of all vertices in T, and then add the edge $e = (u, v)$ to T, where u is the neighbor of v in T. It is easy to verify that the above procedure returns a tree after exactly $n - 1$ iterations. The generated tree T is added to the initial population.

The above algorithm is repeated as long as the number of constructed population is less than $pop - size$.

3.4 Fitness Function

Fitness function is a function used to validate each chromosome. Here, the objective function of the MMST is to minimizes the maximum ratio between all pairs of vertices in the underlying graph G. Formally, the objective function of the MMST is to minimizes $\max_{u,v \in V} \frac{d_T(u,v)}{d_G(u,v)}$, where $d_T(u, v)$ and $d_G(u, v)$ are the distances between u and v in T and G, respectively.

3.5 Selection Process

In this paper, we present three common selection techniques: roulette wheel selection, stochastic universal sampling selection, and tournament selection. All these techniques are called fitness-proportionate selection techniques since they are based on a predefined fitness function used to evaluate the quality of individual chromosomes. Throughout the execution of the proposed algorithm, the reverse of this ratio is used as the fitness function of the corresponding chromosome. We assume that the same selection technique is used throughout the whole algorithm. The rest of this section is devoted to briefly describe these selection techniques.

Roulette Wheel Selection (RWS): [14, 20] Here, the probability of selecting a chromosome is based on its fitness value. More precisely, each chromosome is selected with the probability that equals to its normalized fitness value, i.e., the ratio of its fitness value to the total fitness values of all chromosomes in the set from which it will be selected.

Stochastic Universal Sampling Selection (SUS): [20, 21] Instead of a single selection pointer used in roulette wheel approach, SUS uses h equally spaced pointers, where h is the number of chromosomes to be selected from the underlying population. All chromosomes are represented in number line randomly and a single pointer $ptr \in (0, \frac{1}{h}]$ is generated to indicate the first chromosome to be selected. The remaining $h - 1$ individuals whose fitness spans the positions of the pointers $ptr + i/h, i = 1, 2, \ldots, h - 1$ are then chosen.

Tournament Selection (TRWS): [14, 21] This is a two stages selection technique. We first select a set of $k < pop - size$ chromosomes randomly from the current population. From the selected set, we choose the more fit chromosome by applying the roulette wheel selection approach. Tournament selection is performed according to the required number of chromosomes.

3.6 Crossover Process

In each iteration of the algorithm we repeatedly select a pair of chromosomes (parents) from the current generation and then apply crossover operator with probability p_c to the selected chromosomes to get new chromosomes (offsprings). Simulations and experimental results of the literatures show that a typical crossover probability lies between 0.75 and 0.95. There are two common crossover techniques: single-point crossover and multi-point crossover. Many researchers studied the influence of crossover approach and crossover probability to the efficiency of the whole genetic algorithm, see for example [17, 19] and the references therein. In this paper, we use a multi-point crossover approach by exchanging a randomly selected set of edges between the two parents.

In particular, for each selected pair of chromosomes T_1 and T_2, we generate a random number $s \in (0, 1]$. If $s < p_c$ holds, we apply crossover operator to T_1 and T_2 as follows.

Define the two sets $E_1 = E(T_1) - E(T_2)$ and $E_2 = E(T_2) - E(T_1)$ ($|E_1| = |E_2|$ holds). Let $t = |E_1| = |E_2|$, and generate a random number k from $[1, t]$. We first choose a random subset E_1' of cardinality k from E_1, and then add E_1' to T_2 to get a subgraph T' (i.e., $T' = T_2 \cup E_1'$). Clearly, T' contains k cycles each of which contains a distinct edge from E_1'. For every edge $e = (u, v)$ in E_1', we apply the following procedure to fix a cycle containing e. Let $\widetilde{T}$ be the current subgraph (initially, $\widetilde{T} = T'$). We first find a path $P_{\widetilde{T}}(u, v)$ between u and v in $\widetilde{T} - \{e\}$. We then choose an edge $\widetilde{e}$ in $P_{\widetilde{T}}(u, v)$ randomly and delete it from subgraph $\widetilde{T}$.

Similarly, we apply the above crossover technique by interchanging the roles of T_1 and T_2 one more offspring. Finally, we add each of the resulting spanning trees to the set of generated offsprings.

3.7 Mutation Process

To maintain the diversity among different generations of the population (and hence avoid local minimum), we apply a genetic (mutation) operator to chromosomes of the current generation with predefined (usually small) probability p_m. Namely, for each chromosome T, we generate a random number $s \in (0, 1]$, and then mutate T if $s < p_m$ holds by replacing a random edge (gene) in T with a random edge from $E(G) - E(T)$. Many results analyzed the role of mutation operator in genetic algorithms [15–17].

Formally, a chromosome T is mutated as follows. We first select a random edge $e = (u, v)$ in the graph G but not in the chromosome T, i.e., e is randomly chosen from the set $E(G) - E(T)$ of edges, It is easy to see that the subgraph $T \cup \{e\}$ contains exactly one cycle including e. We then select a random edge e' in the path $P_T(u, v)$ between u and v in T. Let T' denote the offspring obtained from T by exchanging the two edges e and e', i.e., $T' = (T - \{e\}) \cup \{e'\}$. It is easy to see that T' is a spanning tree in G.

A formal description of the proposed genetic algorithm is described in Algorithm 1.

Algorithm 1 Genetic Algorithm for the *Tree t-Spanner* Problem

---

Input: An edge-weighted graph G, a population size $pop - size$, a maximum
    number of generations $maxgen$, a crossover probability p_c, a mutation probability p_m.
Output: A *tree t-spanner* that minimizes t.
1. Compute an initial population I_0 (cf. Sect. 3.3).
2. $gen \leftarrow 1$.
3. **While** ($gen \leq maxgen$) **do**
4.   **For** $i = 1$ to $pop - size$ **do**
5.   Select a pair of chromosomes from I_{gen-1} (Sect. 3.5).
6.   Apply crossover operator with probability p_c to the selected pair of
       chromosomes to get two offsprings (Sect. 3.6).
7.   **Endfor**
8.   For each chromosome in I_{gen-1}, apply mutation operator with
    probability p_m to get an offspring (Sect. 3.7).
9.   Extend I_{gen-1} with valid offsprings output from lines 6 and 8.
10.   Find the chromosome T_{gen-1} with the best fitness value in I_{gen-1}.
11.   If $gen \geq 2$ and the fitness values of T_{gen-2}, T_{gen-1}, and T_{gen} are identical, then **break**.
12.   Select $pop - size$ chromosomes from I_{gen-1} to form I_{gen} (Sect. 3.5).
13. $gen \leftarrow gen + 1$.
14. **Endwhile**
15. Output T_{gen}.

---

4 Experimental Results

In this section, we evaluate the proposed genetic algorithm by applying it to several random edge-weighted graphs. In particular, we generate a random graph G of n nodes by applying Erdos and Renyi [22] approach in which an edge is independently included between each pair of nodes of G with a given probability p. Here, we generate random graphs with sizes 6, 10, 15, and 20, and a randomly chosen probability p. Moreover, all edge weights of the generated graphs are set to random integers from the interval [1, 1000].

For each of the generated graphs, we apply the proposed algorithm with different selection techniques. We set the population size $pop - size = 30$, the maximum number of iterations the genetic algorithm executes $maxgen = 300$, the crossover probability $p_c = 0.9$, and the mutation probability $p_m = 0.2$. All previous parameters are summarized in Table 1. The algorithm terminates if either the number of iterations exceeds $maxgen$ or the solution does not change for three consecutive iterations. All obtained solutions are compared with the corresponding optimal solutions obtained by considering all possibilities of all spanning trees in the underlying graphs.

All results presented in this section were performed in MATLAB R2014b on a computer powered by a core i7 processor and 16 GB RAM.

The results of applying our genetic algorithm to random graphs with sizes $n = 6$, $n = 10$, $n = 15$, and $n = 20$, are shown in Table 2. In particular, Table 2 compare the values of t returned by the algorithm with the corresponding optimal stretch factor. It is seen that the proposed algorithm outputs optimal solution to MMST all the instances the algorithm applies to.

Table 1 Values of algorithm parameters

Parameter	Value
n	6, 10, 15, 20
$pop - size$	30
$maxgen$	300
p_c	0.9
p_m	0.2

Table 2 Values of t corresponding to a random graphs with size n

t n	t-Optimal	t-RWS	t-SUS	t-TRWS
6	1.0833	1.0833	1.0833	1.0833
10	1.0964	1.0964	1.0964	1.0964
15	1.1579	1.1579	1.1739	1.1579
20	1.0132	1.0484	1.0132	1.0132

Fig. 1 The influence of *pop − size* on the running time of the algorithm (n = 6)

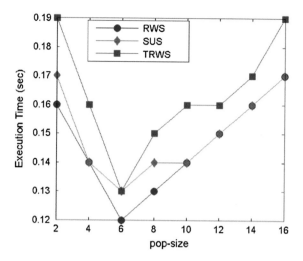

Fig. 2 The influence of *pop − size* on the running time of the algorithm (n = 10)

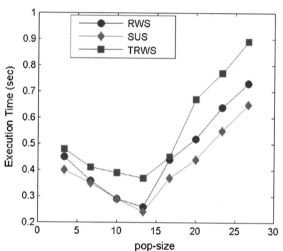

We discuss the effect of the population size *pop − size* on the convergence of the algorithm. Given a random graph with size n, we apply the algorithm with population sizes $n/3$, $2n/3$, n, $4n/3$, $5n/3$, $2n$, $7n/3$ and $8n/3$. Figures 1, 2, 3 and 4 illustrate the running time of the algorithm applied to graphs of sizes $n = 6$, $n = 10$, $n = 15$ and $n = 20$, respectively. The algorithm attains the least running time when the population size is set to a constant fraction of the graph size n.

Fig. 3 The influence of *pop − size* on the running time of the algorithm (n = 15)

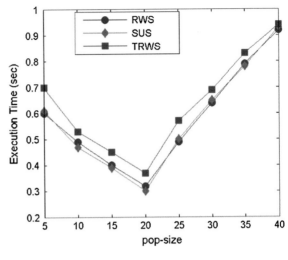

Fig. 4 The influence of *pop − size* on the running time of the algorithm (n = 20)

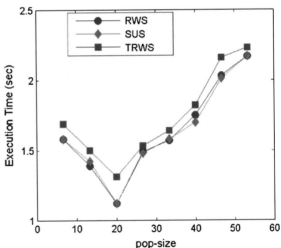

5 Conclusion

In this paper, we have studied the problem of finding the Minimum Maximum Stretch Spanning Tree (MMST) that aims to find a spanning tree T in a given graph G such that the maximum ratio of the distances between every pair of vertices in T to the shortest distance between the two vertices in G is minimized. We have designed a genetic algorithm for the MMST problem which have been evaluated by applying it to random instances of the problem. Experimental results have shown that the proposed algorithm outputs a high quality solutions to the MMST problem. It will be interesting to adapt our algorithm to be applied to the *subgraph t-spanner* problem: the problem of finding a minimum weight subgraph in G such that the distance between any two vertices in this subgraph is at most a given $t \geq 1$ times the shortest distance between the two vertices in G. See [23–26] and the references therein.

References

1. Peleg, D., Ulman, J.D.: An optimal sychronizer for the hypercube. SIAM J. Comput. 740–747 (1989)
2. Emek, Y., Peleg, D.: Approximating minimum max-stretch spanning trees on unweighted graphs. SIAM J. Comput. 1761–1781 (2008)
3. Demmer, M.J., Herlihy, M.P.: The arrow distributed directory protocol. In: Proceeding of the 12th International Symposium on Distributed Computing (DISC), pp. 119–133. Springer (1998)
4. Peleg, D., Reshef, E.: Low complexity variants of the arrow distributed directory. J. Comput.. Syst. Sci, pp. 474–485 (2001)
5. Kuhn, F., Wattenhofer, R.: Dynamic analysis of the arrow distributed protocol. Theory Comput. Syst. pp. 875–901 (2006)
6. Herlihy, M., Tirthapura, S., Wattenhofer, R.: Competitive concurrent distributed queuing. In: Proceedings of the 20th Annual ACM Symposium on Princibles of Distributed Computing, pp. 127–133 (2001)
7. Cai. L., Corneil, D., Tree Spanners. SIAM J. Discret. Math. 359–387 (1995)
8. Brandstädt, A., Dragan, F.F., Le, H.-O., Le, V.B.: Tree spanners on chordal graphs: complexity and algorithms. Theor. Comput. Sci. 329–354 (2004)
9. Brandstädt, A., Dragan, F.F., Le, H.-O., Le, V.B., Uehara, R.: Tree spanners for bipartite graphs and probe interval graphs. Algorithmica 27–51 (2007)
10. Peleg, D., Tendler, D.: Low stretch spanning trees for planar graphs, Technical Report. MCS01-14, Weizmann Science Press of Israel (2001)
11. Fekete, S.P., Kremer, J.: Tree spanners in planar graphs, Discret. Appl. Math, pp. 85–103 (2001)
12. Dragan, F.F., Fomin, F.V., Golovach, P.A.: Spanners in sparse graphs. J. Comput. Syst. Sci. pp. 1108–1119 (2010)
13. Dragan, F.F., Köhler, E.: An approximation algorithm for the tree t-spanner problem on unweighted graphs via generlized chordal graphs. Algorithmica 884–905 (2014)
14. Engelbrecht, A.P.: Computational Intelligence: An Introduction. Wiley, New York (2007)
15. Abdoun, O., Abouchabaka, J., Tajani, C.: Analyzing the Performance of Mutation Operators to Solve the Travelling Salesman Problem, CoRR abs/1203.3099 (2012)
16. Hesser, J., Manner, R.: Towards an optimal mutation probability for genetic algorithms. In: Proceedings of 1st Workshop in Parallel Problem Solving From Nature, pp. 2332 (1991)
17. LIN, W-Y., LEE, W-Y., Hong, T-P.: Adapting crossover and mutation rates in genetic algorithms. In: The Sixth Conference on Artificial Intelligence and Applications, Kaohsiung, Taiwan (2001)
18. Roeva, O., Fidanova, S., Paprzycki, M.: Influence of the population size on the genetic algorithm performance in case of cultivation process modelling. In: Proceedings of the Federated Conference on Computer Science and Information Systems, pp. 371–376 (2013)
19. Vekaria, K., Clack, C.: Selective crossover in genetic algorithms: an empirical study.Lecture Notes in Computer Science, vol. 1498, pp. 438–447 (1998)
20. Chipperfield, A., Fleming, P., Pohlheim, H., Fonseca, C.: The Matlab Genetic Algorithm User's Guide, UK SERC (1994)
21. Blickle, T., Thiele, L.: A Comparison of Selection Schemes used in Genetic Algorithms, Zurich (1995)

22. Erdos, P., Renyi, A.: On random graphs. Publ. Math. **290** (1959)
23. Sigurd, M., Zachariasen, M.: Construction of Minimum-Weight Spanners, pp. 797–808. Springer, Berlin (2004)
24. Farley, A.M., Zappala, D., Proskurowski, A., Windisch, K.: Spanners and message distribution in networks. Dicret. Appl. Math. 159–171 (2004)
25. Gudmundsson, J., Levcopoulos, C., Narasimhan, G.: Fast greedy algorithms for constructing sparse geometric spanners. SIAM J. Comput. pp. 1479–1500 (2002)
26. Navarro, G., Paredes, R., Chavez, E.: t-Spanners as a data structure for metric space searching. In: International Symposium on String Processing and Information Retrieval, SPIRE, LNCS 2476, pp. 298–309 (2002)

Performance Evaluation of Arabic Optical Character Recognition Engines for Noisy Inputs

Shimaa Saber, Ali Ahmed, Ashraf Elsisi and Mohy Hadhoud

Abstract This paper investigates the problem of the effectiveness of input quality on the performance evaluation of Arabic OCR systems. The experimental results show that performance for Arabic OCR systems gives accepted error rate for low noisy images and gives high error rate for images with high noises and Arabic OCR accuracy can be increased by filtering the noise images. Robust word-based and character-based accuracy metrics are used to show the performance evaluation of different Arabic OCR engines using different samples such as newspapers, books, regular text are used.

Keywords Accuracy metrics · Arabic OCR engines · Noisy images

1 Introduction

Optical character recognition (OCR) is a popular research topic in pattern recognition. OCR is aimed to enable computers to recognize optical characters without human intervention. The applications of OCR have been found in automated guided vehicles (AGV), object recognitions, digital libraries, invoice and receipt processing, and recently on personal digital assistants (PDAs) [1].

S. Saber (✉) · A. Ahmed · A. Elsisi · M. Hadhoud
Faculty of Computers and Information, Menofia University, Shebeen El-Kom, Egypt
e-mail: shimaa.saber@ci.menofia.edu.eg

A. Ahmed
e-mail: ali.ahmed@ci.menofia.edu.eg

A. Elsisi
e-mail: ashrafelsisim@yahoo.com

M. Hadhoud
e-mail: mmhadhoud@yahoo.com

© Springer International Publishing Switzerland 2016
T. Gaber et al. (eds.), *The 1st International Conference on Advanced Intelligent System and Informatics (AISI2015), November 28–30, 2015, Beni Suef, Egypt,*
Advances in Intelligent Systems and Computing 407, DOI 10.1007/978-3-319-26690-9_40

449

The Arabic language is one of the world's major languages [2]. Arabic language is widely spoken by roughly 250 million people around the world and has been used since the fifth century when written forms were stimulated by Islam [3]. Arabic OCR lacks research studies when compared to Latin character OCR [2]. The recognition accuracy of Latin character OCR is far beyond that of Arabic OCR because the Arabic language is different in its cursive script and its letter shape [4]. Performance evaluation and characterization of OCR systems is crucial for many reasons [5]: to predict performance, to monitor process, to provide scientific explanation, to identify open problem for research [5], and to get the weakness and the strong of product [6]. The accuracy of the English OCR has become very high, but unfortunately, Arabic OCR outputs accuracy are not satisfying. There have been several studies on Arabic OCR [7].

The performance evaluation paradigm requires comparing an observed variable with a reference variable under controlled conditions. In OCR, the observed variable is the output of an Arabic OCR device; the reference variable is its desired ("Ground truth") output [8].

The performance of evaluation has been programed to calculate the accuracy matrices automated. The Arabic OCR products must be tested under programed for avoiding the human error, speed, precision and repetition.

Many real-world sources of images are "noisy" and require regard to developing techniques that are strong in the presence of such noise. The outputs from OCR contain various degrees of errors. To exploit these inputs, we need to improve techniques to deal with noise. Whether we can successfully handle noise will greatly influence the final usefulness of the information extracted from such images.

Jing, Lopresti, and Shih [5] studied the problem of summarizing textual documents that had undergone OCR and hence suffered from typical OCR errors. This work focus on analyzing how the quality of summaries was affected by the level of noise in the input document, and how each stage in summarization was impacted by the noise, and suggested possible ways of improving the performance of automatic summarization systems for noisy documents.

Brian Kjersten [9] create a post-OCR processing module for noisy Arabic documents which can correct OCR errors before passing the resulting Arabic text to a translation system. To this end, we are evaluating an Arabic-script OCR engine on documents with the same content but varying levels of image quality. The previous work has focused on clean images and well-formatted images [1, 2, 4, 6]. This work shows the effect of noise on the accuracy of Arabic OCR systems and comparing between this accuracy and accuracy after cleaning module. The organization of this paper as follows: input data and Arabic OCR products in Sect. 2. The performance evaluations of Arabic OCR systems are discussed in Sect. 3. The experiments using test data and results of the performance of Arabic OCR systems are examined in Sect. 4. A conclusion of our study is presented in Sect. 5.

2 Input Data and Arabic OCR Products

The automated test system for Arabic OCR needs to the output of Arabic OCR devices, this observed output came from samples of real word scanned images like news-paper, magazine, books or regular sample. And also need the ground truth representation for each sample.

Our work focus on the noisy inputs for images, also focus on making module to clean noise to improve the accuracy, the possibility that inputs of images don't contain on noise. So it needs to simulate this sample. In this case, it is usual to mix up the scanned images, which are ideal, by adding noise.

There are different kind of noise that can be added on the Arabic image and impact it such as salt-and-pepper noise, speckle noise and Gaussian noise. This kind of noise can change letter because some Arabic letter have the same shape but different number of dots so salt-and-pepper has been choose.

The salt-and-pepper noise are also called shot noise, impulse noise or spike noise that is caused by faulty memory locations,malfunctioning pixel elements in the camera sensors, or there can be timing errors in the process of digitization [10]. Salt-and-pepper noise is damaging to Arabic text, because many letters are distinguished only by the presence and number of dots. In addition, image can have blur, pixel shift, or bleed-through, which is nonlinearly dependent on the content of the document [10].

Filtering in an image processing is a basis function that is used to achieve many tasks such as noise reduction, interpolation, and re-sampling [10]. Median Filter is a simple and powerful non-linear filter and Used for de-noising different types of noises.

In this work, four Arabic OCR products are used:

1. Automatic Reader 11.0 Gold "Sakhr": supports the main Arabic based characters languages: Arabic, Farsi, and other international languages. It combines two main technologies: Omni Technology, which depends on highly advanced research in artificial intelligence, and Training Technology, which increases the accuracy of character recognition.
2. Abbyy FineReader V.11: enhances the speed and accuracy of recognition. It creates editable and searchable files from scans. Supports Multilanguage documents with 189 supported language including Arabic.
3. Readiris Pro 11: supports more than 120 languages including the Arabic language. It can convert PDF files into editable text.
4. VajehShenas 1st Edition: is a multi-language OCR that supports Farsi, Arabic, English, Pashtu, and 17 other languages and download license.

3 The Performance Evaluation of Arabic OCR Systems

OCR evaluation can be categorized into two types [6]: (a) blackbox evaluation and (b) whitebox evaluation. Blackbox evaluation is treated as an indivisible unit, it didn't know what happen inside the system, it deal only with the output. Whitebox evaluation, on the other hand of blackbox evaluation, it parts the system to sub-modules and get the evaluation of each submodules alone. This work uses the blackbox evaluation.

3.1 Arabic OCR Evaluation System

A general flowchart of the performance evaluation of Arabic OCR system is shown in Fig. 1. The proposed block diagram compare between the accuracy for noisy image and cleaning one to show the effect of the cleaning filter and to show if the Arabic OCR systems affected by noise or not.

The performance evaluation algorithm can be divided into the following steps:

1. Scan the noisy image and enter the image to the Arabic OCR engine and then get the result to calculate the accuracy metrics.
2. Applying the filter to the noisy image to produce the cleaned one and enter the cleaned image to the Arabic OCR engine and then calculate the accuracy metrics.
3. Scan the groundtruth image and enter the image to the Arabic OCR engine and then calculate the accuracy metrics.
4. Compare between the three outputs of accuracy metrics and show the effect of the performance evaluation of the different Arabic OCR using the noisy images.

Fig. 1 A block diagram shows steps of performance evaluation of Arabic OCR engines

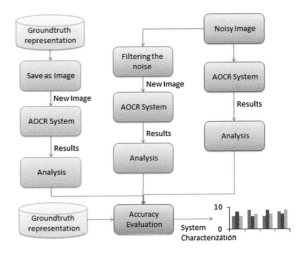

3.2 Accuracy Metrics for Evaluating Arabic OCR Systems

This work uses different metrics for comparing between the different Arabic OCR engines that are character accuracy, word accuracy, non-stop-word accuracy, word precision and character accuracy by character class [11]. These metrics will be discussed in details in following sub-section.

Character Accuracy.
Character Accuracy (CA) calculated as the number of character correctly recognized (CC) by Arabic OCR product by the total number of characters in the groundtruth (TC).

$$CA = \frac{\#CC}{\#TC} = \frac{\#TC - error}{\#TC} \tag{1}$$

Character Accuracy by Character Class.
To measure this metric, the characters in ground truth are divided into two categories. Each category has a number of classes.

Category one.
In this category, the characters in the ground truth are divided into four classes according to the location of the Arabic character in the word. The classes are "Isolated, Initial, Medial, and Final" as seen in Table 1. To calculate the accuracy for each class, we use Eq. (1).
Category two.
The characters in this category are divided into five classes dependent on the number of dots and zigzag in Arabic character. The classes are "1 dot, 2 dot, 3 dot, zigzag, other characters" as seen in Table 2. To measure this metric, we use Eq. (1).

Word Accuracy.
Word accuracy (WA) calculate using the number of words correctly recognized (CW) using Arabic OCR product by the total number of words in the groundtruth (TW) "reference file".

$$WA = \frac{\#CW}{\#TW} = \frac{\#TW - \#errorword}{\#TW} \tag{2}$$

Table 1 Different shapes of Arabic letter 'ع' called 'ayn'

عـ	ـهـ	ـح	ع
(a) Initial	(b) Medial	(c) Final	(d) Isolated

Table 2 Classifying Arabic letter into category two

Class	1 dot	2 dot	3 dot	Zigzag	Other
Letter	ب ج خ ذ ز ض ظ غ ف ن	ة ي ي ت ق ي	ث ش	آ إ أ ئـ ئ ء ؤ ك	ا ح د ر س ص ط ع ل م ه و ك ى

Word Precision.
To calculate the word precision (WP) metric of Arabic OCR system, the number of correctly recognized words divided by the total number of words in the output file (TOW).

$$WP = \frac{\#CW}{\#TOW} = \frac{\#TOW - \#errorword}{\#TOW} \tag{3}$$

Non-Stop Word Accuracy.
We need to calculate the accuracy of the important word "Non-Stop Word" (NWA), and it calculated by the percentage of non-stop words that are correctly recognized (C_NW). To do this, we need a lookup table of Arabic stop words to ignore them from the calculations.

$$NWA = \frac{\#C\_NW}{\#TW} \tag{4}$$

Text Line Accuracy.
Touching and broken characters are difficult problems and are responsible for a significant fraction of all OCR errors. By using text line metric to evaluate the performance of Arabic OCR, there are two important issues. First, they allow the testing of character (and word) level segmentation. Second, they are useful for morphological and lexical techniques can be used to improve accuracy. The textline accuracy calculated the number of character error and determines types of this error and detects the number of error in words and can use Eqs. (1) and (2).

4 Experiments and Results

This section shows the results of the previously selected Arabic OCR systems using the suggested accuracy metrics. The experiments were conducted using Intel(R) Core(TM) i5 CPU device with 2.53 GHZ and 3 GB of RAM and Window 7 operation system, and Matlab R2012a are used.

The four Arabic OCR products were tested using different type of samples such as newspapers, books and regular sample. Calculating the accuracies of four products is used by different metrics that have been suggested in pervious. Our work presents an application that has been done to compare between products. This application takes the text output of four products and groundtruth text, then calculates all accuracy for all metrics, then show what user need to see from the comparing between the metrics.

The first experiment tests the four products using noisy newspaper sample that published in Alahly journal in November 2005 and show in Fig. 2. The comparing accuracy results of the four products using the first experiment are shown in Fig. 3 this figure shows Vajeh OCR is the better comparing to other products. Readiris is

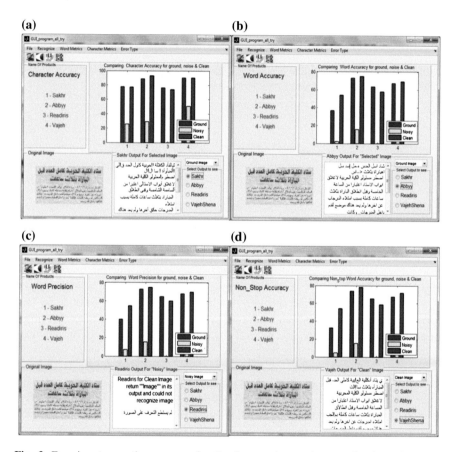

Fig. 2 Experiment one: a journal sample

Fig. 3 Experiment one, the accuracy for the four products using a noisy journal sample. a Character accuracy, b word accuracy, c word precision, d non-stop word accuracy

the most effect product; it couldn't recognize the noisy input image and return the same image for its output. After cleaning, four products improve their output and recognize the image with better output; four products increase their accuracy and

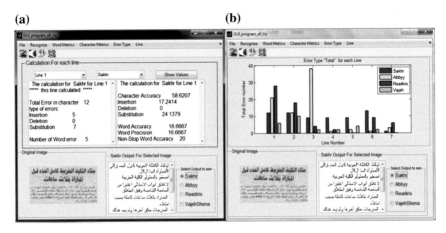

Fig. 4 Experiment one: the textline accuracy. **a** Show details for line 1 of sakhr product, **b** show comparing between of percentage of error for each line

بسم الله الرحمن الرحيم

محمد رسول الله و الذين معه أشداء على الكفار رحماء بينهم تراهم ركعا سجدا يبتغون فضلا من الله و رضوانا سيماهم فى

وجوههم من أثر السجود ذلك مثلهم فى التوراة و مثلهم فى الإنجيل كزرع أخرج شطأه فآزره فاستغلظ فاستوى على سوقه يعجب

الزراع ليغيظ بهم الكفار وعد الله الذين آمنوا و عملوا الصالحات منهم مغفرة و أجرا عظيما

صدق الله العظيم

Fig. 5 Experiment two: a regular sample used with salt and pepper by 0.1

gives values near to the groundtruth. This experiment show that the four products affected by noise. Figure 4 show the textline accuracy of a noisy journal sample. Figure 4a shows the textline accuracy for line one using Sakhr product. In this figure, the different character and word accuracies are calculated. Figure 4b shows the comparison between the four products by calculating the total error of each line.

In the second experiment, testing of the four products is done using an Arabic regular text sample and adding salt and pepper with different percentage (0.02, 0.05, 0.1 and 0.15). Figure 5 shows an Arabic regular sample with salt and pepper by 10 % noise. This sample has been used because it contains all Arabic letters and most cases of different positions like letter ' ت ' that came in three positions; which are isolated in "الصالحات", initial in "تراهم" and in the middle as "فاستوى". Figures 6, 7 and 8 show the comparisons of the character-based accuracy and word-base accuracy of the four Arabic OCRs products. As shown from the character accuracy Fig. 6a, the performance evaluations for all products are affected by the noise and results for all other metrics prove that when the percentages of noise have been increased, the performance evaluations of the products are decreased. The accuracy performance of Readiris OCR is the high affected by noise while comparing to the other products, because in most cases, Readiris OCR couldn't recognize the images

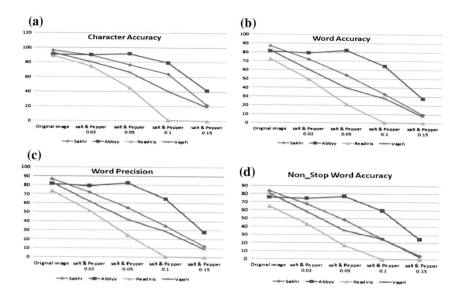

Fig. 6 Experiment two, comparison between the four products using different noise percentage. **a** Character accuracy, **b** word accuracy, **c** word precision, **d** non-stop word accuracy

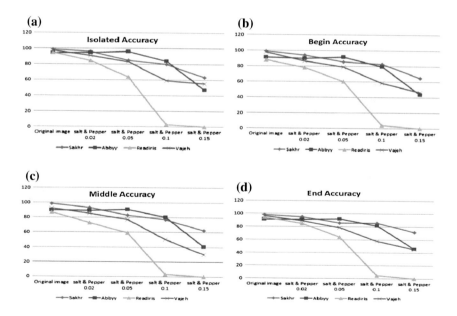

Fig. 7 Experiment two, comparison between the four products for category one metrics

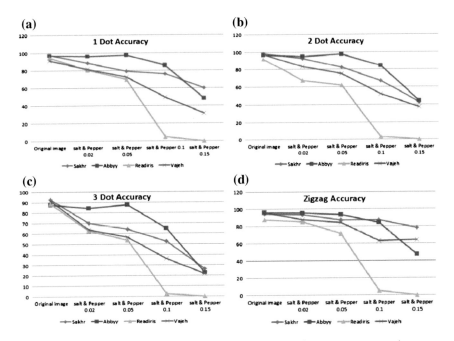

Fig. 8 Experiment two, comparison between the four products for category two metrics

and return the original image as its output. The performance of Abbyy OCR is the best in noisy images while comparing to the other products and its accuracy in noisy is accepted then Sakhr that is also gave accepted accuracy for its output, then Vajeh.

5 Conclusion

The quality of scanned document image is important for characterize the performance evaluation of different Arabic OCR engines. The main contribution of this work is to show the effect of noisy scanned input images on the performance of Arabic OCR engines. Various evaluation metrics to measure the accuracy of the different well known Arabic OCR engines using different samples such as newspapers, books, regular text are used. The experimental results show that the performance evaluation of Abbyy Arabic OCR and Sakhr Arabic OCR are superior in recognizing the noisy image than VajehShenas and Readiris. Readiris couldn't recognize the noisy image and return the input image as its output in most cases. The performance evaluations of the different Arabic OCRs are increased by decreasing the percentages of noises in the image. So this work proved that the image quality is the largest factor of evaluating and comparing between different Arabic OCRs systems.

References

1. Chen, Q.: Evaluation of OCR algorithms for images with different spatial resolutions and noises. Ph.D. thesis, University of Ottawa, Canada 2003. http://bohr.wlu.ca/hfan/cp467/08/nodes/master_thesis.pdf. Accessed Oct 2015
2. Soudi, A.: Arabic computational morphology knowledge-based and empirical methods. Knowledge-Based and Empirical Methods, pp. 3–14. Springer, Amsterdam (2007)
3. AbdelRaouf, A., Higgins, C.A., Pridmore, T., Khalil, M.: Building a multi-modal Arabic corpus (MMAC). IJDAR **13.4**, 285–302 (2010), (Springer)
4. AbdelRaouf, A.: Offline printed Arabic character recognition. Ph.D. thesis, The University of Nottingham for the Degree of Doctor of Philosophy, May 2012
5. Jing, H., Lopresti, D., Shih, C.: Summarizing noisy documents. In: Proceedings of the Symposium on Document Image Understanding Technology, pp. 111–119. Apr 2003
6. Kanungo, T., Marton, G. A., Bulbul, O.: Performance evaluation of two Arabic OCR products. In: The 27th AIPR workshop: Advances in computer-assisted recognition (pp. 76–83). International Society for Optics and Photonics (1999).
7. Albidewi, A.: The use of object-oriented approach for Arabic documents recognition. Int. J. Comput. Sci. Netw. Secur. (IJCSNS) **8.4**, 341 (2008)
8. Nartker, T.A., Rice, S., Nagy, G.: Performance metrics for document understanding systems. In: Proceedings of the Second International Conference on Document Analysis and Recognition, pp. 424–427 (1993)
9. Kjersten, B.: Arabic optical character recognition (OCR) evaluation in order to develop a post-OCR module. Computational and Information Sciences Directorate, ARL, Sept 2011
10. Kamboj, P., Rani, V.: A brief study of various noise model and filtering techniques. J. Glob. Res. Comput. Sci. **4.4**, 166–171 (2013)
11. Saber, S., Ahmed, A., Hadhoud, M.: Robust metrics for evaluating Arabic OCR systems. In: IEEE IPAS'14, pp. 1–6. University of sfax, Tunisia, Nov 2014

Plants Identification Using Feature Fusion Technique and Bagging Classifier

Alaa Tharwat, Tarek Gaber, Yasser M. Awad, Nilanjan Dey
and Aboul Ella Hassanien

Abstract In this paper, a plant identification approach using 2D digital images of leaves is proposed. This approach will be used to develop an expert system to identify plant species by processing colored images of its leaf. The approach made use of feature fusion technique and the Bagging classifier. Feature fusion technique is used to combine color, shape, and texture features. Color moments, invariant moments, and Scale Invariant Feature Transform (SIFT) are used to extract the color, shape, and texture features, respectively. Linear Discriminant Analysis (LDA) is used to reduce the number of features and Bagging ensemble is used to match the unknown image and the training or labeled images. The proposed approach was tested using Flavia dataset which consists of 1907 colored images of leaves. The experimental results showed that the accuracy of feature fusion approach was much better than all other single features. Moreover, a comparison with the most related work showed that our approach achieved better accuracy under the same dataset and same experimental setup.

A. Tharwat
Faculty of Engineering, Suez Canal University, Ismailia, Egypt

T. Gaber (✉)
Faculty of Computers and Informatics, Suez Canal University, Ismailia, Egypt
e-mail: tmgaber@gmail.com

Y.M. Awad
Faculty of Agriculture, Suez Canal University, Ismailia, Egypt

N. Dey
Bengal College of Engineering and Technology, Durgapur, India

A.E. Hassanien
Faculty of Computers and Information, Cairo University, Giza, Egypt

A.E. Hassanien
Faculty of Computers and Information, Beni Suef University, Beni Suef, Egypt

A. Tharwat · T. Gaber · Y.M. Awad · A.E. Hassanien
Scientific Research Group in Egypt (SRGE), Cairo, Egypt
URL: http://www.egyptscience.net

© Springer International Publishing Switzerland 2016
T. Gaber et al. (eds.), *The 1st International Conference on Advanced Intelligent System and Informatics (AISI2015), November 28–30, 2015, Beni Suef, Egypt,*
Advances in Intelligent Systems and Computing 407, DOI 10.1007/978-3-319-26690-9_41

461

Keywords Plant identification · Linear discriminant analysis (LDA) · Bagging classifier · Feature fusion · Color features · Shape features · Texture features · Scale invariant feature transform (SIFT)

1 Introduction

Plants play a vital role in the life and it is important for different purposes such as air, climate, and food security. Plants have different species, which are subject to the danger of extinction. Therefore, classify numerous plant species is needed in order to protect and save economic plant resources. Thus, plant identification techniques have become a cutting-edge research in the recent years [1]. Traditional plant identification is time-consuming and requires more efforts from labors, and experts; however, it is not an easy task and depends mainly on the experts' decisions. On the other hand, automatic plant identification based on information technology is a very vital task for different parties: agriculture, pharmacological, forestry science. Automatic plant identification process will achieve fast, cheap, and accurate systems, which provides a great help to medicine, industry, and foodstuff production, as well as to biologists, chemists, and environmentalists. Plant leaf features are one of the most important morphological evidence taxonomically, contributing significantly to plant classification and conservation. Compared with advanced identification methods such as molecular biology, leaf digital image is a promising technology due to low-cost and convenient [2].

There are many studies have been done for the plant identification based on digital images. Satti et al. [3], has used Flavia image dataset and applied many preprocessing steps on the leaf images. They achieved accuracy ranged from 85.9 to 93.3 % using k-Nearest Neighbor (k-NN) and Artificial Neural Networks (ANN) classifiers, respectively. Gaber et al. used one and two dimensional Linear Discriminant Analysis (LDA) and Principal Component Analysis (PCA) to extract the features from the 2D images of Flavia dataset. The accuracy of the two-dimensional approach (i.e. 2D-LDA and 2D-PCA) achieved accuracy up to 90 % while the one-dimensional approach (i.e. 1D-PCA and 1D-LDA) achieved accuracy ranged from 69 to 75 % [1]. Caglayan et al. [4], have used color and shape features of the leaf images to classify 32 different kinds of plants. They used Support vector Machine (SVM), k-NN, Random Forest, and Naive Bayes classifiers and the best accuracy achieved reached to 96 %. In [5], Arun et al. have used the PCA method to extract the features and k-NN classifiers. They have used Flavia dataset and achieved accuracy ranged from 78 to 81.3 %.

This paper describes an approach addressing the plant identification problem by combining the color, shape, and texture features of the 2D images of leaves. Increasing the number of features may lead to dimensionality problem, thus LDA is used to reduce the dimension and increase the discrimination between different classes. The Bagging classifier is then used to match the testing and training images.

The rest of the paper is organized as follows. Section 2, explains the preliminaries of our research. Section 3, presents the proposed approach. Experimental results and discussion are presented in Sect. 4. Finally, conclusions are summarized in Sect. 5.

2 Preliminaries

2.1 Feature Extraction Method

The main goal of the feature extract methods is to measure some properties of an object and transform these properties into numeric values. There are many types of features such as shape, texture, and color features; these features used to describe the content of the image or any object.

Color Features: Color features are widely used in image retrieval. The main idea of this method is to transform the image's colors into numeric values. Color features are robust to some transformation such as rotation and scaling. Moreover, it is easy to implement, fast and it needs low storage requirements. There are many color features such as color moments, color histograms, Color Coherence Vectors (CCV), and color correlogram [4]. In this paper, color moments will be used as color features.

Color Moments: Color moments method is widely used in information retrieval systems. Color moments features consist of the first order (mean), second order (variance), and third order (skewness). Thus, color moments are represented by nine numbers that represent three moments for each color components [4]. The first color moment of the kth color component ($k = 1, 2, 3$) is defined by, $M_k^1 = \frac{1}{XY} \sum_{x=1}^{X} \sum_{y=1}^{Y} f_k(x, y)$, where XY represents the total number of pixels of the image and $f_k(x, y)$ represents the color value of the kth color component of the image pixel (x, y). The hth moment of kth color component can be calculated as follows, $M_k^h = \frac{h}{XY} \sum_{x=1}^{X} \sum_{y=1}^{Y} (f_k(x, y) - M_k^1)^{\frac{1}{h}}$, $h = 2, 3, \dots$.

Shape Features: The main idea of the shape features is to measure the similarity between shapes. These similarity measurements or shape features can be classified into two main categories: region-based and contour-based methods. Region-Based methods use the whole area of an image or object to extract the features. While contour-based methods use the information that represent the contour of the shape of an object such as shape descriptors (e.g. circularity, length irregularity, and complexity) [4]. Shape features are easy to implement and relatively robust against rotation and transformation. In this paper, invariant moments technique will be used as a shape feature extraction method.

Invariant Moments: Invariant moments was first reported in [6]. Invariant moments are robust features because it is invariant against rotation and translation. However, invariant moments are sensitive to noise, which limits the use of moments in many applications. The set of seven moments (ϕ_n) can be calculated as reported in [6].

Texture Features: The main idea of texture feature extraction methods depends mainly on describing the texture of each region in the image. There are two types of textures, namely, sparse descriptor and dense descriptor. In the sparse descriptor method, the interest points (keypoints) in a given image are first detected and then a local patch around these keypoints are constructed then invariant features are extracted. *Scale Invariant Feature Transformation* (SIFT) [7] is considered the most well-known algorithm in this type. In the dense descriptor-based method, local features are extracted from every pixel (pixel by pixel) over the input image. Examples of the dense descriptor method include *Local Binary Pattern* (LBP) [8] and *Weber's Local Descriptor* (WLD) [9]. In this paper, SIFT technique is used as a texture feature extraction method.

Scale Invariant Feature Transform (SIFT): The idea of SIFT algorithm is to transform the image into a collection of local feature vectors. Each of these feature vectors is supposed to be invariant to any scaling, rotation or translation of the image.

The first step in SIFT is to construct a Gaussian "scale-space" function from the input image. This is formed by convolution (filtering) of the original image with Gaussian functions of varying scales. The *Difference of Gaussian* (DoG) is calculated as the difference between two nearby scales separated by a constant multiplicative factor k as follows, $D(x, y, \sigma) = L(x, y, k\sigma) - L(x, y, \sigma)$, where $L(x, y, \sigma) = G(x, y, \sigma) * I(x, y)$ and $L(x, y, k\sigma) = G(x, y, k\sigma) * I(x, y)$ are two images that produced from the convolution of Gaussian functions with an input image $I(x, y)$ with σ and $k\sigma$, respectively, and $G(x, y, \sigma) = \frac{1}{2\pi\sigma^2} exp[-\frac{x^2+y^2}{\sigma^2}]$ represents the Gaussian function. In the second step, interest points in DoG pyramids which are called keypoints are detected. The keypoints represent the local maxima or minima of $D(x, y, \sigma)$ by comparing each point with the pixels of all its 26 neighbors. In the third step, a low contrast (i.e. sensitive to noise) poorly localized keypoints are eliminated. The fourth step is to assign one or more orientation to the keypoints based on local image properties. An orientation histogram is formed from the gradient orientations of sample points within a region around the keypoint. Peaks in the orientation histogram correspond to dominant directions of the local gradients. The highest peak in the histogram is located and this peak and any other local peaks are used to create a keypoint with that orientation. The final step is to create a descriptor for the patch that is compact and highly distinctive to be robust against changes in illumination and camera viewpoint [7, 10].

There are many parameters that control the efficiency of SIFT algorithm, the length of a feature vector of each keypoint, and the number of keypoints. The first parameter is Peak Threshold *PeakThr*, which represents the amount of contrast to accept a keypoint. Increasing the value of *PeakThr* parameter, decreases the number of features by removing keypoints which are lower than *PeakThr*, therefore, the robustness of feature matching will be decreased. The second parameter is the Patch Size. Increasing the patch size above an extent may extract only global features. On the other hand, decreasing the patch size may not extract more features that enough for identification. The third parameter is the Number of Angles (orientation) and Bins. Increasing number of angles and bins collects features in different angles (orientations) and increases the number of features, hence improves the accuracy of the system. While decreasing the number of angles and bins the features became variant against image rotation [7].

2.2 Linear Discriminant Analysis (LDA)

LDA is one of the supervised dimensionality reduction methods. The main goal of LDA is to transform the data from a high dimensional space to a lower dimensional space by preserving the most discriminative data. The LDA searches for space, which maximizes the ratio of between-class variance (S_b) to the within-class variance (S_w). The within-class variance represents the variance between the samples of the same class while the between-class variance represents the variance or distance between different classes [1, 11].

2.3 Feature Fusion

The aim of the feature fusion technique is to combine many independent (or approximately independent) features to give a more representative features for the objects or patterns. The features are combined by concatenating it into one feature vector.

Feature fusion technique has two problems. The first problem is the compatibility of different features; i.e. the features may be in different ranges of numbers. Thus, the features must be common by normalizing it. There are different normalization methods such as Z_{Score}, Min-Max, and Decimal Scaling [12–15]. Z_{score} normalization method is the most common and simplest method and it is used to map the input scores to distribution with mean of *zero* and standard deviation of *one* as follows, $\acute{f}_i = \frac{f_i - \mu_i}{\sigma_i}$, where f_i represents the ith feature vector, μ_i and σ_i are the mean and standard deviation of the ith vector, respectively, $\acute{f}_i$ is the ith normalized feature vector [13]. The second problem of the feature fusion technique is a high dimensionality, which may lead to high computation time and needing more storage space. Thus, dimensionality reduction technique such as LDA or PCA are used to reduce the largest set of features [13, 16].

2.4 Bagging Classifiers

Bagging is an acronym from *Bootstrap AGGregatING*. Bagging is an ensemble method that creates its ensemble by training different classifiers on a random distribution of the training set ($S = \{s_1, s_2, \ldots, s_N\}$), where s_i represents the ith sample and N represents the total number of samples. Many samples from the training set are drawn randomly with replacement and each sample consists of many patterns. Each training sample (s_i) is used to train or learn one of the learners or classifiers. All classifiers in the ensemble are used to classify the test or unknown pattern by combining the outputs of all classifiers using uniform averaging or voting over class labels. Due to randomness in selecting the samples from the training set in Bagging classifier, some patterns are chosen multiple times while the others are left out. Bagging is effective on unstable learning algorithms such as neural networks and decision trees [1, 17].

3 Proposed Approach

The proposed approach consists of two phases, namely, training and testing phases. In the training phase, the images of leaves (i.e. training images) are collected and then the features are extracted from each image. In our proposed approach we used three different features as follows: (1) invariant moments to extract shape features, (2) color moments as color features, (3) SIFT to extract texture features. The feature vectors are then normalized and combined into one feature vector as shown in Fig. 1. LDA technique is then used to reduce the dimension of feature vectors to avoid the high dimensionality problem. Project the feature vectors (i.e. after fusion) of the training images onto the LDA space to form the training set (S). A number of samples (N) are then drawn randomly with replacement from the training set. For all single classifiers in the Bagging ensemble, each classifier is trained separately using one sample.

In the testing phase, the features are extracted from an unknown image using the same three types of features, which are used in the training phase and then combines the three feature vectors into one feature vector and then project the combined feature vector onto the LDA space. Match the projection of the test image with the training images using the trained classifiers $(C_i, i = 1, 2 \ldots, N)$ and combine the outputs of all classifiers to predict the decision.

4 Experimental Results and Discussion

In this section, two experimental scenarios are performed. The aim of these experiments is to identify 2D leaf plant images using single feature extraction method, while the aim of the second experiment is to identify leaf images using the proposed feature fusion approach.

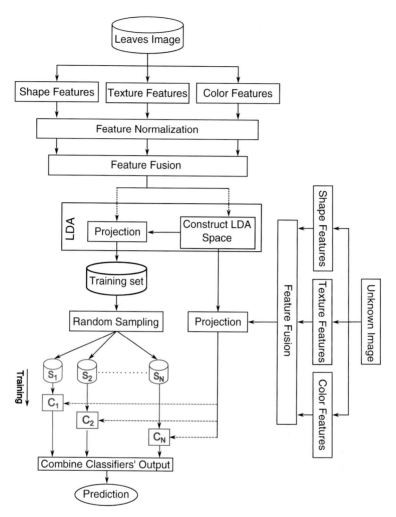

Fig. 1 A block diagram of plant identification system using leaves' images based on feature fusion and Bagging classifier (proposed approach)

4.1 Experimental Setup

In the two experiments, color moments, invariant moments, and SIFT feature extraction methods are used to extract the color, shape, and texture features, respectively. SIFT method has different parameters. In this experiment the values of SIFT parameters are as follows, $PeakThr = 0$, patch size 16×16, eight angles, and each descriptor contains 4×4 array of histograms around the keypoints. Bagging classifier ensemble is used to match between the testing and training images. Different numbers of classifiers (i.e. 5, 15, 51, 101) are used in the Bagging ensemble.

Fig. 2 Sample of different leaves' images (one sample from each class or plant)

Moreover, different numbers of training images ($N = 1, 2, \ldots, 5$) are selected from each class to run our experiment and to show how the influence of the number of training images on the accuracy, while the remaining images are used as a testing images. The accuracy rate (the percentage of the total number of predictions that were correct) is used to evaluate the performance of our proposed method.

Flavia leaves images dataset is used in our experiments. The dataset consists of 120 colored leaves images with size 1600×1200 collected from 20 different plants. We have selected images of 20 different plants (each plant has six images). The images are in different orientation, illumination and quality as shown in Fig. 2.

4.2 Experimental Scenarios

Single Feature Extraction Method: This experiment is designed to evaluate each single feature extraction method, i.e. color moments, invariant moments, and SIFT features, respectively, and to show the influence of changing the number of training images and the size of the ensemble on the accuracy rate. Figure 3a–c summarize the results obtained from this scenario.

Feature Fusion: The aim of this experiment is to identify leaf images using feature fusion approach. In this experiment, the color, shape, and texture features are fused into one feature vector and the accuracy is examined to show how it improves the accuracy through using many different features and how it is powerful than single feature extraction methods. Figure 3d summarizes the results obtained from this scenario.

4.3 Discussion

From the results in Fig. 3a–c many notices can be seen. First, color moments achieved accuracy rate better than invariant moments and SIFT. Second, the accuracy rates proportional to the number of training images of each class. Moreover, the best accuracy achieved when the number of training images was five while the minimum accuracy achieved when only one training image from each class is used. Third, Increasing the size of the ensemble increases the accuracy rate and the best accuracy achieved when the number of classifiers was more than 15 classifiers.

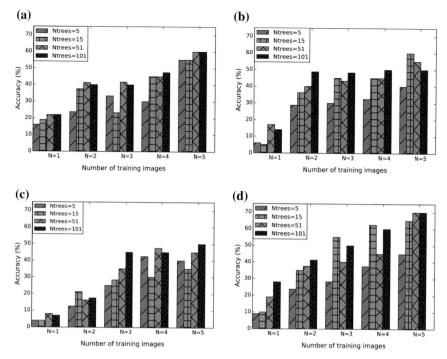

Fig. 3 Accuracy of the proposed approach using single feature extraction and feature fusion methods; **a** accuracy of color moments features, **b** accuracy of invariant moments features, **c** accuracy of SIFT features, **d** accuracy of feature fusion method

Figure 3d shows the accuracy rate of using feature fusion method. As shown from the figure, different findings can be noticed. The feature fusion method achieved results better than all other single feature extraction methods. Moreover, the accuracy rate proportional to the number of training images and the size of the ensemble. As shown in Fig. 3d, the maximum accuracy achieved (i.e. 70 %) using more than 15 classifiers and five training images from each class.

To conclude, the feature fusion method achieved results better than all other single feature extraction methods.

5 Conclusions

In this paper, we have proposed a new approach to classify and identify plant species based on a feature extraction of leaf from its digital images. This approach uses color, shape and texture features in individual experiments. In another experiment, all features are combined into one feature vector, which is robust and invariant to many challenges such as rotation, noise, and illumination. LDA has been used to reduce the dimensions of feature vectors and to increase the discrimination between different classes. Hence, improving the overall recognition rate of our proposed system. Bagging classifier with a different number of learners (i.e. decision tree learner) is used. The experimental results have showed that color feature extraction method achieved the best recognition rate compared with texture and shape features. Moreover, the feature fusion method achieved accuracy (70 %) better than all single feature extraction methods. Furthermore, the accuracy increased when the number of learners increased. In the future work, we are going to test our approach against a large database of leaves' images. This would allow us to evaluate whether our approach will give the same good results and improve it in the future in order to develop an expert system for plant identification.

References

1. Gaber, T., Tharwat, A., Snasel, V., Hassanien, A.E.: Plant identification: two dimensional-based versus one dimensional-based feature extraction methods. In: 10th International Conference on Soft Computing Models in Industrial and Environmental Applications, 375–385. Springer (2015)
2. Liu, J., Zhang, S., Deng, S.: A method of plant classification based on wavelet transforms and support vector machines. In: Proceedings of the 5th international conference on Emerging intelligent computing technology and applications (pp. 253–260). Springer-Verlag (2009)
3. Satti, V., Satya, A., Sharma, S.: An automatic leaf recognition system for plant identification using machine vision technology. Int. J. Eng. Sci. Technol. 5(4), 874–879 (2013)
4. Caglayan, A., Guclu, O., Can, A.B.: A plant recognition approach using shape and color features in leaf images. In: International Conference on Image Analysis and Processing (ICIAP), pp. 161–170. Springer (2013)
5. Arun Priya, C., Balasaravanan, T., Thanamani, A.S.: An efficient leaf recognition algorithm for plant classification using support vector machine. In: International Conference on Pattern Recognition, Informatics and Medical Engineering (PRIME), pp. 328–432. IEEE (2012)
6. Hu, M.K.: Visual pattern recognition by moment invariants. IRE Trans. Inf. Theory 8(2), 179–187 (1962)
7. Tharwat, A., Gaber, T., Hassanien, A.E., Shahin, M., Refaat, B.: Sift-based arabicsign language recognition system. In: Afro-European Conference for Industrial Advancement, pp. 359–370. Springer (2015)
8. Tharwat, A., Gaber, T., Hassanien, A.E., Hassanien, H.A., Tolba, M.F.: Cattle identification using muzzle print images based on texture features approach. In: Proceedings of the Fifth International Conference on Innovations in Bio-Inspired Computing and Applications IBICA 2014, pp. 217–227. Springer (2014)
9. Chen, J., Shan, S., He, C., Zhao, G., Pietikainen, M., Chen, X., Gao, W.: Wld: a robust local image descriptor. IEEE Trans. Pattern Anal. Mach. Intell. 32(9), 1705–1720 (2010)
10. Cheung, W., Hamarneh, G.: n-SIFT: n-dimensional scale invariant feature transform. IEEE Trans. Image Process. 18(9), 2012–2021 (2009)
11. Tharwat, A., Gaber, T., Hassanien, A. E.: Cattle identification based on muzzle images using gabor features and SVM classifier. In: Advanced machine learning technologies and applications: second international conference, AMLTA 2014, Cairo, Egypt, November 28–30, 2014. Proceedings (Vol. 488, p. 236). Springer (2014)
12. Ibrahim, A., Tharwat, A.: Biometric authenticationmethods based on ear and finger knuckle images. Int. J. Comput. Sci. Issues (IJCSI) 11(3), 134–138 (2014)

13. Semary, N.A., Tharwat, A., Elhariri, E., Hassanien, A.E.: Fruit-based tomato grading system using features fusion and support vector machine. In: Intelligent Systems' 2014, pp. 401–410. Springer (2015)
14. Tharwat, A., Ibrahim, A.F., Ali, H., et al.: Multimodal biometric authentication algorithm using ear and finger knuckle images. In: Seventh International Conference on Computer Engineering and Systems (ICCES), pp. 176–179. IEEE (2012)
15. Tharwat, A., Ibrahim, A., Hassanien, A. E., Schaefer, G.: Ear recognition using block-based principal component analysis and decision fusion. In: Proceedings of the 6th international conference, PReMI 2015, Warsaw, Poland. Lecture Notes in Computer Science, Vol. 9124, pp. 246–254 (2015)
16. Tharwat, A., Ibrahim, A., Ali, H.A.: Personal identification using ear images based on fast and accurate principal component analysis. In: 8th International Conference on Informatics and Systems (INFOS), pp. 56–59 (2012)
17. Nath, S.S., Mishra, G., Kar, J., Chakraborty, S., Dey, N.: A survey of image classification methods and techniques. In: International Conference on Control, Instrumentation, Communication and Computational Technologies (ICCICCT), pp. 554–557. IEEE (2014)

Hexart: Smart Merged Touch Tables

Ahmed Samir, Alaa Essam, Esraa Mohamed, Saleh Ahmed, Abdallah M. Zakzouk, Moustafa Attia and Ayman Atia

Abstract Hexart is a multi-touch tables system based on FTIR utilizing hexagon shaped tables, We designed Hexart for people needs in a group activities. Our paper illustrates the proposed system to eliminate common challenges in facing group activities problems such as time wasting, Lack of communication and the most crucially was the inability to integrate different work parts correctly. It also introduces two major cooperation concepts supported which are the merging and splitting. The Merging concept allows users to merge tables together in order to form a large multi-touch table, Whilst the Splitting concept allows users to split a table display into several private areas where each user has their own personal work space. Users can interact together through private areas or a public area which is accessible to all users. Finally, An experiment conducted to measure users' satisfaction and usability of hexagonal multi-touch tables. It showed that such an approach can yield an implementation that was extremely competitive.

Keywords Multi-touch surfaces · FTIR tables · Merging · Splitting · Hexagon · Private and public area · Gestures · Large display

1 Introduction

Recently, Multi-Touch surfaces became commercially available in several forms [5] such as tables [7, 23], Touch-screens [3], IPads and IPods. Therefore, Multi-Touch technology is used in different activities and places, For example; Browsing data in museums, Galleries, Medical visualization, Education, Entertainment, Consumer electronics domains [18] and power station control rooms. There are variety of tech-

A. Samir (✉) · A. Essam · E. Mohamed · S. Ahmed · A.M. Zakzouk · M. Attia · A. Atia
HCI-LAB, Department of CS Faculty of Computers and Information,
Helwan University, Cairo, Egypt
e-mail: ahmedsamirabdelkarim@gmail.com

A. Atia
e-mail: ayman@fci.helwan.edu.eg

© Springer International Publishing Switzerland 2016
T. Gaber et al. (eds.), *The 1st International Conference on Advanced Intelligent System and Informatics (AISI2015), November 28–30, 2015, Beni Suef, Egypt,*
Advances in Intelligent Systems and Computing 407, DOI 10.1007/978-3-319-26690-9_42

nologies for multi-touch surface architectures such as Sensitive Technology [14], Capacitive Technology [17] , Optical-Based Technology [2] and Acoustic-Based Technology [12]. Optical-Based technology is known for its cheap setup and high performance. It includes different techniques such as DI [18, 21], DSI [6] and FTIR [8]. One of these techniques is FTIR which characterized by its low cost, High accuracy, Robustness, Multi-finger recognition.

In the field of Computer Supported Cooperative Work (CSCW), Continuous researches have reached radical solutions in an attempt to enrich group activities and try to help people in their workplaces. For example, Interactive Room (IRoom) [7] presented a set of applications to serve the cooperators in meeting rooms. Classroom 2000 [1] supported teachers and learners in lecture environments by enabling students to take notes during live lecture. Blue Board [16] was a large interactive display system for groups facilitating the exchange of information in an informal collaborative way. These examples among others showed that large interactive surfaces concept is a technology which can help multiple users interactions in group activities. However, The limitation of users number may introduce unusable system specially if we created traditional implementation for large interactive surfaces as one big surface, Users exceed the maximum number supported or additional functionality is required. Furthermore, Problems such as over crowdedness and limitations in the number of private areas and tables will be resulted From an insurmountable amount of wasted space and difficulty in interacting between users.

In Hexart, We implemented the concept of large surfaces by merging small multitouch surfaces to form a larger surface to improve collaborative work between users, Collaborative design is very important in group activities to save time and cost in companies, Multiple users can interact with merged surfaces simultaneously to form a single group, Subgroups or even individual work space (Fig. 1). Merging small surfaces increases users flexibility to merge or break-up any number of surfaces to suit a varying number of users and functionalities, Mergeing between two private areas will happen if they are collided. The hexagon shape of Hexart tables and areas also helped in resolving various problems that have challenged in previous research groups and enriched Hexart with more interactive features as we demonstrated the hexagon usability section in our paper. Hexart also supports Private/Public areas

Fig. 1 Hexagon tables with public and private areas

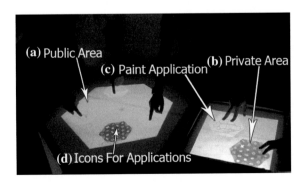

implementation using the Splitting concept, Which could be used in group activities effectively by allowing each individual user to have their own separated work area containing several applications to interact with the system without affecting or being affected by other work-spaces and then facilitate the interaction between all users through a public area that are regarded as shared space that can be accessed by all users and enabled them to share data and files between their private areas [10].

Our research aims to enrich collaborative activities so we used Multi-Touch tables to increase the effectiveness of group activities based on these papers, Isenberg et al. [9] presented collaborative Activities on multi touch tables, They increased the shared awareness information in group activities, They allowed users to work in common or different work spaces.

To help large number of interactions to maintain shared or separated tasks, The merging allowed connecting spaces from six sides. We created private area for each user to interact with separated tasks and the public area was to interact with common tasks. Hexart presents the following contributions: We introduce a hexagonal hardware and software interfaces that ensure the best spaces consuming. We present a concept of merging and splitting that is implemented physically and virtually to improve, enrich and facilitate group activities.

2 Related Work

2.1 Groupware Activity

Verma et al. [22] created a collaborative environment to support content creation and sharing in meetings for small groups. The System consisted of a shared wall-mounted workspace and all users can interact together using mouse and keyboard or using pen and paper. It enabled group members to interact with same shared workspace simultaneously. They performed an experiment which consisted of groups, who were asked to complete murder mystery task which designed by Stasser et al. [20], Group members should use collaborative environment to create and share content. The experiment was separated into two sets, First one was regards to the collaboration between group members, Second one considered as the groupware usage during the task. They classify groups into three kinds based on: Time when they were created, Visual differences and the immediate purpose of the content served. The results indicated that there existed a mapping between input device affordances and knowledge representation which were the task demands.

Educational activities are most promising domains for collaborative work and group participation so Meredith Ringel Morris et al. [13] created CollabDraw cooperative gesturing multi-user interaction system for collaborative art and photo manipulation. CollabDraw evaluation experiment consisted of six female participants, The mean age was 25 years . The experiment was that each group should recreate a drawing target then complete a questionnaire. The goal of evaluation was to measure the usability of cooperative gestures and the questions enabled people to find their intuitive or confusing, if tasks were Fun or tedious, Easy or difficult to use. The evaluation result was CollabDraw which is easy to use and the gestures is easy to learn. Participants took 28.8 min on average Time (stdev = 6.2 min) to learn all 15 gestures and all seven pairs were able to accurately recreate the target drawing with a mean time of 8.2 min (stdev = 1.2 min).

2.2 Private and Public Areas

Scott et al. [19] presented an experiment which prove collaborative benefits of digital tabletop displays. Three tables activities were set up at a local university; Tasks on tables which represent activities of collaborators actions on table such as manipulating items (puzzle), sharing, discussing items and ideas with each other. The results revealed that collaborators used three types of tabletop territories to help coordinate their interactions within shared tabletop workspace: personal, group, and storage territories. Personal territories appeared to provide each person with dedicated space on table for performing independent activities. Collaborators used their personal territories to customize items that were added to Floor Plan layout. When it was necessary to modify an item, They typically removed the item from its position on the Floor Plan and it was added to specific Group or Storage territories. The group territory was primarily used for arranging Items (puzzle) in Floor Plan. It was also used for discussing layout ideas and for assisting others to create or modify arrangements. Storage territories were at various locations on table. These territories were relocated in the workspace at different stages of the task, Depending on where participants were currently working and what resources they currently needed, so we used personal and group area in Hexart whereas each user has private area and all users can share items on public area.

2.3 Hexagon and Merging

Willis et al. [24] presented a new way for merging portable surface multi-user interaction. Their system consists of hardware components such as IR cameras, Handheld projectors and sensor, Software was consists of: Synthesizing IR markers and visible images, Detection of projected images from fiducial markers and information transferring between devices. The handheld projector displayed application content in different places. Invisible fiducial markers were tracked using cameras to detect the collision of virtual objects which were projected on the environment. They developed different scenarios for merging surfaces applications that used anywhere. Hexart enriched the idea of Collaborative merging of surfaces by interacting over merged multi-touch tables using gestures. Implementation of merging concept over two levels: Virtual and physical. This concept helped several users to work separately and to work as a group. They can combine their works or share the data with each other.

Rooke et al. [15] presented a way to make scenarios using hexagonal tiles, where each side in these tiles contains an infrared (IR) light source that was tracked using a camera with IR pass filter. The displayed scenarios on tiles are controlled by a projector according to the tracked infrared light. Scenarios could be images, videos or integrating information into the interactive workspaces. Their idea inspired us to use the hexagon shape over the multi-touch table. We built Hexart tables and created private areas as hexagon shape. While Hexart targets serving multiple users using private/public areas. We used hexagon shape to solve crowdedness problem as discussed in hexagon section.

3 Hexart Architecture

The Software of Hexart is divided into four main parts as (Fig. 2): Detection part, Tracking part, Gestures classifier and End-user application. Blob detection part is responsible for capturing frames of the table surface and detecting blobs which resulted from touches, Then it sends the detected blobs to be tracked in tracking part using TUIO protocol [11]. Tracking gives more details containing blob ID and its motion path of points to be translated into meaningful action. Blob will be translated into a direct interaction or to a specific gesture and finally, The client application receives details about tracked blobs and executive actions with visual displayed

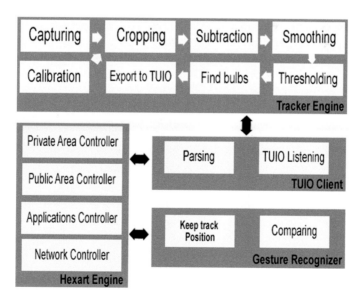

Fig. 2 Hexart architecture

feedback on the table surface. If any finger touches the surface; Some of internally reflected infrared light refract and spread inside the table. The camera that is provided with infrared pass filter captures refracted infrared beam Then the captured frames are passed to Core Community Vision (CCV) engine [2]. CCV is a software library developed by NUI Group used some algorithms for multi-touch systems development. Hexart engine which tracks blobs to get all path points and determine if blob is direct interaction or gesture. In case of gesture, It will be sent with its all details to gesture classifier to be recognized. Finally blobs transferred to the Hexart end-user application which translates them to specific actions according to the received blobs details.

3.1 Splitting and Merging

Virtual Level One of the main concepts of Hexart is splitting surfaces into smaller displayed areas which called private area. As shown in (Fig. 3a) Private areas contain icons for applications that are available to use. This concept enabled several users to work on the same table simultaneously, Each private area was considered as a miniature display. Around the private areas there is a free space called Public area, which is available to be accessed by all users, see (Fig. 3b). Users can use the public area to share files, transfer data or merge private areas in order to combine their works. One of the advantages of using the private and public areas is to help users to work independently. We implemented Paint application to collaborate work with allowing each user to draw separately in his area. In this case the public screen was considered as the palette for all users over the table to show the drawing tools.

Private areas could be merged to form a new larger area that contains collective data from each merged private area. For example, If users use the paint application and want to combine their paints shapes, The system collects all painting details from

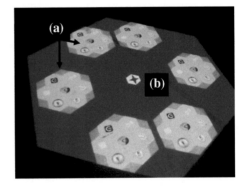

Fig. 3 Hexagon table, **a** the private area **b** public area private areas

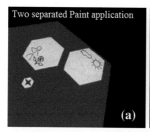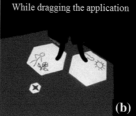

Fig. 4 Merging of two private areas

each private area and displays them in the new larger area as a complete painting. Virtual merging is useful for user Groupware activities that needs to combine the works together as shown in (Fig. 4). Users can create a tile by tip on a new button in his private area. Hence, He can share any object like (image, document, etc) by dragging it into tile then drag tile outside this private area. The public area will move those tiles automatically to the central position between the users areas. This tile can be gotten by dragging his private area. In other word, The tile was used as Object holder to pass any object from or into the public area. Users allowed to create (n) number of tiles where (n) is the size of private area over the tile size which depends on the table size, they can destroy the tiles by tipping close button on it.

Physical Level Another main concept of Hexart is merging, To increase the number of users we have to expand the interaction area in optimized spaces, For example: We have a team consists of two sub teams, Each one has a specific task, How to integrate these tasks? We can merge these separated hexagon tables to complete a scenario, The physical merging will cause integrating separated sub group activities tasks, We can utilize the room area spaces depending on the number of users and tables, Using the dynamic interactional model, we can resize the interaction area rather than the interaction in fixed tables size which waste room spaces, In Hexart each table works independently, They will draw a straight line to integrate tasks gesture starting from the first table to the second one, Each table recognizes the direction of gesture.

The starting point will be as a server (Fig. 5a). The ending point in another table will represent the clients (Fig. 5b). After that each table will send a unique identifier to other tables, working as one table each user will be able to drag his private area from table to another, After all, Users can split the merged tables again using closing gesture as shown in (Fig. 5c). In order to make merging more efficient, we have to avoid circular paths in merging where circular path will cause one central table that cannot be used by any user for example, if we need to merge seven tables, circular path will result one central and six tables around so the central table is useless in this case otherwise, non-circular path of merging will avoid this problem.

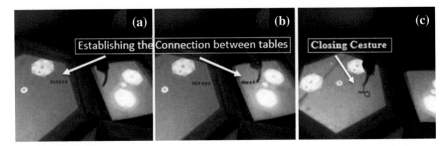

Fig. 5 **a** and **b** show how the user will draw a straight line to merge tables. **c** shows how to close the merging using closing gesture

3.2 GESTURES

Gestures increase the number of functionalities that provided for users without causing any confusion in interaction. There are a lot of classifiers that were used to classify and recognize gestures like Hidden Markov Models (HMM) and Dynamic Data
wrapping (DTW). In Hexart, we used Dynamic Time Warping (DTW) classifier
because its more faster in recognition than HMM [4]. DTW is a classifier algorithm
to estimate the optimal alignment of two different time dependent sequences. DTW
was used in gesture recognition; The algorithm calculates the characteristic sequence
of time based on warping the time iterative and matches the optimal results. The
algorithm decision was depending on the extracted features of the gestures. It needs
to track the start end point to calculate the distance and compare the tracked gestures with the predefined ones to take the true decision. Hexart has two types of
gestures: The first one is Direct gestures, That do not need a classifier to be recognized, The Second is symbolic gestures that has different shapes with different functions. It divided into two orientations: System gestures is reserved for the system
functionalities such as merging or break up tables. Unlike direct gestures, Symbolic
gestures need to be recognized before translating it into meaningful action and the
other orientation is application gestures that can be developed by Hexart applications
developers.

4 Evaluation

4.1 Experiment of Using Hexagon and Rectangle

We have performed a user study to evaluate our system. This study depends on experiment to evaluate the usability of the rectangular shape compared with the hexagonal
shape in crowdedness. We have conducted an experiment to measure the needed time
to sort a set of tiles in two ways, The first way was sorting rectangle tiles sce (Fig. 6).

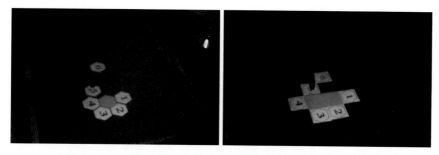

Fig. 6 Two experiments of hexagon and rectangle usability

Fig. 7 Experiment paths

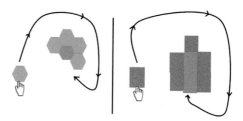

We chose rectangle which used in most UI interfaces used in computers or mobile OS. The second way was sorting Hexagon tiles. Participants had to move and sort tiles in a circular path clock-wise around a fixed place (Fig. 7). The goal of this experiment was to determine which shape was more usable in crowded areas. There were 10 participants, who were computer science students. They were 6 males and 4 females. Aged between 19 and 23 years old. Participants had to avoid tiles collision. Each participant performed the experiment of hexagon and rectangle two times. After that we calculated the time for each two tries.

4.2 Results

We observed that participants were significantly more different in their opinions (Fig. 8) then we used one way Anova test, The results of the average time, st. of deviation per seconds for the rectangle was ($\mu = 68.45$, $\alpha = 24.00$) and for hexagon was ($\mu = 61.65$, $\alpha = 16.26$) and the $F(1, 18) = 0.55$, $p < 0.001$ when the F-critical value was Equal to 4.414 which assured that the time of sorting hexagon tiles was less than rectangle tiles. Which proved that we fail to reject the null hypothesis and the hexagon shape is more usable than the rectangle in crowded spaces.

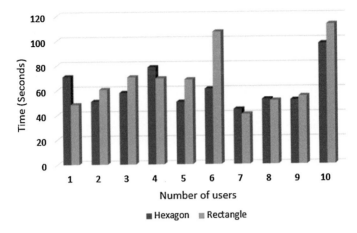

Fig. 8 Experiment results

4.3 Questionnaire About Smart Merged Tables

The second experiment was a Questionnaire to help us in evaluating the system (Fig. 9) According to the questionnaire results, 90 % of users preferred touch devices rather than regular devices. About 70 % of users thought that hexagon shape is more interested and engaging more than regular rectangle. And 90 % approved that merging and splitting are useful in group activities environments.

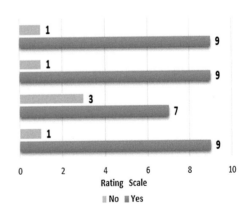

Fig. 9 Survey result

5 Conclusions and Future Work

We developed Hexart system for integrating multi-touch tables to enrich groupware activities, which are based on FTIR technique. Splitting and merging enabled the users to create a larger table from connecting number of private areas, Our system supports serving multiple users in one table by displaying a private area for each user, And supports grouping activity by providing a public area that accessed by all users. In our future work, we will use Hexart in implementing a virtual system for laboratories to help students in virtual learning. We will change also the usage constraints depending on users identities.

References

1. Abowd, G.D.: Classroom, 2000: an experiment with the instrumentation of a living educational environment. IBM Syst. J. **38**(4), 508–530 (1999)
2. Teiche, A., Rai, A. K., Yanc, C., Moore, C., Solms, D., Cetin, G., et al.: Multi-Touch Technologies, 1st edn. NUI Group, (2009)
3. Alt, F., Schmidt, A., Schmidt, A.: Advertising on public display networks. Computer **45**(5), 50–56 (2012)
4. Carmona, J., Climent, J.: A Performance Evaluation of HMM and DTW for Gesture Recognition. Springer, Heidelberg (2012)
5. Chang, R., Wang, F., You, P.: A survey on the development of multi-touch technology. In: 2010 Asia-Pacific Conference on Wearable Computing Systems (APWCS), pp. 363–366 (2010)
6. Dietz, P., Leigh, D.: Diamondtouch: a multi-user touch technology. In: Proceedings of the 14th Annual ACM Symposium on User Interface Software and Technology, UIST '01, pp. 219–226. ACM (2001)
7. Fox, A., Johanson, B., Hanrahan, P., Winograd, T.: Integrating information appliances into an interactive workspace. IEEE Comput. Graph. Appl. **20**(3), 54–65 (2000)
8. Han, J.Y.: Low-cost multi-touch sensing through frustrated total internal reflection. In: Proceedings of the 18th Annual ACM Symposium on User Interface Software and Technology, UIST '05, pp. 115–118. ACM (2005)
9. Isenberg, P., Fisher, D.: Collaborative brushing and linking for co-located visual analytics of document collections. In: Proceedings of the 11th Eurographics/IEEE—VGTC Conference on Visualization, EuroVis'09, The Eurographs Association, pp. 1031–1038. John Wiley Sons, Ltd. (2009)
10. Jin, Chunming, Takahashi, Shin, Tanaka, Jiro: Interaction between small size device and large screen in public space. In: Gabrys, Bogdan, Howlett, Robert J., Jain, Lakhmi C. (eds.) KES 2006. LNCS (LNAI), vol. 4253, pp. 197–204. Springer, Heidelberg (2006)
11. Kaltenbrunner, M., Bovermann, T., Bencina, R., Costanza, E.: Tuio—a protocol for table based tangible user interfaces (2005)
12. Lopes, P., Jota, R., Jorge, J.A.: Augmenting touch interaction through acoustic sensing. In: Proceedings of the ACM International Conference on Interactive Tabletops and Surfaces, ITS '11, pp. 53–56. ACM, New York, USA (2011)
13. Morris, M.R., Huang, A., Paepcke, A., Winograd, T.: Cooperative gestures: multi-user gestural interactions for co-located groupware. In: Proceedings of the SIGCHI Conference on Human Factors in Computing Systems, CHI '06, pp. 1201–1210. ACM (2006)
14. Pickering, J.A.: Touch-sensitive screens: the technologies and their application. Int. J. Man-Mach. Stud. **25**(3), 249–269 (1986)

15. Rooke, M., Vertegaal, R.: Physics on display: tangible graphics on hexagonal bezel-less screens. In: Proceedings of the fourth international conference on Tangible, embedded, and embodied interaction, TEI '10, pp. 233–236. ACM, New York, USA (2010)

16. Russell, D.M., Trimble, J.P., Dieberger, A.: The use patterns of large, interactive display surfaces: Case studies of media design and use for blueboard and merboard. In: Proceedings of the Proceedings of the 37th Annual Hawaii International Conference on System Sciences (HICSS'04)—Track 4—Volume. 4, HICSS '04, p. 40098.2. IEEE Computer Society (2004)

17. Schöning, J., Brandl, P., Daiber, F., Echtler, F., Hilliges, O., Hook, J., Löchtefeld, M., Motamedi, N., Muller, L., Olivier, P., Roth, T., von Zadow, U.: A technical guide. Technical report, Multi-touch surfaces (2008)

18. Schning, J., Hook, J., Bartindale, T., Schmidt, D., Oliver, P., Echtler, F., Motamedi, N., Brandl, P., Zadow, U.: Building interactive multi-touch surfaces. In: Mller-Tomfelde, C. (ed.) Tabletops—Horizontal Interactive Displays. Human-Computer Interaction Series, pp. 27–49. Springer, London (2010)

19. Scott, S.D., Carpendale, M.S.T., Inkpen, K.M.: Territoriality in collaborative tabletop workspaces. In: Proceedings of the 2004 ACM Conference on Computer Supported Cooperative Work, CSCW '04, pp. 294–303. ACM (2004)

20. Stasser, G.: Information salience and the discovery of hidden profiles by decision-making groups: a thought experiment. Organizational Behavior and Human Decision Processes, Group Decision Making 52(1), 156–181 (1992)

21. Tse, E., Shen, C., Greenberg, S., and Forlines, C.: Enabling interaction with single user applications through speech and gestures on a multi-user tabletop. In: Proceedings of the working conference on Advanced visual interfaces, AVI '06, pp. 336–343. ACM (2006)

22. Verma, H., Roman, F., Magrelli, S., Jermann, P., Dillenbourg, P.: Complementarity of input devices to achieve knowledge sharing in meetings. In: Proceedings of the 2013 Conference on Computer Supported Cooperative Work, CSCW '13, pp. 701–714. ACM (2013)

23. Wall, J.: Demo i microsoft surface and the single view platform. In: Proceedings of the 2009 International Symposium on Collaborative Technologies and Systems, CTS '09, IEEE Computer Society, Washington, DC, USA (2009)

24. Willis, K.D., Poupyrev, I., Hudson, S.E., Mahler, M.: Sidebyside: ad-hoc multi-user interaction with handheld projectors. In: Proceedings of the 24th annual ACM symposium on User interface software and technology, UIST '11, pp. 431–440. ACM (2011)

Malware-Defense Secure Routing in Intelligent Device-to-Device Communications

Hadeer Elsemary and Dieter Hogrefe

Abstract Device-to-Device (D2D) communications are foreseen to be an essential part of internet of things in the near future. Multi-hop D2D communication plays an essential role for exchanging messages and sharing information from disconnected areas as well as highly congested network. In fact, many research works are focusing on the connectivity while the security issue remains a significant challenge. Today, mobile devices are capable of initiating advanced security attacks without passing through a powerful centralized entity, especially malware attacks. Malware attacks formalize security risk that threatens the mobile network. However, the existing security schemes for malware attacks are not efficient as well as the tracking and isolating of infected devices remains very challenging. A major concern is that the malware attacks are happening at rate far exceeding the evolution of security techniques. As a first step toward thwarting the success of the malware attacks, we seek to mitigate the mobile infection. In this paper, we propose Malware-defense Secure Routing (MSR) based on zero-sum static Bayesian game. The protocol aims at mitigating the mobile infection by detecting malware attached to message before it infects the device. As this game achieves Nash equilibrium, and leads to an optimal infection-defense strategy for the network. Through simulation we evaluate the proposed routing protocol in terms of the percentage of the detected malicious messages. Results show that our protocol significantly increases the chance of success in infection-defense strategy for D2D network.

Keywords Game theory · Malware · Security · Device-to-Device

H. Elsemary (✉) · D. Hogrefe
Faculty of Mathematics and Computer Science,
Georg-August-University, Gottingen, Germany
e-mail: hadeer.el-semary@informatik.uni-goettingen.de

D. Hogrefe
e-mail: hogrefee@informatik.uni-goettingen.de

© Springer International Publishing Switzerland 2016
T. Gaber et al. (eds.), *The 1st International Conference on Advanced Intelligent
System and Informatics (AISI2015), November 28–30, 2015, Beni Suef, Egypt,*
Advances in Intelligent Systems and Computing 407, DOI 10.1007/978-3-319-26690-9_43

1 Introduction

Device-to-Device communication has been widely recognized as a promising and innovative feature of the next generation 5G cellular networks [1]. Due to D2D communication manifold advantages, it becomes the popular approach that gains much interest nowadays. D2D communication provides high bit-rate, low communication delay and locally computational offloading as well as high throughput in the cell area [2]. Furthermore, it enables direct communication between two mobile devices in cellular network without passing through base station or core network. The communication can occur on the cellular spectrum (e.g. LTE) or unlicensed (e.g. IEEE 802.11) [3]. Therefore, D2D communication is expected to be an essential part of internet of things (IoT) [4].

D2D communication in the IoT will typically be a multiple hop in nature, the devices will communicate with each other independently without any centralized control. The devices cooperate to share, collect and relay information in multi-hop manner thus performing routing functions [4]. The importance of multi-hop D2D communication is realized in the disaster scenarios where the communication infrastructure were physically damaged. However, relay by mobile terminals (smart phones, laptops, tablet PCs) could deliver messages through multi-hop D2D communication. Furthermore, the decentralized infrastructure-less multi-hop communication plays an essential role in the disaster salvage and emergency cases [5].

Relay by mobile terminals can be the unique option for emergency situations and in designing systems for disaster recovery, where there is no communication infrastructure [5]. Additionally, it can be applied for commercial purposes (e.g., advertisement, coupons and flyer distribution) by delivering advertisements to subscribers when they are in surrounding area instead of traditional methods such as email. Another important application is the sharing of information (i.e., exchange private message, document) among groups in places outside the cellular coverage (e.g., mountains, island, military domain). Additionally, D2D communication facilitates the sharing of information among groups of people where the cellular communication is highly congested.

Due to the explosive growth in demand of D2D communication in large areas, it has become an attractive target for attacks. However, the security requirements for multi-hop D2D communication depend on the type of the application. While some applications may need little security, other applications may need more security (i.e., private messages exchange, distributing important documents). As a result, the security concerns in the multi-hop D2D communication should be addressed to support all the possible applications.

It is worth mentioning that cyber security is moving from infrastructure to advanced mobile infrastructure-less threats. In fact, the growth in computation, sensor and communication capabilities of mobile devices makes us move towards advanced mobile security threats. Additionally, mobile devices are capable of initiating advanced security attacks without passing through a powerful centralized entity, especially malware attacks. Mobile malware attacks are becoming a significant threat to mobile wireless network [6]. Mobile malware attacks formalize a serious security risk that threatens to retard the large-scale reproduction of wireless applications. Additionally, mobile malware can disseminate offensive content, or provide unauthorized access to personal and financial information (e.g., mobile banking, private data, and or SMS). Furthermore, these attacks sometimes attempt to disrupt normal functions of the devices [7], alter the network traffic, or even kill the device or launch epidemic attacks.

Researchers have recognized the security threat of these attacks in mobile wireless network. Thus, they have been studying the maximum damage of malware attacks taking the dynamic behavior of the malware and the evolution of future malware into consideration [6, 8].

In order to accomplish secure intelligent D2D communication in the future, research issues need to be addressed. Researches focus much more on the connectivity, however, the security issue has to receive more attention for practical applications. Researchers have been studying the malware spread and propagation within wireless network and cellular networks. However, no studies are conducted so far on disconnected distributed mobile networks such as Mobile Device Cloud [9, 10]. On the other hand, the mobility of the devices can increase the malware infection and propagation rate. As a result, the tracking and the isolation of the malware attacks are very challenging. Motivated by all previous trends, lightweight efficient countermeasures to hinder the malware infection are highly required. As a result, the mitigation of the malware infection is considered a first step toward thwarting the success of the malware attacks.

In this paper, we are proposing a Malware-defense Secure Routing Protocol (MSR). The MSR protocol attempts to mitigate the mobile infection by detecting the malware before it infects the device. The main objective of our protocol is to choose the optimal secure route to send a message to a targeted device in multi-hop D2D network. Thus, anti-malware controls are residing on all mobile devices to detect the mobile malware attached to message. Additionally, due to the energy constraints of mobile devices, some devices may not collaborate to relay other device's traffic. Therefore, our protocol considers the energy level of the devices, involved in multi-hop D2D communication, in the routing decision. Thus, our protocol ensures the cooperation and the participation of the devices in the routing process. It is worth mentioning that all the mobile terminals should be able to achieve the multi-hop D2D communication. Therefore, our protocol considers different operating systems (iOS, Android, Windows, Symbian).

We formulated Malware-defense Secure Routing Game to derive the optimal infection-defense strategy for the network. This Bayesian game models the interaction between the D2D network (defender) and the attacker. The motivation behind the formulation of Bayesian game is that defender-attacker is incomplete information game because the defender is uncertain about his adversary's type. The latter's objective is to primarily launch malware attacks by injecting message with malware into the network. On the other hand, the defender has certain statistics (i.e., beliefs) about different malware types of different operating systems. Therefore, the defender's objective is to select the most secure route in terms of malware detection and route availability to send the message to the targeted device. The utility of the defender is affected by route detection rate of different malware types and the route availability. The game achieves Nash Equilibrium, thus leading to infection-defense strategy for defender then it selects the route with mobile devices, that maximizes his payoff.

The rest of this paper is organized as the following: In Sect. 2, we review related work. Section 3 we present the proposed framework, including the system model and game model. Section 4 describes the MSR protocol. We present some preliminary results in Sect. 5. Section 6 concludes this paper and mentions the future work.

2 Related Work

There are some secure routing protocols that have used the game theory in wireless network. In [11], the authors proposed a secure protocol to prevent the passive denial of service attack in wireless sensor networks (WSN) as a repeated game between an intrusion detection system (IDS) and nodes. This framework enforces cooperation among the nodes and punishment for non-cooperative behavior. The authors in [12] proposed a secure mechanism that prevents passive denial of service attacks and enforces node cooperation. This mechanism based on a Bayesian game between monitoring devices and nodes of WSN. Malicious nodes are isolated from the network and IDSs monitor the nodes in each stage of the game, using the updated beliefs about the nodes, and providing secure routing in WSN. In [13], the authors introduced stochastic routing based on game theory that mitigates the effects of interception, eavesdropping, and improve fault tolerance. They presented two techniques to compute multipath routing tables and select among these paths randomly to forward the packets.

In [14], the authors introduced a repeated game theoretical model to examine equilibrium conditions of packet forwarding strategies with Tit-For-Tat punishment strategy. In [15], an information theoretic framework is proposed to model trust propagation in ad hoc networks for secure routing and detection of the malicious nodes. Therefore, by examining the nodes' trustworthiness, a most trustworthy route is selected to send a message to a destination. The authors in [16] investigated the interactions between good nodes and malicious nodes and formulated secure routing and packet forwarding games in mobile adhoc network. The optimal defense strategies have been derived and the maximum possible loss that the attackers can cause have been examined. In [17], the packet forwarding game are investigated for independent ad hoc networks in hostile and noisy environments and derive a set of attack-resistant cooperation stimulation strategies. In [18], the authors addressed the cooperation stimulation in Mobile Ad-Hoc Networks. A secure routing and packet forwarding game is formulated which modeled the interactions among nodes under noise and imperfect observation. In [19], the authors constructed a malware propagation model of seven states. They formulated a malware-defense differential game and explored optimal defense strategies for preventing malware propagation in WSNs.

In [8], the authors studied the dynamic behavior of the malware in response to the mobility of the devices in the network over time. A zero-sum dynamic game model is proposed and a network defense strategy against the spread of malware in mobile ad-hoc networks is derived. Although the amount of game theoretic for security in wireless network, few studies are conducted for malware attacks. As we have seen in the related work, no studies have been conducted to address the mitigation of the malware infection.

3 Proposed Framework

3.1 System Model

In this paper, we consider a disconnected distributed mobile wireless network consisting of two D2D clusters as shown at Fig. 1. Each cluster is consisting of a set of $\mathcal{N}$ different mobile devices denoted by the set $[\mathcal{N}]$, where all the devices trusted each other. The multi-hop D2D communication is enabled among these mobile devices. The mobile devices in each cluster are sharing private messages using short range technologies (e.g., WiFi) with another devices in the other cluster. Any open wireless environment becomes an attractive target to attackers. The attacker attempts to infect one or more targeted mobile devices by injecting malware with message into the network. We assume that for each message request to certain targeted device, there is a set of all routes $[R]$ from the source device to targeted device. Therefore, source device selects $r_j \in [R]$ to send the message, where $[S_j]$ is the set of devices along the route r_j.

We consider Ω different mobile operating systems, expressed by the finite set $[\Omega]$. We denote by $[\mathcal{M}_\omega]$ the set of $\mathcal{M}_\omega$ as a different malware available to the attacker to infect devices that run the mobile operating system ω.

For each $\omega \in [\Omega]$, we assume $\mathcal{A}_\omega$ anti-malware controls (i.e., Resources) expressed by the finite set $[\mathcal{A}_\omega]$. Anti-malware control is residing on each mobile device s_i and each anti-malware control has its detection rate to detect successfully certain malware type. Since the routing is a cooperative process where the messages are relayed among devices. Every device is responsible for inspecting the received message using its detection capability. Therefore, we denote by $\psi(a_k^i, \mathcal{M}_m; \omega)$, the probability that the device s_i that runs the anti-malware control a_k successfully detects malware $\mathcal{M}_m \in [\mathcal{M}_\omega]$. As a result, for the route r_j with number of hops $\mathcal{H}$, the average probability of $\mathcal{M}_m \in [\mathcal{M}_\omega]$ to be detected before it reaches the targeted device that runs ω is given by:

$$\psi(r_j, \mathcal{M}_m; \omega) := \frac{\sum_{s_i \in S_j} \psi(a_k^i, \mathcal{M}_m; \omega)}{\mathcal{H}} \tag{1}$$

Fig. 1 Mobile wireless
network

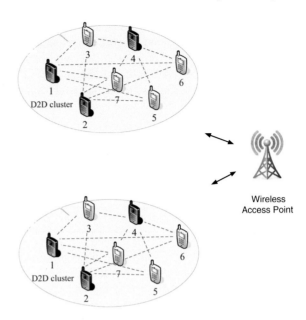

On the other hand, the multi-hop D2D communication and malware detection will necessitate cooperation between devices. Some devices may not collaborate to relay other device's traffic because of their limited available energy. Therefore, our protocol ensures the route availability during the routing process. It considers the energy level of the devices in the routing decision. It chooses the routes with highest energy devices on the basis of residual energy of the device. Formally the energy level of device s_i, $s_i \in [\mathcal{N}]$ is given by:

$\mathcal{E}(s_i) = \frac{\mathcal{E}_r}{\mathcal{E}_{max}}$, such that $\mathcal{E}_r$ is the remaining energy and $\mathcal{E}_{max}$ is the maximum energy available for the device.

Therefore the route energy level on r_j is derived by multiplying the energy level of all the devices along the route as follows:

$$\mathcal{E}(r_j) := \prod_{s_i \in S_j} \mathcal{E}(s_i) \tag{2}$$

3.2 Malware-Defense Secure Routing Games (MSRG)

We formulate a finite zero-sum Bayesian game (MSRG) to characterize the interactions between the defender (Network) and attacker in order to derive the infection-defense strategy for the defender. In the Bayesian game, the defender faces uncertainty about the multiple types of adversaries, denoted by $\omega \in \Omega$. In other words, we consider a finite number of possible attacker types with different payoffs. The defender has a subjective priori probability distribution $\rho(\omega)$ on Ω.

The defender has the statistics (i.e., $\rho(\omega)$) about different malware types for each mobile operating system ω. Furthermore, the defender uses the game history to perform better. We assume that the attacker exploits the mobile operating system vulnerabilities. Therefore, the attacker selects the malware that targets the vulnerability of certain operating system of the targeted device.

Strategy Set: The strategy set of a player refers to all available moves the player is able to take. We consider that the defender's pure strategies is a set of all possible routes $r_j \in [R]$ from the source device to the target device. An attacker's pure strategies are solely determined by his/her type ω. The attacker's pure strategy is a set of different malware $\mathcal{M}_m \in [\mathcal{M}_\omega]$, $\omega \in \Omega$, from which the attacker selects to send to the targeted device aiming its infection.

Payoff: we define the $\mathcal{U}_\Theta$ as the payoff of the defender. The payoff of the defender depends on the route detection rate and the route availability. We define $\mathcal{U}_\psi$ as the payoff of the attacker of type $\omega \in \Omega$, where the attacker's payoff is opposite to defender's payoff (i.e. zero sum game).

Bayesian Nash Equilibrium (BNE)

We consider the defender is the row player in the payoff matrix and the attacker types are the columns player. For a given pure strategy profile $(r_j, \mathcal{M}_m)$, $r_j \in [R]$, $\mathcal{M}_m \in [M_\omega]$ where $\omega \in \Omega$, the payoff of the defender is given by

$$U_\Theta(r_j, \mathcal{M}_m; \omega) = \sum_{\omega \in \Omega} \rho(\omega)[\psi(r_j, \mathcal{M}_m; \omega)\mathcal{V} + \mathcal{E}(r_j)] \quad (3)$$

We assume the $\mathcal{V}$ is the defender's security gain value (monetary), where $\mathcal{V} > 0$. The defender's payoff is the expected gain of detecting the malware before infecting the targeted device depends on the route detection rate in Eq. (1) summed up the route energy level in Eq. (2).

The defender's mixed strategy $X = [x_1, x_2, \ldots, x_\xi]$ is the probability distribution over different routes in $[R]$ (i.e. Pure strategies) from the source device to the targeted device. Where x_j is probability that the defender will choose its jth route to send the message.

For each $\omega \in \Omega$, the attacker's mixed strategies $Y^\omega = [y_1^\omega, y_2^\omega, \ldots, y_\eta^\omega]$ is the probability distribution over different malware (i.e. Pure strategies) against targeted devices that run ω. Where y_l^ω is probability that the attacker will choose its lth malware to infect device that runs ω.

In two players zero-sum game with finite number of actions for both players, there is at least a Nash equilibrium in mixed strategy [10].

The MSRG game consists of mixed strategy profile (X, Y^ω), therefore the expected payoff of the defender is denoted by:

$$\mathcal{U}_\Theta \equiv U_\Theta(X, Y^\omega) = \sum_{x=1}^{\xi} \sum_{l=1}^{\eta} x_j y_l^\omega U_\Theta(r_j, \mathcal{M}_m; \omega) \quad (4)$$

For zero sum game, $U_\Psi = -U_\Theta$, This means that the defender's gain is considered the attacker's loss.

Theorem The MSRG has mixed strategy Nash Equilibrium denoted by $(X^*, Y^{\omega*})$ with the property that

$$X^* = arg\ max_X\ min_{Y^\omega}\ U_\Theta,\ \forall Y^\omega$$
$$Y^{\omega*} = arg\ max_{Y^\omega}\ min_X\ U_\Psi,\ \forall X \tag{5}$$

4 Malware-Defense Secure Routing Protocol

Our proposed protocol MSR is strengthened version of Dynamic Source Routing (DSR) and described as follows:

Route Discovery procedure: First the source device broadcasts a Route Request message, all the devices receive this message should broadcast it to their neighbors. If the receiving device is the targeted device, it sends back the Route Reply message containing the full source route in reverse order. We have appended two new fields in the Route Reply message (route detection rate, route energy level). Each intermediate device on receiving the Route Reply, sums up its detection rate to the value in the route detection rate field and multiplying its energy level to the value in the route energy level field.

Route selection by the source device: After the source device receives several routes, then stores its routing table. It uses its routing table to solve the MSRG and deriving the optimal defense X^*. The source device selects the best route randomly according to X^* to send the message.

5 Performance Evaluation

In this section, we study the properties of the BNE in MSRG through the simulation and present the results of our study. The simulation is implemented in Omnet++ simulator, and fixed parameters are showed in Table 1. All the devices are equipped

Table 1 Simulation parameter values

Parameter	Value
Number of nodes	20
Mobility model	Linear mobility
Mobility speed	10 mps
Mobility update interval	0.1s
Packet size	512 bytes
Packet generation rate	2 packets/s

with the anti-malware systems with some known signatures attack. We assume that there is one attacker sending a sequence of malicious messages to a certain device. We consider two ad-hoc routing protocols, DSR and AODV, and obtain the performance of infection-defense strategy for MSR in terms of the percentage of the detected malicious messages comparing with other nonstrategic protocols. In this paper, we consider two distinct attacker profiles: Nash and Uniform. For Nash profile, the attacker plays the attack mixed strategy given by BNE of the game. For Uniform profile, the attacker plays the attack with the same probability.

In the Fig. 2 shows the results of the percentage of the detected malicious messages, varying the pause times in case of nash attacker profile and uniform attacker profile. In case of nash attacker, it can be seen that in DSR and AODV protocols, a small percentage of detected malicious messages. While the MSR protocol still keeps this percentage high even in a hostile environment. In case of Uniform profile, we can notice that MSR protocol outperforms the other two ad-hoc protocols and still has high percentage of detected malicious messages. Additionally, MSR performs better in the presence of a Uniform attacker profile rather than a Nash attacker profile. We can notice the average values of detected malicious messages in case of Uniform attacker within the range [65–80 %]. Whilst in case of Nash attacker the average values of detected malicious messages within the range [60–70 %]. This is

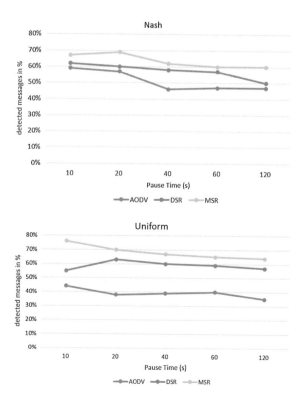

Fig. 2 Percentage of detected malicious messages verus pause time

because the Uniform attacker chooses any of his strategies with the same probability therefore this affects the performance of the malware detection of the network rather than Nash attacker.

6 Conclusion and Future Work

In this paper, we model a static Bayesian security game. We show that the game has the Nash equilibrium leading to optimal infection-defense strategy. Results show that using prevention mechanisms alone are not sufficient to handle the secure networks and our protocol based on strategic plan outperforms the other nonstrategic protocols. The preliminary results are promising and we have plans for further investigations. Future work is to model this static game as a dynamic Bayesian game where the defender updates his beliefs at each stage of game. The other work is using another signature database for the intrusion detection.

References

1. Mumtaz, S., Rodriguez, J. (eds.): Smart Device to Smart Device Communication, pp. 1–22. Springer International Publishing, Switzerland (2014)
2. Doppler, K., Rinne, M., Wijting, C., Ribeiro, C.B., Hugl, K.: Device-to-Device communication as an underlay to LTE-advanced networks. IEEE Commun. Mag. 47(12), 42–49 (2009)
3. Asadi, Arash, Wang, Qing, Mancuso, Vincenzo: A survey on Device-to-Device communication in cellular networks. IEEE Communication Surv. Tutor. 16(4), 1801–1819 (2014)
4. Bello, O., Zeadally, S.: Intelligent Device-to-Device communication in the internet of things. IEEE Syst. J. PP(99), 1–11 (2014)
5. Nishiyama, H., Ito, M., Kato, N.: Relay-by-smartphone: realizing multihop Device-to-Device communications. IEEE Commun. Mag. 52(4), 56–65 (2014)
6. Khouzani, M.H.R., Sarkar, S., Altman, E.: Maximum damage malware attack in mobile wireless networks. IEEE/ACM Trans. Netw. 20(5), 1347–1360 (2014). Dreese Labs., Ohio State University, Columbus
7. Karlof, Chris, Wagner, David: Secure routing in wireless sensor networks: attacks and countermeasures. Ad Hoc Netw. 1(23), 293–315 (2003)
8. Khouzani, M.H.R., Sarkar, S., Altman, E.: Saddle-point strategies in malware attack. IEEE J. Sel. Areas Commun. 30(1), 31–43 (2011)
9. Mtibaa, A., Harrasy, K.A., Alnuweiri, H.: Malicious attacks in mobile device clouds: a data driven risk assessment. In: 23rd International Conference on Computer Communication and Networks (IC-CCN), pp. 1–8 (2014)
10. Equilibrium Points in N-Person Games. http://www.ncbi.nlm.nih.gov/pmc/articles/PMC1063129/?page=2
11. Agah, Afrand, Das, Sajal K.: Preventing DoS attacks in wireless sensor networks: a repeated game theory approach. Int. J. Netw. Secur. 5(2), 145153 (2007)
12. Mohi, M., Movaghar, A., Zadeh, P.M.: A Bayesian game approach for preventing DoS attacks in wireless sensor networks. In: International Conference on Communications and Mobile Computing, vol. 3, pp. 507–511 (2009)
13. Bohacek, S., Hespanha, J.P., Lee, J., Lim, C., Obraczka, K.: Game theoretic stochastic routing for fault tolerance and security in computer networks. IEEE Trans. Parallel Distrib. Syst. 18(9), 1227–1240 (2007)

14. Felegyhazi, M., Buttyan, L., Hubaux, J.-P.: Nash equilibria of packet forwarding strategies in wireless ad hoc networks. IEEE Trans. Mob. Comput. **5**(5), 463–476 (2006)
15. Sun, Y.L., Yu, W., Han, Z., Liu, K.J.R.: Information theoretic framework of trust modeling and evaluation for ad hoc networks. IEEE J. Sel. Areas Commun. **24**(2), 305–317 (2006)
16. Yu, W., Ji, Z., Liu, K.J.R.: Securing cooperative ad-hoc networks under noise and imperfect monitoring: strategies and game theoretic analysis. IEEE Trans. Inf. Forensics Secur. **2**(2), 240–253 (2007)
17. Yu, W., Liu, K.J.R.: Game theoretic analysis of cooperation stimulation and security in autonomous mobile ad hoc networks. IEEE Trans. Mob. Comput. **6**(5), 507–521 (2007)
18. Yu, W., Liu, K.J.R.: Secure cooperation in autonomous mobile ad-hoc networks under noise and imperfect monitoring: a game-theoretic approach. IEEE Trans. Inf. Forensics Secur. **3**(2), 317–330 (2008)
19. Shen, S., Li, H., Han, R., Vasilakos, A.V., Wang, Y., Cao, Q.: Differential game-based strategies for preventing malware propagation in wireless sensor networks information forensics and security. IEEE Trans. Inf. Forensics Secur. **9**(11), 1962–1973 (2014)

Compiling a Dialectal Arabic Lexicon Using Latent Topic Models

Rana Elmarakshy and M.A. Ismail

Abstract This paper introduces DALex, a Dialectal Arabic Lexicon that comprises 4 major regional Arabic dialects: Maghrebi, Egyptian, Levantine and Peninsular or Gulf Arabic, covering about 75 % of the Arabic dialects spoken sub-regionally. We describe the approach used to compile the lexicon and the inference algorithms that extract semantically similar words from the different dialects lists of the word-topic distribution generated by Probabilistic Topic Models sampled on dialectal comparable documents that are harvested from web blogs and forums. The power of this Lexicon lies in its automatic, dialect independent way of construction that makes it easily created and expanded. Furthermore, the wide coverage of multiple dialectal variants, we believe, will be a great resource for future Computational Linguistics research.

Keywords Arabic dialects · Arabic lexicon · Latent topic models

1 Introduction

Arabic is the largest member of the Semitic language family spoken by approximately 206 million native speakers and 246 million non-native speakers[1] in the 22 countries of what is called the Arab-World. When it comes to the Arabic speaking communities, we

---

http://looklex.com/e.o/arabic_l.dialects.htm

---

[1]https://www.ethnologue.com/language/arb.

R. Elmarakshy (✉) · M.A. Ismail
Computer and Systems Engineering Department, Alexandria University, Alexandria, Egypt
e-mail: rana.elmarakshy@gmail.com

M.A. Ismail
e-mail: maismail@alexu.edu.eg

© Springer International Publishing Switzerland 2016
T. Gaber et al. (eds.), *The 1st International Conference on Advanced Intelligent System and Informatics (AISI2015), November 28–30, 2015, Beni Suef, Egypt*,
Advances in Intelligent Systems and Computing 407, DOI 10.1007/978-3-319-26690-9_44

497

find the concept of *diglossia* which is defined as a situation in which two languages, or dialects, are used under different conditions within the same language community [1]. In case of Arabic, we find that Modern Standard Arabic, or MSA, is used nearly exclusively in written context and formal situations while the Arabic dialects occupy the daily communications, mostly in spoken form but also in informal written contexts.

With the introduction of web 2.0, and the increasing use of emails, text messages, blogs and chats the Dialectal Arabic, or DA, started taking over more space in written contexts. Nowadays, it is completely normal to write in a dialect, however, it is not limited to the private communication. In fact, it is becoming a trend to write and publish in local dialects e.g. Kelmetna[2] is a youth magazine, among many others, that is published in the Egyptian dialect. Specially after the Arab Spring revolution, we noticed an increasing number of authors publishing books in their local dialects. Dialectal poetry has become even more common.

In the last decade, the increasing content of DA on the web has motivated researchers to give more interest to DA text understanding, specially in the tracks of Sentiment Analysis and Reviews Mining. However, written dialects have relatively scarce resources and the written materials usually involve informal conversations or traditional folk literature, and with the absence of linguistic rules and writing standards, understanding such documents represents a big challenge for researchers.

To cope with DA processing challenges, researchers have followed two main approaches: the first is to explicitly create dialectal corpora and tools such as morphological analyzers and the second is to take advantage of the closeness between MSA and dialects in order to adapt MSA resources and tools to dialects. Although the first approach is more linguistically accurate because it takes into account specificities of each dialect, building resources from scratch is very expensive in terms of time and effort.

In this paper, we will present an approach to automatically compile a Dialectal Arabic Lexicon of 4 dialects, beside the MSA. First, we review the related work in Sect. 2. Then, we overview the variations between the Arabic dialects in Sect. 3. Section 4 presents the proposed algorithm to extract semantically similar words for building the lexicon. Section 5 explains the process of harvesting data to create a comparable dialectal corpus for training the algorithm. Section 6 discusses the evaluation of the lexicon. Section 7 highlights the potential applications of DALex in Arabic NLP tasks, while Sect. 8 concludes the work and suggests possible future extensions.

2   Related Work

Unlike MSA, Dialectal Arabic text understanding is a relatively new research field that has a smaller number of linguistic resources, corpora and tools, which makes of it a more challenging task for Arabic NLP researchers.

---

[2]https://www.kelmetnamag.com/.

A previous approach to build a Multi-Dialectal corpus was to scrap dialectal content from users' comments on newspaper websites and manually annotate them. A corpus of 3 dialects (Levantine, Gulf and Egyptian) was built for the task of dialect identification [2]. This work was later extended to support 2 more dialects (Iraqi and Maghrebi) and to include Twitter as an additional source of data [3]. This is considered the most diverse corpus of Dialectal Arabic known so far in terms of both the source of the content and the number of dialects.

Bilingual dictionaries for English speaking learners of Moroccan, Syrian and Iraqi Arabic [4] were manually developed by human natives.

Tharwa [5] is a Egyptian Arabic, MSA and English three-way lexicon compiled from existing resources such as paper dictionaries and electronic corpus based dictionaries. It is a static resource whose maintenance and update require a big manual effort.

Few efforts were made to automatically build dialectal lexicons, each of which comprised single minor Sub-Regional dialect, e.g. the Tunisian Dialect [6, 7] and the Cairene Dialect [8]. The latter was automatically compiled where word synonyms are induced from unrelated documents based on correlations between word co-occurrence patterns.

Our approach follows this line of thinking, with the difference of extending the lexicon to support Multi-Dialects and using Latent Topic Models to infer the topics-per-document and words-per-topic distributions and get a list of similar context occurring words across different dialects, which are further processed for the extraction of semantically similar terms.

3 Arabic Dialects Variations

Arabic dialects can be classified regionally as Maghrebi, Egyptian, Levantine, Gulf, and Yemeni or Sub-Regionally as Moroccan, Tunisian, Syrian, Lebanese, Kuwaiti, Qatari, etc. Other parameters that cause further differences in dialects are ethnicities, religious groups, social classes, education levels, gender and age. In this work we cover the four main regional dialects; Maghrebi, Egyptian, Levantine and Gulf.

Arabic is a morphologically rich language that has an ambiguous orthography which represents a big challenge for Arabic NLP researchers [9]. Arabic dialects vary phonologically, lexically, and morphologically from one another and from MSA [10].

3.1 Phonology

The biggest example of phonological differences can be observed in the dialectal mitigation of MSA hard sounds. Since a dialect is the language of everyday communication, speakers frequently mitigate the hard sounds of MSA, leading to

Table 1 Prominent sound mitigation relations among MSA and colloquial dialects

MSA sound	MAR	EGY	LEV	GULF
ذ		ز، د	ز، د	
ض		د	د	
ظ		ض	ض	
ق	ق، أ، ج	ا	ا	ج، ق
ك	ك	ك	ك	ك، ج
ج	ق	ج	ق	ق
ث	ت	س، ت	س، ت	ث، ت
ئ		ي	ي	

new words that are used in spoken and sometimes even in written speech. Table 1 shows a list of MSA hard sounds and their common mitigations across the dialects.

3.2 Orthography

While MSA has a standard orthography, the dialects do not. People often write a word the way they say it. Regardless how it should be written as an original Arabic word. Slang words and loanwords don't have a standard orthography since they are never written in a formal context, DA can even be, sometimes, written in Roman script [11].

3.3 Morphology

Morphological differences are very common. Table 2 presents a clarifying example that shows the variations in the prefixes added to a stem verb to make the future form in different dialects.[3]

3.4 Lexicon

A common feature of the human natural language is being influenced by the social aspects of its speakers. In case of Arabic dialects, we notice that the major

---

[3]Arabic transliteration is presented in the Habash-Soudi-Buckwalter scheme [22]

ا	ب	ت	ث	ج	ح	خ	د	ذ	ر	ز	س	ش	ص	ض	ط	ظ	ع	غ	ف	ق	ك	ل	م	ن	ه	و	ي
A	b	t	θ	j	H	x	d	ð	r	z	s	š	S	D	T	Ď	ς	γ	f	q	k	l	m	n	h	w	y

.

Table 2 Future form variations in dialects

Dialect	MAR	EGY	LEV
MSA	+ s ا س	سأشرب	saÂašrabu
EGY	+ ha ه ها	هشرب	hašrab
LEV	+ raH ر رح	رح أشرب	raH Âašrab
TUN	+ bAš باش	باش نشرب	bAš nišrab

Table 3 Foreign languages most influencing each dialect

Dialect	Source of loanwords
Maghrebi	Tamazight, French, Spanish, English
Egyptian	English, French, Greek, Italian, Turkish
Levantine	French, English
Gulf	English, Turkish, Farsi

differences in lexicons across the dialects return to historical and geographic aspects. Table 3 lists the foreign languages that are the source of most of the loanwords in each dialect.

4 The Proposed Algorithm

4.1 Building a Translation Graph

Our lexicon is internally represented as a weighted undirected graph where each vertex $v \in V$ in the graph is a pair (w, d) where w is a stem word of dialect d. Undirected edges in the graph denote translation relations between words: an edge $e \in E$ linking (w_1, d_1) and (w_2, d_2) represents the probability that w_1 and w_2 are semantically similar.

Construction. The translation graph is initialized with the stem words of the Aralex MSA lexicon [12]. It is particularly useful to take advantage of the huge corpus already built for MSA to build the first dimension of our graph which will serve as a reference to compare and locate new dialectal words before adding them to the graph. The Lexicon Extraction algorithm will be described in more details in Sect. 4.2.

Inference. The translation graph supports three query modes; first, it can be searched for the meaning of a word without restricting the target dialect, in which case we return a set of the words, from all dialects, that have the highest similarity value with this term. Secondly, the graph can be searched for the closest word in meaning for a source word specifying both source and target dialects; this is solved by a doing a graph BFT until we reach the closest node in the defined target dialect. Finally, a user can search for the similarity degree of a word-pair in the graph regardless of their dialects; in which case we return the shortest path from all the occurrences of the source word to the target word or an error if the 2 words have significantly dissimilar meanings. A case study of the system performance in those three query modes will be presented in Sect. 6.3.

4.2   Lexicon Extraction Algorithm

The problem of mining the web for Dialectal Arabic lexicons can be considered a special case of Cross-Language Lexicon Extraction even a harder one due to the lack of dialectal resources and almost rarity of parallel documents. Therefore, several studies attempted to make use of the similarity between DA and MSA to overcome the problem of the under-resourced dialects. Our approach follows this line of thinking, we implement a customized version of the edit distance algorithm that discovers the mutations happened to a dialectal term and links it to its MSA original term. In this section, we present the topic model used to infer a list of contextually similar words and the matching algorithm used to get potential word synonyms.

Bilingual LDA. The topic model that we use is the bilingual extension of the standard LDA model which was first presented in several works [13–15].

Outline of the method. This topic model assumes that each tuple is a set of documents that are broadly parallel to each other, but are written in different languages. Those don't have to be strictly parallel (sentence-aligned) texts, as long as the documents in a tuple have comparable (document-aligned) content i.e. having the same tuple distribution over topics.

Generative Model. Since we apply the topic model to a collection of comparable (document-aligned) texts which share their topic distributions over the documents, we can safely use a single variable θ to contain the topic distribution of the paired comparable documents. Furthermore, the vocabulary words for each of the source dialect S and the target dialect T are sampled on the distribution φ and ψ respectively. The Generative process for the BiLDA is described as follows:

```
for each document pair dⱼ in S and T
    for each word position i ∈ dⱼ in S
        sample z_{ji}^S ~ Multinomial(θ)
        sample w_{ji}^S ~ Multinomial(φ, z_{ji}^S)
    for each word position i ∈ dⱼ in T
        sample z_{ji}^T ~ Multinomial(θ)
        sample w_{ji}^T ~ Multinomial(ψ, z_{ji}^T)
```

Parameter Estimation. To train the model we use Gibbs sampling with 1000 Gibbs iterations and a value of 10/k and 0.01 to the parameters α and β respectively, where k is the number of topics. After the training, we end up with a set of φ and ψ word-topic probability distributions that are used for the calculations of potential word synonyms.

Matching algorithm.

The second part of the algorithm denotes extracting the potential synonymous word pairs from the set of word-topic distributions generated by the topic model. The matching process happens in 3 steps.

DA to MSA word mapping using Edit Distance. We derive a special substitution penalty to the known edit distance algorithm to encode deeper morphological and phonological knowledge based on each dialect's MSA mitigated letters defined in Table 1.

For example: We give a lower penalty for subtracting the letter ت t for the letter ث θ, which results in giving a higher value for the EGY word تمن taman 'price' to match the MSA word ثمن θaman 'price' rather than matching the MSA word يمن yaman 'Yemen' which both would have had equal values in a normal edit distance function that considers only their 1-letter difference. Moreover, we give a low penalty for the addition and deletion of the letters ا a, و w, ي y, since Arabic is written in an Abjad script that doesn't write short vowels, it is very common that, in an unstandardized dialect orthography, short and long vowels are confused.

Named Entity Recognition. We make use of the fact that Named Entities are normally written in Roman script then followed by their Arabic transliteration. So when an English NE is encountered, a window of 2 words preceding and following that term is scanned to detect the Arabic transliterated name of that Entity. If found, this word pair is saved for future detection of Named Entities written in Arabic script only.

Loanwords detection. If the edit distance calculated at the first step didn't return a similarity value defined within an accepted range, it is more likely that this word is derived from a foreign language. Therefore, we try to infer the origin of that loanword by generating all possible transliterations of the word and looking them up in bilingual dictionaries of the foreign languages, mentioned in Table 3, that are potential sources of loanwords of that dialect. In case of a successful lookup, the loanword inherits the similarity value of its Arabic translation when it is added to the graph.

Unmatched words. When the 3 previous matching steps fail to locate the new word in the graph, we fall back to the contextual information learned from the BiLDA. Moreover, we insert unmatched words as outliers to the cluster formed by the words distributed over this topic to explicitly show that this word is similar in context to the topic represented by this set of words but it is not specifically a synonym for any of them.

5   Data Collection

The process of building a dialectal corpus, to train the topic model and generate the translation candidate words, was done in a semi-automatic way. As previously mentioned, the BiLDA model needs a large collection of comparable texts which are thematically-aligned but not necessarily sentence-aligned. Therefore, dialectal micro-posts such as tweets and users' comments were not the best fit for our model. Instead, we went for harvesting blog posts and forum articles for all the dialects using 2 approaches.

First, querying documents about shared topics of interest among the Arab-World, an example can be the literature works or religious stories like *Alf leila w leila* 'One thousand and one nights' and *kesas elanbeyaa* 'Tales of the prophets' that although being written in local dialects by different authors, they still share the same content which guarantees that we have loosely comparable texts. We tried to cover as many topics as possible in the genres of Art, Literature, Culture, Religion, and Child stories.

Second, crawling specialized web blogs written in local dialects and aligning documents that share tags or keywords. For example, an article reviewing the security issues in the new version of Android on an Egyptian technical blog is much likely to be having the same word-topic distribution of an article having the keywords *Android*, *Security,* and *Lollipop* in a Moroccan technical blog. This approach helped us to cover more topics including Technology, Politics, Sports and Travel.

6 Evaluation and Discussion

We cannot make a comparative evaluation with other compiled lexicons or with previous algorithms applied to DA Lexicon Extraction since, to the best of our knowledge, this is the first study done to automatically compile a lexicon of multiple dialects.

Instead, we investigate three key questions: (1) how does the coverage of our Lexicon compare to the largest existing Multi-Dialectal Lexicon? (2) what is the precision of our Lexicon Extraction algorithm? (3) how accurately does our translation inference algorithm perform in extreme cases?

6.1 Coverage

We measure the coverage of a lexicon according to two parameters; the number of dialects that it comprises and the number of entries in the dictionary, and the breadth of topics that they cover. As mentioned earlier, our Lexicon covers 4 regional dialects which make about 75 % of the Arabic native speakers.[4] The Lexicon's terms are extracted from 9 genres of documents covering the terminology of most of the topics encountered in a native speaker's daily life. A detailed

---

[4]http://looklex.com/e.o/arabic_l.dialects.htm.

Table 4 Comparison of previous attempts of building Arabic dialect lexicons

Lexicon name	Languages	Vocabulary	Number of words
LDC bilingual dictionaries [4]	EN—Iraqi	NA	4368
	EN—Syrian	NA	3323
	EN—Morrocon	NA	3014
Tunisian Dialect lexicon [7]	MSA—Tunisian	NA	Verbs: 1500 Nouns: NA
Tunisian lexicon for deverbal nouns [6]	MSA—Tunisian	Deverbal Nouns	Root lexicon: 1,357 Verbal lexicon: NA
ECA lexicon [8]	MSA—Cairene	NA	1,000
Tharwa [5]	MSA—EN—EGY	NA	73,000 EGY words
AIPSeLEX [16]	MSA—EGY	Idioms and proverbs	3632

example of the number of compiled terms will be discussed in the next subsection. Table 4 presents a comparison of the coverage of the earlier efforts of building colloquial Arabic lexicons.

6.2 Lexicons Extraction Evaluation

We present a detailed example of the system performance during the Lexicon Extraction process of the "Mobile" section of the technical corpus. A word is defined as the stem word of an occurring word after removing its prefixes and suffixes. Table 5 shows the total number of unique words resulting from the word segmentation process of every dialect's corpus. The list of unique Arabic words is further processed using the Lexicon Extraction algorithm discussed in Sect. 4.2. Table 6 shows the number and percentage of identified lexicons at every step of the matching algorithm.

Even if DALex doesn't precisely identify or match infrequent words, it still has the power of estimating their context and getting a live example of how these words are used by natives.

Table 5 Unique word counts of the technical corpus

	Maghrebi	Levantine	Egyptian	Gulf
Total unique words	1116	1138	1888	1517
Unique words in Arabic script	989	1045	1654	1448
Unique words in Roman script	127	93	234	69

Table 6 The numbers of identified lexicons at every step of the matching algorithm

Technique	Accuracy	Maghrebi	Levantine	Egyptian	Gulf
DA to MSA mapping	Words	880	971	1463	1333
	Percentage (%)	88.98	92.91	88.45	92.06
Named entity Recognition	Words	19	29	25	53
	Percentage (%)	1.92	2.77	1.51	3.66
Loanwords detection	Words	27	23	110	24
	Percentage (%)	2.73	2.21	6.65	1.66
Unmatched words	Words	63	22	56	38
	Percentage (%)	6.37	2.11	3.39	2.62

6.3 Lexicon Lookup Evaluation

In this section, we present a case study of the system performance under extreme cases for each of the three query modes presented in the inference subsection of Sect. 4.1.

Searching for all synonyms of a word. In this query mode a word in a source dialect is looked up without defining a target dialect so the closest translation synonyms are returned in all dialects. Figure 1 illustrates a subgraph of the lexicon that represents the nodes returned from the look up of the Gulf word الدوام, AldwAm 'work'. We notice an error in the Levantine synonym of the word, as the lexicon gives a higher similarity value to the word 'issue' than the actual synonym of the word 'work'. We return this to the fact that the Levantine word الشغلة, Alšaγlh 'issue' has a big orthographic similarity to the Egyptian word الشغل, Alšoγl 'work' and that both words appeared in an article taking about safety issues at work.

Searching for the closest word in meaning for a source word specifying both source and target dialects. In this experiment, we try to verify the case that the similarity measure of 2 words is not dominated by their orthographic similarity. To test that, we use the Egyptian query word مبسوط, mbswT 'happy' and request to retrieve its synonym in Gulf, we notice that the returned word is رايق, rAyq 'having peace of mind' although the Gulf lexicon includes the word مبسوط, mbswT 'stroke' that has a similar orthography but a completely different meaning.

This experiment proves that word similarity is only considered when words in different dialects share the same contextual co-occurrence.

Searching for the similarity degree of a word-pair regardless of their dialects. The critical point in this search mode lies in finding the source and the target nodes that are closest in meaning when any of the query terms is polysemous, hence is represented in multiple nodes in the graph. A good example of a source word is the MSA word 3 lm علم; a word that without having its diacritics can be interpreted by Native speakers as either 3 ,ilm علم 'science' or 3 ,alam علم 'flag'. Another example

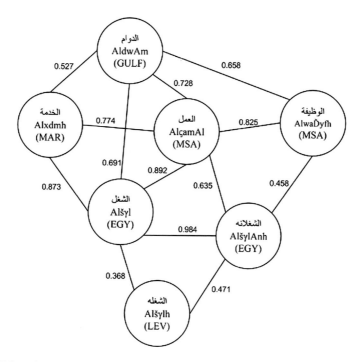

Fig. 1 Subgraph result of the query word الدوام, AldwAm 'work'

of a target query is the word مسيد, masYd which has the meaning 'school' in Maghrebi and 'mosque' in Gulf. The query returns the path between the MSA node, holding the word science, and the Maghrebi node, holding the word school, as the shortest path between the occurrences of the 2 words that have the highest semantic similarity value.

7 Applications of DALex in Arabic NLP

The existence of a dialectal lexicon that have such a wide coverage of words over many dialects can boost many Arabic Natural Language Processing tasks such as Arabic Dialect Identification [2, 17], Machine Translation [18] and Arabic Sentiment Analysis [16, 19–21] by providing a standard Arabic thesaurus that can be used in a wide range of Computational Linguistics tasks instead of re-implementing a customized Lexicon for every project instance of a task.

8   Conclusion

In this work, we have presented an unsupervised approach to compile a Dialectal
Arabic Lexicon. First, a Latent Topic Model, Bilingual LDA, is trained to generate
sets of word-topic distributions, then word synonyms are extracted from those lists
using a customized similarity matching algorithm that exploits the closeness of the
Arabic dialects to MSA, and to each other.

Possible future extensions are to further divide the dialects to a Sub-Regional,
rather than regional, basis to open the door for more accurate synonym words from
every region, country, or community. Another future direction is to extend the
algorithm to support processing Micro-Posts like tweets or users' comments which
will allow more terms to be added to the Lexicon specially slang words and chat
acronyms which don't normally appear in blog posts and other semi-formal articles.

References

1. Ferguson, C.A. Diglossia. Word (1959)
2. Zaidan, O.F., Callison-Burch, C.: The arabic online commentary dataset: an annotated dataset
   of informal arabic with high dialectal content. In: Proceedings of the 49th Annual Meeting of
   the Association for Computational Linguistics: Human Language Technologies: Short Papers
   (2011)
3. Cotterell, R., Callison-Burch, C., A multi-dialect, multi-genre corpus of informal written
   Arabic. In: Proceedings of the Language Resources and Evaluation Conference (LREC)
   (2014)
4. Graff, D., Maamouri, M., Developing LMF-XML bilingual dictionaries for colloquial Arabic
   dialects. In: LREC (2012)
5. Diab, M., Al-Badrashiny, M., Aminian, M., Attia, M., Dasigi, P., Elfardy, H., Eskander, R.,
   Habash, N., Hawwari, A., Salloum, W.: Tharwa: a large scale dialectal arabic-standard
   Arabic-English lexicon. In: Proceedings of the Language Resources and Evaluation
   Conference (LREC) (2014)
6. Hamdi, A., Gala, N., Nasr, A., Automatically building a Tunisian lexicon for deverbal nouns.
   In: COLING *2014* (2014)
7. Boujelbane, R., Khemekhem, M.E., Belguith, L.H.: Mapping rules for building a tunisian
   dialect lexicon and generating corpora. In: Proceedings of the International Joint Conference
   on Natural Language Processing (2013)
8. Al-Sabbagh, R., Girju, R., Mining the web for the induction of a dialectical Arabic lexicon. In:
   LREC (2010)
9. Habash, N.Y.: Introduction to Arabic Natural Language Processing, vol. 3. Morgan &
   Claypool Publishers, Dordrecht (2010)
10. Watson, J.C.: The Phonology and Morphology of Arabic. Oxford University Press, Oxford
    (2007)
11. Darwish, K.: Arabizi detection and conversion to Arabic. In: CoRR (2013)
12. Boudelaa, S., Marslen-Wilson, W.D.: Aralex: a lexical database for modern standard Arabic.
    Behav. Res. Methods **42**(2), 481–487 (2010)
13. Mimno, D., Wallach, H.M., Naradowsky, J., Smith, D.A., McCallum, A.: Polylingual topic
    models. In: Proceedings of the 2009 Conference on Empirical Methods in Natural Language
    Processing (2009)

14. De Smet, W., Moens, M.-F., Cross-language linking of news stories on the Web using interlingual topic modeling. In: Proceedings of the CIKM 2009 Workshop on Social Web Search and Mining (2009)
15. De Smet, W., Tang, J., Moens, M.-F.: Knowledge transfer across multilingual corpora via latent topics. In: Proceedings of the PAKDD: the 15th Pacific-Asia Conference on Knowledge Discovery and Data Mining (2011)
16. Ibrahim, H.S., Abdou, S.M., Gheith, M.: Idioms-proverbs lexicon for modern standard Arabic and colloquial sentiment analysis. Int. J. Comput. Appl. **118**, 26–31 (2015)
17. Malmasi, S., Refaee, E., Dras, M.: Arabic dialect identification using a parallel multidialectal corpus. In: Proceedings of the 14th Conference of the Pacific Association for Computational Linguistics (PACLING 2015), Bali, Indonesia (2015)
18. Jeblee, S., Feely, W., Bouamor, H., Lavie, A., Habash, N., Oflazer, K.: Domain and dialect adaptation for machine translation into Egyptian Arabic. In ANLP 2014 (2014)
19. Ibrahim, H.S., Abdou, S.M., Gheith, M.: Sentiment analysis for modern standard arabic and colloquial. Int. J. Nat. Lang. Comput. (IJNLC) **4**(2) (2015)
20. Duwairi, R.M.: Sentiment analysis for dialectical Arabic. In: 6th International Conference on Information and Communication Systems (ICICS), IEEE (2015)
21. El-Makky, N., Nagi, K., El-Ebshihy, A., Apady, E., Hafez, O., Mostafa, S., Ibrahim, S.: Sentiment analysis of colloquial Arabic tweets. In: ASE BigData/SocialInformatics/PASSAT/BioMedCom 2014 Conference, Harvard University, Dec 2014
22. Habash, N., Soudi, A., Buckwalter, T.: On Arabic transliteration. Arabic Computational Morphology, pp. 15–22. Springer (2007)

Access Control Models for Pervasive Environments: A Survey

Walaa Elsayed, Tarek Gaber, Ning Zhang and M. Ibrahim Moussa

Abstract Pervasive computing is a concept in computer science where computing appear everywhere and anywhere, the devices are heterogeneous and may belong to different domains, and they interact with each other to provide smart services. Traditional access control models such as DAC, MAC and RBAC are not suitable to this environment. Therefore, we need more flexible, dynamic and generic access control models for controlling access to such environments. There have been many proposals proposed to address the new security requirements in the pervasive computing environments. The goal of this paper is to review and analyze the existing proposals seen in literature and to compare the approaches taken in these proposals based on the security requirements. This comparison will lead to the identification of some research gaps that require further investigation.

1 Introduction

Computer security is protection afforded to automated computer systems to detect and prevent unauthorized users from accessing secured data and using computer resources. There are three main security requirements in any computer system namely, *confidentiality*, *integrity*, and *availability* [1]:

W. Elsayed · T. Gaber
Faculty of Computers and Informatics, Suez Canal University, Ismailia, Egypt

N. Zhang
School of Computer Science, University of Manchester, Manchester, UK

M. Ibrahim Moussa
Faculty of Computers and Information, Benha University, Benha, Egypt

T. Gaber (✉)
Scientific Research Group in Egypt, (SRGE), Cairo, Egypt
e-mail: tmgaber@gmail.com
URL: http://www.egyptscience.net

© Springer International Publishing Switzerland 2016
T. Gaber et al. (eds.), *The 1st International Conference on Advanced Intelligent System and Informatics (AISI2015), November 28–30, 2015, Beni Suef, Egypt,*
Advances in Intelligent Systems and Computing 407, DOI 10.1007/978-3-319-26690-9_45

511

Confidentiality: Data confidentiality assures that confidential data is available only to authorized users and the user's privacy which ensures that individuals' information is collected and stored by authorized users and determines who has the privilege to access them [1].

Integrity: Data integrity ensures that data can be modified or edited only by authorized user. A system integrity is making sure that the system performs its proposed function according to the designed way without any unauthorized manipulation [1].

Availability guarantees that the system and the data are available to the authorized users when needed [1].

In computing, access control security mechanisms are used to ensure that unauthorized users are not allowed to access the system and that authorized users cannot make inappropriate modifications. According to [2], an access control system consists of four main components; subject, object permission and credential (Fig. 1).

The subject is an active entity that needs permission to access an object or the data within an object. The object is a passive entity that being accessed. The permission is an access right to a subject over an object. The credential is a piece of information that is used to prove the identity of a subject [2]. Based on the classification in [2], there are three types of access control models.

- Discretionary Access Control (DAC) is a class of access control models where the access control decision is based on the identity of a user. The model is flexible in supporting commercial solutions where no strict information flow is required. However, it is not suitable for military applications that need a precise control of information flows.
- Mandatory Access Control (MAC) is another class of access control models. It enables the access rights to be determined by a manager or a central authority of the system. MAC is suitable for governmental and military applications, but it is too precise for other applications, e.g. commercial domains.
- Role-Based Access Control (RBAC) is a third class of access control models which governs access to a resource object based on a subject's organizational role. In RBAC, access rights are defined for roles instead of individual users. Privileges are assigned to different roles which are then assigned to users [3].

Fig. 1 The security requirements tried [1]

However, the major weakness of the RBAC model is that it cannot capture any security-relevant information from its environment [4].

Traditional access control methods are suitable to the centralized and relatively static environment in which subjects and objects are known and relatively static and the permissions are rarely changed. Furthermore, traditional access control methods face some new challenges in open and distributed environments, and they can hardly meet the open and dynamic requirements. Access control in distributed environments must adapt to the dynamic addition and deletion of entities [5].

The objectives of this paper are reviewing existing access control models and determining the requirements needed for building an effective access control model suitable for pervasive systems and dynamic environments. The rest of this paper is organized as the follows: Sect. 2 presents the related work; Sect. 3 conducts a comparison among some of the existing access control models. Finally, Sect. 4 concludes the work done on this paper.

2 Related Work

Most of the existing access control models make their decision based on the context, user behavior, attributes or relation. Therefore, we classified the literature into the following four classes: context-aware, attribute-based, user behavior-based and relation-based access control models.

2.1 Context-Aware Based Access Control Models

The traditional access control models did not take into account how to use the surrounding information about the user, the object and the environment when making the access decision; that concept is called context-awareness. A context represents any information that can be used to describe the state of an entity. An entity such as a user, environment, or object that is related to the interaction between a user and application, including the user and applications themselves [3].

- Zhang et al. [3] presented the problem of how the traditional access control models are not suitable for pervasive applications. Therefore, they proposed Dynamic Role Based Access Control (DRBAC) model which extends the role-based one. In this model, the access decision depends on the required credentials of users and the context such as location and the state of the system such as the load on resources. They defined the components of DRBAC and described the operation by examples. However, they did not implement the model but gave some issues that should be considered to implement it successfully in real applications. The major strength of the DRBAC model is its ability to make access control decisions dynamically based on context information.

- Kim et al. [4] proposed a more flexible context-aware access control model that depends on different context attributes such as location and time. In addition to that, they presented the State Checking Matrix (SCM) that specifies the relationship between the user and the object. In their proposed model, user's role is only active when all elements of the context information are active; for example when location and time are active. Compared with [3], this model is more flexible as it presents how to accommodate with different context-aware attributes.

- Diep et al. [5] proposed risk as an idea that is used for decision-making process which makes the model more precise and dynamic. They presented access control framework and explained how the risk value can be estimated and compared with threshold to determine the access control decision. This model is dynamic because the access control decision is based on the risk value and flexible in using more attributes to measure the risk value such as context attributes and cost of outcomes.

- Ahmed and Zhang [6] designed a Context Risk-Aware Access Control (CRAAC) model for ubiquitous computing and presented the major requirements of this model. The requirements are flexible, generic, adaptive, extensible and low cost performance.

- In this model, the access control decision depends on the risk assessment and the Level of Assurance (LoA) to make the model more flexible, generic, adaptive and extensible. They classified resources and services into object groups each with a distinctive object level of assurance (OLoA). When an object access request is received, the decision engine will compare the OLoA with the request level of assurance (RLoA) and the request is granted iif RLoA $\geq$ OLoA. This model is adaptive as it adapts its decision to the surrounding information. The model is flexible enough to accommodate different contextual attributes. Moreover, it is extensible because it allows easy addition of new and removal of obsolete, contextual attributes and any alterations imposed to the architecture. Finally, the model is precise as the access control decision depends on the sensitivity of the resources and level of assurance of the user. The experiments conducted in [7] showed that the basic RBAC model takes 2.16 min to process 100,000 access request while the RBAC takes 5.57 h to process 100,000 access request. However, CRAAC model takes a large proportion of the average access delay to access and evaluate all policy rules on policy files.

- Priya et al. [8] noticed that web service have been widely used in the field of distributed application systems. Therefore, they proposed a context-aware architecture for user access control model (CAUAC) that is suitable for web services. They securely transmitted the context information between the requester (**client**) and the provider (**server**). The proposed model did not explain how to accommodate with different contextual attributes. It also did not explain how to add or remove the contextual attributes and did not take into account **the trust** and the **behavior** of the subject. However, it is **dynamic** because the access decision is based on contextual attributes.

2.2   Attribute-Based Access Control Models

In ABAC, the access control decisions are based on the attributes of the requester, the service, the resource, and the environment.

- Cha et al. [9] presented Attributes-Based Access Control (ABAC) model which makes access control decision based on the attributes. This model achieves the scalability and the flexibility so it is more suitable for distributed systems such as cloud computing. In this model, the access decision depends on the attributes of the requester, the service, the resource, and the environment by using policy evaluation function that permit or deny the request according to these attributes. This model is flexible because it can accommodate with different attributes and scalable because it can add or remove more attributes and policies.
- Sun and Yin [10] showed that the traditional access control models are not suitable for the Internet of Things (IoT). They proposed an access control model that relies on attributes. In this model, firstly, the policy enforcement point (PEP) receives natural access request (NAR) which includes all kinds of users attributes, such as age, location, identity etc., and the way to access the resource. Secondly, PEP extracts the required attribute values and formats it as one unified form. The attribute-based access request (AAR) represents the attribute values as four-tuples (Subject, Object, Behavior attribute, Environment). Thirdly, the PEP transfers AAR to the policy decision point (PDP). Finally, PDP makes a decision according to the policy administration point (PAP) and sends the decision to PEP. PAP stores the default access policies which need to be optimized and the conflicts need to be detected. This model solves the scenarios of large-scale dynamic users, and improves the policy management and permissions assignment compared with RBAC.

2.3   User Behavior-Based Access Control Models

Using a user behavior is another way to design a dynamic and precise access control model. In the behavior-based access control models, the access control decision depends on the evaluation of the user's behavior.

- Ma et al. [11] designed an access control model based on trust quantification (TQBAC). This model depends on evaluating the user behavior or user information after the interaction between the subject and the object completes. Consequently, they used grey fuzzy comprehensive evaluation method to assess the trust value. The proposed model is dynamic; however, it sometimes incorrectly calculates the trust value as it depends on the whole history and not just the recent one.
- Ahmed and Inajem [12] used the past behavior and the history of the entity to predict the future behavior with a certain assurance. The authors designed an

access control model based on three variables: Recency of the transaction, Frequency of the transaction and monetary of the transaction (RFM model). The values of these variables are used to quantify the trust value that used to determine the access control decision. This model is dynamic because the access control decision depends on the trust value and more precise because it cares about the recency of transaction, frequency and monetary. However this model needs to be implemented and evaluated.

- Zerkouk et al. [13] presented the problem of how the pervasive system may provide an assistive environment which allows dependent people to perform their required services in their living spaces. Therefore, the access control model should be suitable to this smart environment. In this model, the access control decision depends on the user behavior and capability. The authors designed a new user behavior and capability based access control model in four steps. Firstly, they constructed the user behavior profile model by assigning behavior classes to users using a classifier process based on historic data. Secondly, they represented the data on standard format using ontologies. Thirdly, they used the currently captured data and the inference rules to deduce new knowledge and to check the consistency of the ontology. Finally, they updated the behavior classes by learning the data provided over the time from different sources. Theoretically, the model is adaptable because the access control decision depends on the user behavior, capability and context aware data, and the model is also dynamic in updating the history file.

- Compared to [13], Mhamed et al. [14] added the trust to the model and presented the main features of this trust modeling. Firstly, it combines many of the useful features introduced in other models, like trust recommendations, trust distribution, and service-dependent trust. Secondly, it introduces the new concept of Judgment to imitate rational human thinking on evaluating. Thirdly, it provides different levels of trust based on requested services. Finally, it provides a user rating that allows taking feedback from users and integrating it into the trust evaluation process. This model is more precise because it uses trust value as the attribute to determine the access control decision.

- Zerkouk et al. [15] developed a personalized and adaptive health care model while taking into account: context awareness, personalization, smartness, adaptability and extensibility. The authors proposed a context aware access control model and its related architecture based on user behavior and capabilities. This model takes the context aware feature to deal with dynamic data by using the most significant sensors. It is reactive by making the decisions based on the basis of the current contextual data and the past behavior. The model takes into account the personalization because it is the main security policy feature which requires richer user context. It is required to the identified sensitive user features (profile, capability, preference, and habit and health state) to provide the most adaptive service. It is extensible because it can integrate dynamically more devices and services. The model takes into account the trustworthy because it used the past behavior and present contextual data. Moreover, it cares about the privacy by taking into account the security of dependent people's personal life. It is smart and this feature is defined by means of the insuring of delivering the suitable service for the specific user.

- Burnett et al. [16] proposed a trust and risk-aware access control mechanism (TRAAC) and a sparse zone-based policy model. In this model, the access control decision is based on the basis of the requester's trustworthiness and the completion of obligations designed to mitigate risk. The model is dynamic and precise because the access control decision depends on the change in the update of the trust value and the risk value. However, it did not explain how to accommodate with different context attributes and did not explain how to dynamically add or remove attributes.

- Smari et al. [17] explained the problem of using standard attribute-based access control model (ABAC) which lacks provisions for trust and privacy issues. The authors presented an extended access control model based on attributes and it incorporates trust and privacy issues in order to make access control decisions sensitive to the cross-organizational collaboration context. The ABAC model is composed of two aspects: the policy model (PBAC: Policy Based AC) and architecture model which applies the policy. In extended ABAC model, the subject attempts to access an object attribute in order to perform an operation. Before authorizing or denying the operation, the request is assessed using the extended model: ABAC rules and contexts are evaluated using the attributes of the subject and the object then the trust algorithm. The model represents the trust by monitoring the subject's behavior. The proposed model is flexible because it can accommodate with different contextual attributes. The model is dynamic because the decision is based on some of the attributes about the requester and the object. Furthermore, the model is extensible because it can add or remove more contextual attributes. Finally, it is precise as it takes into account the trust and the privacy preserving.

- Li et al. [18] proposed access control model using the role mapping relationship between the domains. In the same domain, each time a user requests access to cloud services or cloud resources the authentication and authorization center (AAC) will examine cloud users' trust degree and ensure that the trust degree reaches its threshold. In Cross-Domain Access Control Policy, users often need to access different cloud services or cloud resources in different domains. The authors proposed a RBAC model and presented a new role mapping to achieve resource sharing. The role mapping may cause the problem of permission penetration and privilege escalation. In order to avoid the problem, they presented the mirror role which is based on role mapping. The model is dynamic because the access control decision depends on the domain trust degree. They did not explain how the model can accommodate with different context attributes and how it can dynamically add or remove attributes.

2.4 Relation-Based Access Control Models

The relation-based access control model considers the relation between the subjects or between the subject and the object which makes the access control model decision flexible, precise and generic.

- Zhang et al. [19] proposed a relation-based access control (RelBAC) model for context-aware environment with a domain specific description logic. The description logic is a family of formal knowledge representation languages used to solve decision problem. In this model, the authors represented the permissions of a subject onto an object as a relation that is formalized using the description logic. The model is dynamic because the access control decision depends on context attributes such as location and time. Moreover, it is flexible in dealing with more than one context attribute and it is extensible as it can add or remove attributes.
- Jung and Park [20] proposed an access control model that is different from the traditional research in two ways: it regards the relationship among employees as contextual information and it designs access control architecture using Near Frequency Communication (NFC) technique. This model consists of two main components namely, certification server and control administrator. The certification server publishes certification value (certification id) to each user access. The control administrator makes the access control decision based on the access control policy, user's certification id, requested permission and relationship information. The authors proposed two protocols for permission assignment and permission delegation. They demonstrated two scenarios; the first one shows how to assign permissions in hospital environment, while the second delegate permissions in a company. The authors compared their access control model with the existing RBAC and context-based access control models. The results show that suggested technique can satisfy access control requirements including relationship based access control model.

3 Comparison of Access Control Models

Through the solutions that addressed previously can determine the requirements of the effective access control model:

- Dynamic: the access control decision depends on change of different attributes.
- Flexible: the access control model can accommodate different context attributes.
- Personalization: the access control model can make use of a richer set of user context (profile, capability, preference, habit, health situation).
- Adaptability: the access control model can identify sensitive user features to provide the most adaptive service.

- Extensibility: the access control model can take into account of more constraints generated dynamically as the result of integration of devices and/or services with the underlying smart environment.
- Context aware: This feature is required to make sure that access control decision depending on the surrounding information.
- Trustworthy: Ensuring the system take into account the past behavior and present contextual data.
- Privacy: by taking into account the privacy of user's personal life.
- Smartness: This feature is defined by means of the insuring of delivering the suitable service for the specific user.
- Relation: the access control model can consider user's relationship to decide permission assignment and delegation.

3.1 Context-Aware Based Access Control Models

- Symbol Y: Means that model did achieve this probability.
- Symbol X: Means that model did not achieve this probability.

We can understand from Table 1 that the context-aware access control model can provide a flexible, dynamic, adaptive and extensible model. However, we also need a more precise model that takes into account some issues such as privacy and trust. Moreover, a more generic model is required to take into account the relation between the users themselves and between the users and the objects.

3.2 Attribute-Based Access Control Models

Table 2 shows that the attribute-based access control model can provide a flexible, dynamic, adaptive and extensible model. However, this model also needs to take

Table 1 Context-aware access control models

	Dynamic	Flexible	Personalize	Adaptive	Extensible	Context aware	Trust	Privacy	Smart	Relation
Zhang et al. [3]	Y	Y	X	Y	Y	Y	X	X	X	X
Kim et al. [4]	Y	Y	X	Y	Y	Y	X	X	X	X
Diep et al. [5]	Y	Y	X	Y	Y	Y	Y	X	X	X
Ahmed and Zhang [6]	Y	Y	X	Y	Y	Y	Y	X	X	X
Priya et al. [8]	Y	X	X	Y	X	Y	X	X	X	X

Table 2 Attributes-based access control models

	Dynamic	Flexible	Personalize	Adaptive	Extensible	Context aware	Trust	Privacy	Smart	Relation
Cha et al. [9]	Y	Y	X	Y	Y	X	X	X	X	X
Sun and Yin [10]	Y	Y	X	Y	Y	X	X	X	X	X

into account the context-aware attributes to be more flexible and adaptive. Moreover, a precise model is required by bearing in mind some issues such as privacy and trust. Finally, this model is needed to be more generic by considering the relation between the users themselves and between the users and the objects.

3.3 User Behavior-Based Access Control Models

Table 3 shows that the user behavior-based access control model can provide a precise, flexible, dynamic, adaptive and extensible model. This model is taking into account the capability of the user, so it provides smartness and personate model. However, it needs to be more generic by bearing in mind the relation between the users themselves and between the users and the objects.

3.4 Relation-Based Access Control Model

Table 4 shows that the relation-based access control model can provide a generic model. Moreover, this model is flexible, dynamic, extensible and adaptive because

Table 3 User behavior-based access control models

	Dynamic	Flexible	Personalize	Adaptive	Extensible	Context aware	Trust	Privacy	Smart	Relation
Ma et al. [11]	Y	Y	X	Y	X	X	Y	X	X	X
Ahmed and Inajem [12]	Y	Y	X	Y	X	X	Y	X	X	X
Zerkouk et al. [13]	Y	Y	Y	Y	Y	Y	X	X	Y	X
Mhamed et al. [14]	Y	Y	X	Y	Y	Y	Y	X	Y	X
Zerkouk et al. [15]	Y	Y	Y	Y	Y	Y	Y	Y	Y	X
Burnett et al. [16]	Y	Y	X	Y	X	X	Y	X	X	X
Smari et al. [17]	Y	Y	X	Y	Y	Y	Y	X	X	X
Li et al. [18]	Y	Y	X	Y	X	X	Y	X	X	X

Table 4 Relation-based access control models

	Dynamic	Flexible	Personalize	Adaptive	Extensible	Context aware	Trust	Privacy	Smart	Relation
Zhang et al. [19]	Y	Y	X	Y	Y	Y	X	X	X	Y
Jung and Park [20]	Y	Y	X	Y	Y	Y	X	Y	X	Y

it takes into account the contextual attributes. However, it needs to be more precise by bearing in account the user behavior.

4 Conclusion

In this paper, we presented how the access control is critical to the computer system and discussed the traditional access control models and how these models are not suitable to pervasive systems and dynamic environments. Therefore, we presented the solutions that used to address the limitation of traditional access control models then discussed the security requirements to build effective access control model suitable to pervasive environments and finally we compared between these proposed solutions. In the future, we can combine between more than one feature such as between context-aware, user behavior and relation to making more flexible, dynamic, adaptive, generic and precise access control model.

References

1. Forouzan, B.A.: Cryptography and Network Security. McGraw-Hill, Inc., Boston (2007)
2. Ahmed, A.: Context-Aware Access Control in Ubiquitous Computing (CRAAC), pp. 44–62 (2010)
3. Zhang, G., Zhang, G., Parashar, M., Parashar, M.: Context-aware dynamic access control for pervasive applications. In: Proceedings of the Communication Networks and Distributed Systems Modelling Simulation Conference, pp. 21–30 (2004)
4. Kim, Y.G., Mon, C.J., Jeong, D.W., Lee, J.O., Song, C.Y., Baik, D.K.: Context-aware access control mechanism for ubiquitous applications. Adv. Web Intell. Proc. **3528**, 236–242 (2005)
5. Diep, N.N., Hung, L.X., Zhung, Y., Lee, S., Lee, Y., Lee, H.: Enforcing access control using risk assessment. In: Fourth European Conference on Universal Multiservice Networks, pp. 419–424 (2007)
6. Ahmed, A., Zhang, N.Z.N.: A context-risk-aware access control model for Ubiquitous environments. In: 2008 International Multiconference on Computer Science and Information Technology, pp. 775–782 (2008)
7. Ahmed, A., Zhang, N.Z.N.: An access control architecture for context-risk-aware access control: architectural design and performance evaluation. In: 2010 Fourth International Conference on Emerging Security Information, Systems and Technologies (SECURWARE) (2010)

8. Priya, P., Charles, I.I.P.J., Britto, I.I.I.S., Kumar, R.: Context-aware architecture for user access control, vol. 2, no. 3, pp. 201–204 (2014)
9. Cha, B., Seo, J., Kim, J.: Design of attribute-based access control in cloud computing environment. In: Proceedings of the International Conference on IT Convergence and Security 2011, pp. 41–50 (2012)
10. Sun, K., Yin, L.: Attribute-Role-Based Hybrid Access Control, no. 61100181, pp. 333–343 (2014)
11. Ma, S.M.S., He, J.H.J., Shuai, X.S.X., Wang, Z.W.Z.: Access control mechanism based on trust quantification. In: 2010 IEEE Second International Conference on School of Computing (SocialCom) (2010)
12. Ahmed, A., Inajem, A.: Trust-aware access control: how recent is your transaction history?. In: 2012 Second International Conference on Digital Information and Communication Technology and it's Applications (DICTAP), pp. 208–213 (2012)
13. Zerkouk, M., Mhamed, A., Messabih, B.: User behavior and capability based access control model and architecture. In: Computer Networks and Communications (NetCom), pp. 291–299. Springer (2013)
14. Mhamed, A., Zerkouk, M., El Husseini, A., Messabih, B., El Hassan, B.: Towards a context aware modeling of trust and access control based on the user behavior and capabilities. In: Inclusive Society: Health and Wellbeing in the Community, and Care at Home, pp. 69–76. Springer (2013)
15. Zerkouk, M., Cavalcante, P., Mhamed, A., Boudy, J., Messabih, B.: Behavior and capability based access control model for personalized telehealthcare assistance. Mob. Netw. Appl. **19** (3), 392–403 (2014)
16. Burnett, C., Chen, L., Edwards, P., Norman, T.J.: TRAAC: Trust and Risk Aware Access Control, pp. 371–378 (2014)
17. Smari, W.W., Clemente, P., Lalande, J.-F.: An extended attribute based access control model with trust and privacy: application to a collaborative crisis management system. Futur. Gener. Comput. Syst. **31**, 147–168 (2014)
18. Li, B., Tian, M., Zhang, Y., Lv, S.: Strategy of domain and cross-domain access control based on trust in cloud computing environment. In: Computer Engineering and Networking, pp. 791–798. Springer (2014)
19. Zhang, R., Giunchiglia, F., Crispo, B., Song, L.: Relation-based access control: an access control model for context-aware computing environment. Wirel. Pers. Commun. **55**(1), 5–17 (2010)
20. Jung, K., Park, S.: Context-aware role based access control using user relationship. Int. J. Comput. Theory Eng. **5**(3), 533–537 (2013)

Similarity Measures Based Recommender System for Rehabilitation of People with Disabilities

Rehab Mahmoud, Nashwa El-Bendary, Hoda M.O. Mokhtar
and Aboul Ella Hassanien

Abstract This paper proposes a recommender system to predict and suggest a set of rehabilitation methods for patients with spinal cord injuries (SCI). The proposed system automates, stores and monitors the heath conditions of SCI patients. The International Classification of Functioning, Disability and Health classification (ICF) is used to stores and monitors the progress in health status. A set of similarity measures are utilized in order to get the similarity between patients and predict the rehabilitation recommendations. Experimental results showed that the proposed recommender system has obtained an accuracy of 98 % via implementing the cosine similarity measure.

Keywords Recommender system · Collaborative filtering · Cosine similarity · The international classification of functioning · Disability and health classification (ICF) · Rehabilitation · Spinal cord injuries (SCI)

1 Introduction

In 2013, statistics of the world health organization (WHO) showed that as many as 500,000 persons suffer from spinal cord injuries (SCI) each year. People with spinal cord injuries are likely to die with more rates in low and middle income countries

R. Mahmoud
Faculty of Computers and Information, Fayoum University, Fayoum, Egypt

N. El-Bendary (✉)
Arab Academy for Science, Technology, and Maritime Transport, Cairo, Egypt
e-mail: nashwa.elbendary@ieee.org

H.M.O. Mokhtar · A.E. Hassanien
Faculty of Computers and Information, Cairo University, Cairo, Egypt

A.E. Hassanien
Faculty of Computers and Information, Beni-Suef University, Beni-suef, Egypt

R. Mahmoud · N. El-Bendary · A.E. Hassanien
Scientific Research Group in Egypt (SRGE), Cairo, Egypt
URL: http://www.egyptscience.net

© Springer International Publishing Switzerland 2016
T. Gaber et al. (eds.), *The 1st International Conference on Advanced Intelligent System and Informatics (AISI2015), November 28–30, 2015, Beni Suef, Egypt,*
Advances in Intelligent Systems and Computing 407, DOI 10.1007/978-3-319-26690-9_46

[1]. Spinal cord injuries may be caused by sports, violence, falls, vehicular, and other different causes, as since 2010 the ratio of the previous causes are estimated to 9.2, 14.3, 28.5, 36.5, and 11.4 %, respectively [2].

The International Classification of Functioning, Disability and Health classification (ICF) is a standard framework, devolved in 2001 by the WHO, for coding the human health status in conman language. The ICF describes the health status from the view of the body structure, body function, activities and participations, and environmental factors in addition to the interaction between these components.

Recommender systems are software applications that apply machine learning for filtering unseen information and can predict whether a user would like a given resource. There are many approaches of recommender systems such as collaborative, content based filtering, knowledge based, utility-based, ontology based, demographic based, and hybrid recommender systems [1, 3]. Medical recommender systems use the recommender system methods and algorithms on medical information [4, 5].

This paper proposed an ICF based automated system that stores and monitors health status of patient with spinal cord injuries in order to recommend rehabilitation methods for patients based on his/her status.

The proposed system is divided into three phases; namely (1) prepossessing, (2) similarity, and (3) recommendation phases, and has utilized a number of similarity measures; which are cosine, correlation, Jaccard, spearman, and hamming. The remaining of this paper is organized as follows. Section 2 introduces basics of collaborative filtering recommender systems (CFRS). Section 3 describes the details of the proposed system structure. Section 4 depicts experimental results. Section 5 presents conclusions and discusses future work.

2 Related Work

This section reviews a number of current researches related to similarity measure, medical automated systems and ICF coding based rehabilitation systems.

In [6], authors build medical recommender system that protected users privacy of users, maintain data of users and achieve the flexibility of functionality to make recommendations according to individuals conditions. The proposed a privacy friendly framework to reliable collection of medical information of users. The proposed frame work base on two architectures to achieve that goal the Secure Processing Architecture (SPA) and the Anonymous Contributions Architecture (ACA). SPA user submission is done without giving any privacy information and a computations of recommendations processes over the secured data. ACA the user submission is done in clear way where there is no connection between the submission and users data. They made a comparison between the two architectures and found some drawback to the other and they leave the community to decide which architecture is more suit to practical use.

In [7] Authors aimed to get the dependency between health conditions and performed physical activities. They developed a mathematical model of novel algorithm to generate a recommendations and suggestion instead of health care and hospital admission and help the user to improve his own health. Authors grouped the dataset they used based on the similarity of health conditions and the physical activity history using classification algorithms and the similarity measure of recommender system. To evaluate the proposed recommendation model they change the influence of parameter and the experimental showed that the algorithm is increased with the increase of number of activity observed and the amount of data obtained.

In [8], authors introduce a broach to integrate multi-component rating into collaborative filtering recommender system to collect more information on user performance on special item. Multi-rating components aims to rate item on multi-components not one components as the most literature of collaborative filtering do. they used Expectation Maximization (EM) algorithm to estimate the parameter of the algorithm. Also to test the algorithm authors use training data in two tasks 1: predict the Overall rating on an unseen user-item pair 2: retrieve the highest rated items by a user. they found that the multi-rating component model makes a better predication and retrieval with little training data then the sufficient training data.

In this paper we proposed an automated system to store, monitor the health conditions of patients with SCI, Also a recommender system is developed to recommend a rehabilitations methods for SCI Patient using a set of similarity methods. The results of similarity measures is compared to get the similarity with high accuracy used on dataset.

3 Collaborative Filtering Recommender Systems (CFRS)

Collaborative filtering recommender systems (CFRS) are popular recommendation algorithms with predictions and recommendations based on the ratings or behavior of other users in the system [9]. These systems take into account the ratings provided by users on items and build the user-item rating matrix, where each row of the matrix represents a user profile and the column represents an item profile. Collaborative Filtering (CF) can perform two tasks, leading to two kinds of results [10].

- **Rating prediction**: that predicts the rating that a given unseen product will have for the target user.
- **Recommendation task**: that provides top-N recommendations list of unseen relevant items for the target user.

The proposed recommender system has utilized *cosine, correlation, spearman, hamming*, and *Jaccard* similarity measures. Calculation details of each similarity measure is shown as follows.

3.1 Cosine Similarity

Cosine measures the similarity between vectors of an inner product space by measuring the cosine angle between them, as shown in Eq. (1) [4].

$$cos(i,j) = \frac{\sum_{u \in I_{ij}} y_{u,i} y_{u,j}}{\sqrt{\sum_{u \in I_{ij}} y_{u,i}^2 \sum_{u \in I_{ij}} y_{u,j}^2}} \qquad (1)$$

where $cos(i,j)$ is the cosine similarity measure between two vectors i,j, user u are represented as $|I|$-dimensional vectors i,j and $y_{u,i} \rightarrow$ rating by user u for item i.

3.2 Correlation Similarity

This measure computes the statistical correlation (Pearson's r) between two user's common ratings to determine their similarity [3]. The Correlation similarity is calculated by the formula in Eq. (2) [11].

$$sim(u,v) = \frac{\sum_{i \in I_{uv}} (y_{u,i} - \hat{y}_u)(y_{v,i} - \hat{y}_v)}{\sqrt{\sum_{i \in I_{uv}} (y_{u,i} - \hat{y}_u)^2 \sum_{i \in I_{uv}} (y_{v,i} - \hat{y}_v)^2}} \qquad (2)$$

where $y_{v,i} \rightarrow$ rating by user v for item i, and users ratings vector y_u.
The output of correlation similarity is in the domain of -1 to 1.

3.3 Spearman Similarity

Spearman rank correlation coefficient is a similarity method that measures the similarity between users based on rank of item's rating [11]. That is, items with the highest rating takes rank 1, and items with lower rating values takes ranks >1; e.g. $2, 3, \ldots$, etc. Equation (3) shows the spearman rank correlation coefficient formula [11].

$$\rho = 1 - \frac{6 \sum_{i=1}^{n} [R(x_i) - R(y_i)]^2}{n(n^2 - 1)} \qquad (3)$$

where ρ is the spearman similarity between two vectors X, Y, N is the number of attributes in each vector, i is the current attribute, $R(X_i)$ is the rank if i in vector X.

3.4 Hamming Similarity

Hamming similarity calculated the number of positions at which the corresponding symbols are different between two strings of equal length. In another way, it measures the minimum number of substitutions required to change one string into the other, or the minimum number of errors that could have transformed one string into the other [12]. Equation (4) shows the calculations of the hamming similarity measure [13]

$$d^{HAD}(i,j) = \sum_{k=0}^{n-1} [y_{i,k} \neq y_{j,k}] \tag{4}$$

where $d^{HAD}(i,j)$ is the hamming similarity between two vectors i, j, and k is the index of the reading variable y of the total number of variables n.

3.5 Jaccard Similarity

Jaccard similarity measure similarity between users based on rating the users who are more similar if they have more common ratings [9]. The Jaccard similarity is expressed in Eq. (5), also jaccard distance with two different elements in the collection of all elements measured by the ratio of the two sets of discrimination, jaccard distance is expressed in Eq. (6) [4].

$$J(A, B) = \frac{|A \cap B|}{|A \cup B|} \tag{5}$$

$$d_J(A, B) = 1 - J(A, B) = \frac{|A \cup B| - |A \cap B|}{|A \cup B|} \tag{6}$$

where $J(A, B)$ is the jaccard similarity measure between two vectors A and B, and $d_J(A, B)$ is the distance between two vectors.

4 The Proposed Rehabilitation Recommender System

The proposed system consists of three phases; namely *pre-processing, similarity,* and *recommendation* phases. Figure 1 describes the general structure of the proposed system.

Fig. 1 Proposed system
model

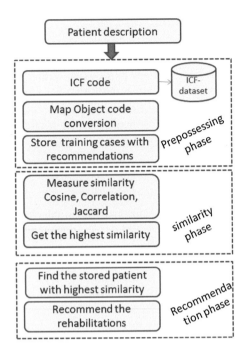

4.1 Pre-processing Phase

In the pre-processing phase, the proposed system converts the entire overall medical expert description of each patient status into its equivalent ICF code. This code is stored for each patient on different period of times in order to evaluate and monitor the progress in patient status. The ICF code for different periods of time are compared to give progress in patient status, as presented in our previous research [14]. The ICF code is developed for unique numeric values using map object code conversion. Map object code conversion is a method that convert each item of ICF code to given numeric value and stored codes in map object. The new codes with its recommendations are stored as training cases.

4.2 Similarity Phase

The similarity phase aims to get similarity between two vectors in input matrix by applying the similarity measures previously mentioned. So, firstly, from the pre-processing phase, the input for the similarity phase is a matrix of two vectors. Both vectors represent the data of two patients. The desired output from this phase is to get the vector with the highest similarity for the input vector via repeating the similarity calculation for input vector against the previously stored ones. Then, the result from similarity process is being considered to get the vector with maximum similarity value. Algorithm 1 illustrates the process of similarity measurement using cosine, correlation, spearmsan, hamming, and Jaccard similarity measures.

Input: Two vectors of ICF code for two patients in text format
Output: Similarity between vectors
 1: ICF code for two patients
 2: Convert the input ICF code into numeric values using map object conversion
 3: Store these data in a matrix as each row is a vector that represents data of one patient
 4: Choose similarity measurement type (cosine, correlation, spearman, hamming or Jaccard) that will be applied on two vectors A and B by using equations (1), (2), (3), (4), and (5) respectively
 5: IF *cosine* similarity selected
 6: Apply it on two vectors, store the the output similarity
 7: Subtract the results from 1 and store the the output similarity value
 8: ELSE
 9: Apply the selected similarity measure on two vectors, store the the output similarity value
 10: Repeat steps (1)-(9) to compare other vectors with the new vector
 11: Get the maximum output value from similarity process
 12: Suggest the rehabilitation based on the highest similarity

Algorithm 1: Calculating cosine, correlation, spearmsan, hamming, and Jaccard similarity measures for two vectors

4.3 Recommendation Phase

The target of this phase is to recommend a set of rehabilitation methods for SCI patients. Based on the output of the previous similarity measure phase, the input of recommendation phase is the vector with the highest similarity for input vector. To get recommendations, we need to get data of similar vector, so search process will be occurred to get the data from the training cases. After finding data, a set of rehabilitations methods will be recommended/returned.

5 Experimental Results and Discussion

The proposed system is built and tested on two categories of dataset; namely *ICF core dataset* and *twenty real cases of patient with spinal cord injuries*. ICF core dataset for spinal cord injuries is developed as a cooperative effort between the ICF Research Branch, the Classification, Assessment and Terminology team and the Disability and Rehabilitation team at the World Health Organization (WHO). The first version was proposed in 2005 and the final version was introduced in 2007 [15]. The twenty case studies of real patients with SCI were gathered by the Swiss Paraplegic Research for the aim of contributing to optimal functioning, social integration, and health and quality of life for people with SCI through clinical and community-oriented research [16].

The proposed system is tested on real case studies for patients with SCI. As previously stated, five similarity measures have been implemented; cosine, correlation, Jaccard, spearman, and hamming similarity measures. Table 1 illustrates and shows comparative analysis for results obtained of measuring similarity for input patient X against previous ten stored patients cases, as a training dataset, using the five similarity measures mentioned. The predicated most similar cases will have the largest common number of diagnosis (i.e. ICF codes) and have the most close number of diagnosis. Also, as shown in Table 1, considering contents in range from 1 to −1 for each similarity measure, the closer values to −1 represent the lower value of similarity and the closer values to 1 represent the higher values of similarity.

Moreover, Table 2 summarizes and highlights the most similar cases from the perspective of cosine, corrections, Jaccard, spearman and hamming similarity measures. Based on results shown in Table 2, the most similar cases from perspectives of cosine, jaccard, and hamming similarity is case 5, and for correlation and spearman similarity is case 7. Based on manual calculations and human expert recommendations, the most similar cases for input patient X is case 5. So, the cosine, jaccard, and hamming similarity measures gives more accurate results than the correlation and spearman similarity measures. That's due to the observation of considering the angle between two vectors and ignoring the number of occurrences in the attributes of vectors. In addition to that the tested dataset has a unique attribute value for each

Table 1 Results of applying five similarity measures on input patient against previous ten stored patients cases

	Cosine	Correlation	Jaccard	Spearman	Hamming
Case 1	0.327	0.143	0.966	0.034	0.104
Case 2	0.219	0.354	1	0.365	1
Case 3	0.373	0.127	0.900	0.032	0.794
Case 4	0.219	0.323	0.918	0.093	0.918
Case 5	0.617	0.395	1	0.599	1
Case 6	0.584	0.363	1	0.650	1
Case 7	0.241	0.427	1	0.706	1
Case 8	0.502	0.163	0.966	0.229	0.630
Case 9	0.463	0.195	1	0.295	0.652
Case 10	0.429	0.128	0.978	0.385	0.978

Table 2 Results of most similar case from perspective of five similarity measures

Similarity measure	Cosine	Correlation	Jaccard	Spearman	Hamming
Most similar case	Case5	Case7	Case5	Case7	Case5
Max similarity value	0.781	0.427	1	0.706	1

vector, so cosine similarity measure gives the most accurate results. Also, Jaccard equation measures the similarity via calculating the intersection of vectors divided by the union of them. Based on the tested data set, the ICF codes in each vector is occurring once and the intersection gives the common numbers or codes, which the two vectors share, regardless the occurring number itself. For hamming similarity measure, it showed more similarity ands gives the common attribute for both vectors via measuring the required number of positions to make one vector equal to another. So, based on the tested dataset, which has unique attributes in each vector, the hamming similarity gives more accurate results.

The correlation similarity measure uses the standard deviation of each vector for measuring the similarity between them. In the situation of the system proposed in this paper, correlation similarity resulted in low accuracy because the number calculated as standard deviation of each vector was far from the mean of each tested pair of vectors. The Spearman was also calculated as the correlation coefficient between the ranked variables.

Accordingly, based on both the expert based calculations and the obtained results from the proposed system, the accuracy of the five similarity measures is calculated using Eq. (7), which calculates by the summation of the obtained values divided by the summation of the correct values multiplied by 100. The accuracy obtained by cosine, jaccard, correlation, spearman, and hamming are 98, 80 %, and 36, 40, and 92 % respectively. Figure 2 shows the accuracy of the five similarity measures.

$$Accuracy = \frac{\sum obtained\ value}{\sum correct\ value} * 100 \tag{7}$$

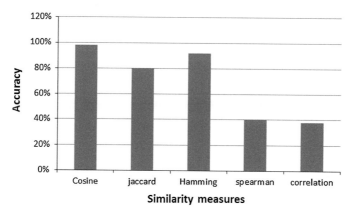

Fig. 2 Accuracy obtained by cosine, jaccard, hamming, spearman, and correlation similarity measures

6 Conclusions and Future Work

As rehabilitation progress monitoring, and specially for spinal cord injuries (SCI) patients, represents a great challenge, this paper presents a rehabilitation recommender system that predicts and suggests a set of rehabilitation methods for patients with SCI. The proposed system automates, stores and monitors the progress in heath conditions of SCI patients using the aids of The International Classification of Functioning, Disability and Health classification (ICF). The proposed recommender system measures similarities on data of patient with spinal cord injuries via applying cosine, Jaccard, correlation, spearman and hamming as similarity measures and the highest similarity accuracy of 98 % has been obtained by the cosine similarity measure. That is because the cosine similarity measure the angle between two vectors, regardless the magnitude, which is the number of occurrences. So, based on the tested patients data, each code in vectors is occurred once and accordingly the similarity is high for the cosine measure. For future work, it's planned to optimize the results obtained in this research by using bio-inspired optimization algorithms and also to consider different datasets and additional similarity measures.

References

1. Joan, L.H.: World health organization releases internship perspective on spinal cord injuries. J. Vent. Assist. Living **28**(1), 1–2 (2014)
2. World Health Orgnization (WHO): Spinal cord injury. http://www.who.int/mediacentre/news/releases/2013/spinal-cord-injury-20131202/en/ (2013). Accessed 11 May 2015
3. Kulev, I., Vlahu-Gjorgievska, E., Trajkovik, V., Koceski, S.: Development of a novel recommendation algorithm for collaborative health: care system model. Comput. Sci. Inform. Syst. **10**(3), 1455–1471 (2013)
4. Su, X., Khoshgoftaar, M.T.: A survey of collaborative filtering techniques. Adv. Artif. Intell. **2009**, 19 (2009)
5. Ghazanfar, M.A., Prugel-Bennett, A.: An improved switching hybrid recommender system using Naive Bayes classifier and collaborative filtering. Proc. Eng. Comput. Sci. Hong Kong 2010 **I**, 978–988 (2010)
6. Hoens, T., Blanton, M., Chawla, N.: Reliable Medical Recommendation Systems with Patient Privacy. IHI'10 Arlington. Virginia, USA (2010)
7. Igor, K., Vlahu-Gjorgievska, E., Trajkovik, V., Koceski, S.: Development of a novel recommendation algorithm for collaborative health: care system model. COMSIS J **10**(3), 1455–1471 (2013)
8. Mazurowski, M.: Estimating confidence of individual rating predictions in collaborative filtering recommender systems. Expert Syst. Appl. **40**(10), 3847–3857 (2013)
9. Shabanpoor, M., Mahdavi, M.: Implementation of a recommender system on medical recognition and treatment. Int. J. e-Educ. e-Bus. e-Manag. e-Learn. **2**(4), 315–318 (2012)
10. Shinde, S., Kulkami, U.: Hybrid personalizad recommender system using centeringbunching based clustering algorithm. Expert Syst. Appl. **39**(1), 1381–1387 (2012)
11. Ekstrand, M.T.D., Riedl, J., Konstan, J.A.: Collaborative filtering recommender systems. Hum. Comput. Interact. **4**(2), 81–173 (2010)
12. Bobadilla, J., Hernando, A., Ortega, F., Bernal, J.: A framework for collaborative filtering recommender systems. Expert Syst. Appl. **38**(12), 14609–14623 (2011)

13. Desoky, A., Ali, H., Abdel-Hamid, N.: Enhancing iris recognition system performance using templates fusion. Ain Shams Eng. J. **2012**(3), 133–140 (2011)
14. Mahmoud, R., Elbendary, N., Mokhtar, H., Hassanien, A.:ICF based automation system for spinal cord injuries rehabilitation. In: ICCES 2014, Cairo, Egypt, pp. 192–197 (2014)
15. Icf-research-branch.org: ICF core set projects, ICF Research Branch—Development of ICF Core Sets for Spinal Cord Injury (SCI). http://www.icf-research-branch.org/icf-core-sets-projects-sp-1641024398/neurological-conditions/development-of-icf-core-sets-for-spinal-cord-injury-sci (2015). Accessed 11 May 2015
16. Jaimie, F., Cieza, A., Hoogland-Eriks, I., Rauch, A., Stucki, G., Wong, V.: ICF case studies, [online] Implementation of the International Classification of Functioning, Disability and Health (ICF) in rehabilitation practice. http://www.icf-casestudies.org/ (2015). Accessed 7 May 2015

Author Index

© Springer International Publishing Switzerland 2016
T. Gaber et al. (eds.), *The 1st International Conference on Advanced Intelligent System and Informatics (AISI2015), November 28–30, 2015, Beni Suef, Egypt,*
Advances in Intelligent Systems and Computing 407, DOI 10.1007/978-3-319-26690-9

Printed in the United States
By Bookmasters